Farming the Woods

Farming the Woods

An Integrated Permaculture Approach to Growing Food and Medicinals in Temperate Forests

KEN MUDGE AND STEVE GABRIEL

FOREWORD BY JOHN F. MUNSELL

CHELSEA GREEN PUBLISHING
WHITE RIVER JUNCTION, VERMONT

Unless otherwise noted, all photographs copyright © 2014 by Steve Gabriel and Ken Mudge.
Unless otherwise noted, all illustrations copyright © 2014 by Carl Whittaker and Travis Bettencourt.

Project Manager: Hillary Gregory
Developmental Editor: Makenna Goodman
Copy Editor: Eileen M. Clawson
Proofreader: Helen Walden
Indexer: Lee Lawton
Cover design: Melissa Jacobson
Cover images: Ken Mudge and Steve Gabriel, except for the second-from-the-right
bottom image by VoDeTan2Dericks-Tan, Wikimedia Commons

Printed in the United States of America.
First printing September, 2014
10 9 8 7 6 5 4 3 2 1   14 15 16 17

Our Commitment to Green Publishing
Chelsea Green sees publishing as a tool for cultural change and ecological stewardship. We strive to align our book manufacturing practices with our editorial mission and to reduce the impact of our business enterprise in the environment. We print our books and catalogs on chlorine-free recycled paper, using vegetable-based inks whenever possible. This book may cost slightly more because it was printed on paper that contains recycled fiber, and we hope you'll agree that it's worth it. Chelsea Green is a member of the Green Press Initiative (www.greenpressinitiative.org), a nonprofit coalition of publishers, manufacturers, and authors working to protect the world's endangered forests and conserve natural resources. *Farming the Woods* was printed on paper supplied by RR Donnelley that contains at least 10% postconsumer recycled fiber.

Library of Congress Cataloging-in-Publication Data
Mudge, Ken, 1949–
  Farming the woods : an integrated permaculture approach to growing food and medicinals in temperate forests / Ken Mudge and Steve Gabriel ; foreword by John F. Munsell.
    pages cm
  Other title: Integrated permaculture approach to growing food and medicinals in temperate forests
  Includes bibliographical references and index.
  ISBN 978-1-60358-507-1 (pbk.) — ISBN 978-1-60358-508-8 (ebook)
  1. Permaculture. 2. Agroforestry. 3. Tree crops. I. Gabriel, Steve, 1982– II. Title. III. Title: Integrated permaculture approach to growing food and medicinals in temperate forests.

S494.5.P47M83 2014
631.5'8—dc23

2014020393

Chelsea Green Publishing
85 North Main Street, Suite 120
White River Junction, VT 05001
(802) 295-6300
www.chelseagreen.com

This book is dedicated to the memory of Chris Dennis,
and to all of the next generation of those interested in
cultivating a working relationship with the forest.

Contents

A few years ago I asked a group of around one-hundred and fifty farmers how many own forested land. About ninety percent raised their hands. I then asked how many actively manage their forests. Two hands remained. So it goes. Farmers generally confine their work to the field. Here and there they may cut trees or turn out livestock to relieve pastures, but forest management within a farming enterprise is rare. At the same time, many of the forests owned by government, nonprofits, investment institutions, industry, and families run the same course. From time to time these owners also use their forests, but likewise contribute to the patchwork penumbra that befalls the modern temperate legacy.

Aside from intensively managed tree plantations, which produce a great deal of wood but account for a relatively small portion of forested land, extensive uses of temperate forests such as outdoor recreation and the occasional timber harvest are most common. Whether acted out in the woodlands of a farm or in the remnant forests of a rolling, rural residential development, these uses provide much-needed material like timber and valued amenities such as hunting. The trouble is most are not the result of an integrated plan. Temperate forests are therefore useful but mostly not well used. Yet opportunities to do better by them remain.

When left to their own devices, forests are thought to constantly change within the bounds of site-specific biomass benchmarks. Their biotic volume and structure are largely functions of available resources and microclimate. In other words, forests generally grow according to how much and what kinds of life a specific place in a given time can support. These place-based assemblies typically consist of a complex web of flora, fauna, and fungi, all of which use, share, and sometimes improve available water, nutrients, and light. It is here that forests offer their lesson, providing structural and functional signposts that inform thoughtful and productive use.

The work of Ken Mudge and Steve Gabriel is situated at this intersection. They are agroforesters, and in this spirit complement rather than co-opt temperate forest biodiversity. Carrying out their trade at the interface between species is part and parcel of their philosophy. Opportunities to interact with the abundance of forest life for betterment of both people and environment are embraced. In so doing, they aim to integrate forest farming and permaculture alongside silviculture within a single framework. As such, theirs is a notable inflection in the dialogue and direction of sustainable food, wood, medicine, and decorative-materials production.

Like any good agroforester, the authors work with complex and sometimes unpredictable production systems. By necessity, therefore, they are adaptive managers, using creative problem solving via permaculture principles and practices to respond to lessons learned. The challenges faced today in terms of food and health security along with questions surrounding energy and climate wholly warrant such an approach. And instead of articulating beliefs and lamenting challenges without offering comprehensible recourse, their book positions forest farming as a practical process and provides a sizable survey of cold temperate production possibilities. Perhaps most impressive is that they supplement this useful content with case studies of successful forest farmers.

The authors begin by sharing their worldview of farming and forestry. In their own way, they challenge divisions in cold temperate land-use regimes, discussing how historical compartmentalization of farming and forestry systems reduced landscapes into parts that subsequently became alienated. The theme is that this boundary disserves contemporary needs. Be it planting trees with crops or livestock in the field or growing non-timber crops alongside timber logs in the forest, finding mutual ground when and where possible is exigent by the authors' account. They argue there likely is no other way if land-management systems are to adequately improve lives and sustain productive environments in the face of such rapid global change.

The points and examples offered by the authors in drawing lines around the status quo are grounds for contemplation, but little requires pondering when it comes to the book's practical content. Expansive and detailed, it offers those who encounter it a variety of options for impactful forest farming. All the while underpinned by personal perspective, the book swirls with information on applications and opportunities for crafting an integrated and multifunctional forest enterprise. From detailed information on mushroom production to an inspirational celebration of a meal created wholly from forest products, the collection of forest farming opportunities, tricks of the trade, and case studies will take the reader on a tour of what could, and by many accounts should, play a larger role in the cultural context of integrated land use.

The practices and principles of agroforestry and permaculture are alive throughout the book. Interestingly, boundaries between the two often are blurred in the face of more important issues like feeding, treating, and sheltering growing populations while sustaining environments that are able to meet needs for generations to come. Ultimately, the authors work toward a strategy wherein the practice of forest farming is positioned relative to the future of cold temperate forests and their place within multifunctional and well-planned landscape management.

When thinking back to the number of farmers who kept their hands up when asked about forest management, it is not altogether surprising that they often do not consider themselves to be active managers. Long an issue for those concerned with landscape health and productivity, it can be said that the segregation of forestry and farming has not only divided land use, but also people. Realizing the need for an inclusive approach, the authors of this book correctly emphasize a variety of opportunities, rather than a single land-management prescription.

That Ken Mudge and Steve Gabriel believe people should be empowered in pursuits of integrated, multifunctional forest management is clear. As a result, the book is better positioned to positively impact forest owners, farmers, policy makers, and general readers alike. I encourage you to take advantage of this resource, because at your fingertips is a useful and inspirational forest farming guide.

—John F. Munsell
Blacksburg, VA
July, 2014

Acknowledgments

The authors would like to thank a number of individuals for their support and assistance while writing this book. Collecting, reviewing, and presenting such a wide swath of information was a tremendous effort, which was truly based in a community of supporters and advocates.

Several people gave their skills and expertise to this writing. First off, our sincere thanks to Marguerite Wells, who provided some initial copy editing to the manuscript and to Patrick Castrenze, who gave expertise on structure and conventions of writing that helped us set a crucial pattern and tone for the book. We are also grateful for the talented drawing and sketching skills of Carl Whittaker, whose mushroom line drawings are both accurate and fantastical, and to Travis Bettencourt, who came through on technical diagrams with his crisp drawings. Finally, we owe a deep appreciation to Michael Burns, who helped set up the Farming the Woods website and conducted marketing and promotions for the Indiegogo fund-raising campaign for the book. Our heart goes out to the family of Chris Dennis, a student of ours who filmed several videos for the book and on forest farming and tragically went missing on a canoe trip on Cayuga Lake in spring 2013.

A number of experts in the fields of agroforestry, permaculture, and horticulture offered their feedback on various sections throughout the book. We are thankful that many of our technical details were reviewed with extensive commentary (exhausting but appreciated!) from John Munsell and Catherine Bukowski of Virginia Tech, Mike Gold from the University of Missouri, Jim Chamberlin of the USDA, Peter Smallidge, Louise Buck, Marvin Pritts, and Jim Ochterski (all from Cornell), and author and permaculture designer Dave Jacke. A special thanks to "Mr. Ginseng," Bob Beyfuss, and to "Mr. Mushroom," Steve Sierigk, who provided resources and feedback for their crops of passion.

The case study sites for the book came from so many wonderful and committed individuals that inspire us by bringing theory to practice in the real world. Each site was responsive and accommodating to our visits, opening their homes and lives to us and taking time from their busy schedules to share, and we are delighted to have made some new friends along the way. Our thanks to Philip Rutter and Brandon Rutter-Daywater of the Badgersett Research Corporation; Mary Ellen Kozak and Joe Krawczyk of Field and Forest Products; Steve and Julie Rockcastle of Green Heron Growers; Johann Bruhn and the great folks at the University of Missouri Center for Agroforestry; Wayne and Kim Lovelace of Forrest-Keeling Nursery; Mark Shepard of New Forest Farm; Jamin and the Uticone family and Swamp Road Baskets; Bonnie Gale of English Basketry Willows; Rodney Webb of Salamander Springs Farm, Bruce Phetteplace in Central New York; the Menominee nation; Chris Chmiel of Integration Acres and the folks who organize and sponsor the Ohio Pawpaw Festival; Nicola and Dan of Ozark Forest Mushrooms; Sean Dembrosky of Edible Acres; Josh Dolan and Sapsquatch Maple Products; and last but not least, two Daves: Dave Carman of Purdue, West Virginia, not to be confused with Dave Cornman of Spring Haven Nursery in Pennsylvania.

Of course, it should be recognized that our work stands on many shoulders, and that we owe what we know and how we know it to those who have written, researched, and advocated for these practices. Our gratitude reaches back in time to Lawrence MacDaniels, for giving us a living example and a playground to study forest farming, and to Robert Hart, J. Russell Smith, and other visionaries who wrote and advocated in a time when agroforestry was even more obscure. Along these lines, we appreciate the individuals who work tirelessly to promote tree agriculture, including the U of Missouri Center for Agroforestry, the USDA

National Agroforestry Center, the Association for Temperate Agroforestry, and innumerable researchers, Extension agents, educators, and activists who are helping to move us all forward.

Steve would like to give endless thanks to his wife Elizabeth and doggies Sadie and Vida for being loving and putting up with his craziness while writing the book. He is thankful to his family, Bob, Susan, Jennifer, and Scott, as well as his grandmother Ruby and all the Bergschneiders for the kind assurances and support along the way. And much appreciation is given to the many amazing teachers he's been blessed with, including Dale Bryner, his grandfather Fred Bergschneider, Julie Kulick, Mike Demunn, Jeff Carter, Chant and Suzanna Thomas, the Findhorn Community in Scotland, and the folks at Empire State College. Finally, immense gratitude goes to the numerous friends who support, students who inspire, and the permaculture community who strengthens.

Ken is lovingly indebted to his wife Beth Mudge for her encouragement, for providing a calm in the middle of a storm (from time to time), and for putting up with the long hours that went into our book.

Visiting as many case study sites as we did would not have been possible but for the 149 contributors to our Indiegogo campaign in February of 2013. We are endlessly grateful for your support.

And finally, we'd like to thank Chelsea Green Publishing and Makenna Goodman for helping to make this book a reality.

Introduction

This book begins, and ends, with our love for the forest. For our temperate climate, the place we know, the forest is the natural ecosystem type that exists, or would exist if we were to stop mowing, cutting, and plowing in a given place. The forest, which is a place where trees of all shapes, sizes, and characters live, is what covers land, protects soil, and harvests rainwater. It is the home for many mammals, birds, reptiles, amphibians, insects, and fungi. It is the story of hundreds to thousands of years of growth, death, and decay. It is also a place where people have gained sustenance for much of our time on Planet Earth.

Spending time in the woods leaves many people feeling calmer, happier, and more peaceful. The Japanese even have a word for this, *Shinrin-yoku*, which can be translated as, "taking in the forest atmosphere or forest bathing."[1] Research has shown that spending time in the woods is good for health and can be therapeutic. A landscape dominated by trees is where we authors like to spend our time. We are fascinated with trees themselves, as well as the complex communities they can create. We are constantly awed and amazed as we work in the woods grafting, pruning, cutting, pollinating, and eating. We are visited by red-tailed hawks or surprised by the discovery of a flush of chicken of the woods mushrooms. The forest is a place where time slows down and surprises emerge.

For many in the modern world the forest is a place to recreate, which means visiting for a time before returning to "town" or to the places we call home. Yet as we spend time grafting trees, moving mushroom logs, and sharing forest farming with students and youth, we realize that the forest is a home, too. Students in our forest farming class look forward to the weekly sessions when we meet in our classroom—the woods—and learn ideas and skills. For many the chance to wield a shovel or simply sit in the forest is a welcome break from lecture halls and library study sessions.

Some of my fondest memories from childhood took place in the woods. (Steve) When I was five we moved to a new neighborhood in New York, a new subdivision, and for a moment in time we were the last house on the street. Next door was a wild woodland of hawthorn, locust, and other varieties of thorny trees and shrubs that emerged when the farmer abandoned his fields. The other neighborhood boys and I spent endless days after school clearing the thorns with sticks we'd fashioned so that we could run through the woodland with ease without poking our eyes out. We built forts and hideouts and secret places where we could observe the neighbors around us without being detected. One day I got off the bus from school and found our beloved forest on fire, lit by the builders who were starting to construct the next lot of homes.

I remember, too, the first time I visited our local nature center (which I ended up working at many years later) and witnessing for the first time the tapping of a maple tree—and the delicious sugary sap that came forth. I was blown away by the fact that sugar could come from a tree: and I was hooked. This was just one of many peak experiences that connected me to the forests of the place where I grew up. Whether it was tromping through an amazing grove of sycamore and lying under the grasp of massive weeping willows at the south end of Cayuga Lake with a high school girlfriend, taking hiking and camping trips to the Adirondack woods with my father, or drying off on a long hemlock tree that had washed ashore near a favorite swimming hole, trees litter my memories of place and define my experience of being alive.

During my last two years of high school I started reading and learning about the long list of things humans have done to damage life systems on Earth, taking note especially of the track record of damage to forests. I couldn't believe that the places I loved to spend time in were actually quite rare and endangered, and that future generations might not be able to enjoy them. I was not convinced that environmental destruction was a necessity of progress for a nation. These convictions led me to explore ideas of how humans could interact more positively with the natural world, including organic farming, permaculture, and sustainable forestry. I hopped around different colleges, traveled abroad to witness the almost total disappearance of any forests in Scotland, and participated in anti-logging activism in the West.

When I moved back to my hometown of Ithaca and decided it was time to wrap up my degree once and for all, I learned about a unique professor at Cornell who taught a class in the woods. The story was that he had found a lost nut tree planting started by a previous professor and had revived it with students. Here was a truly unique place in the world: a ninety-year-old nut forest that was born of human design. I was amazed that such a high-level institute as Cornell was offering a class on forest farming, which was pretty obscure for a university interested more in large-commodity crops and livestock. I ended up taking Ken Mudge's class, and that began a collaboration and a friendship and, though we didn't know it at the time, this book.

We bonded over the concept that forest farming offered a unique opportunity to rethink how we farm in the modern world. It was a concept that many of Ken's colleagues have scoffed at over the years, citing the mantra of any big agricultural school, which is, "That's cool, but you aren't going to make money doing it." Yet Ken's persistence and dedication to forest farming, and specifically mushroom cultivation, over many years proved just the opposite; today more and more farmers are growing mushrooms on logs each year, and they are making money doing it. It's just the beginning.

Humanity has a mixed relationship and a complicated history with the woods. Yet when Ken and I visited forest farmers around the country and met the people who manage them, we saw a mirror of ourselves in them. There are incredible people already doing this stuff out there, in many different ways. The common thread that binds them all is a passion for the woods and a desire to spend as much time in it as possible,

doing things they love. Some of them cultivate beneath stands of old trees, and some plant new ones, leaving a forest in the footsteps of their farming activities.

This book is about many things, but fundamentally it is about a new way to relate to the forest. It offers not only new ways of seeing and valuing forests for both preserving and enhancing forest health but also the potential to make an income. However, the tips, tricks, and techniques found within are no good if readers don't take time to connect to the forested landscape. It is our love for the woods that keeps us going above all, and the reason we wrote this book.

To explore the many facets of forest farming, this book will take you on a journey through the cool shade of a hemlock forest, where Steve and Julie Rockcastle cultivate shiitake mushrooms right alongside wild mushrooms rotting in tree stumps and the devilish red ones with white spots that pop up right out of the ground.

Make a stop at the MacDaniels Nut Grove and view the many hickory trees that have strange bulges on their trunks, indicating that they were grafted over seventy years ago. Climb a ladder to help pollinate the blood-red flowers of the pawpaw with fine-tipped paintbrush, bringing pollen from the male parts of a flower of one tree over to the female part of a flower on another tree. A bit tiresome, but well worth the trouble when the plump aromatic fruits come in, about the same time of year when hickory nuts rain down from the sky, some with their husk nearly the size of a tennis ball. You may want to wear one of the shiny yellow hard hats for protection.

Visit the woods in late February collecting sap from the sugar trees to make maple syrup. Catch a few drops on your finger as it drips out the spile. It barely tastes sweet at all, being many hours of boiling away from finished syrup. As syrup season ends, witness the early spring sun shining down on the forest floor, warming the soil and calling up ramps (wild leeks) from dormancy. Dig one out of the ground, and you'll find that the plump scallion-like bulb smells like garlic.

Visit Dave Carman, an Appalachian forest farmer, at his home in West Virginia, where he grows forest herbs such as spikenard, fairy wand, ginseng, goldenseal, and many more on a wooded hillside beneath a high, green canopy of tulip poplar. Although these are valuable medicinal herbs, he's not so much interested in harvesting the roots, rhizomes, or other structures used as medicinals. Instead he harvests and sells seeds for others to grow the herbs. Over generations enough rich organic soil has accumulated for Dave to grow Virginia snakeroot scattered among the fragrant wild ferns that were there first. The small S-shaped seedpods are reminiscent of Santa's curved pipe.

At a Pennsylvania forest nursery Dave Cornman grows the medicinal herb black cohosh, with its delicate arching spray of fragrant white flowers. Butterflies scatter as you approach it. He sells it to shade gardeners, who think of it as an ornamental. Dave built a ruggedly beautiful stone house before building the nursery, which includes a couple of quaint little storage buildings painted forest green with brown trim; they remind you of the witch's house in "Hansel and Gretel," but without the candy.

These and the other forest farmers profiled in this book offer a vision for how more people can live—*with* and *in* the forest rather than outside it, a foreigner who only visits from time to time. Human civilization is in a time when the decisions we need to make are unlike those any generation has had to make before. With increasing inequality, the collapse of ecosystems around the world, and the uncertain effects of climate change, there is not a better time to consider farming the woods.

— Steve Gabriel and Ken Mudge, 2014

Farming the Woods

Figure 1.1. Forest farming of sugar maple, mushrooms, and ducks at Wellspring Forest Farm, Mecklenburg, New York

1   What Is Forest Farming?

In the eyes of many people, the practices of forestry and farming are at odds, because in the modern world it's often the case that agriculture involves open fields, straight rows, and machinery to grow crops, while forestry is primarily reserved for timber and firewood harvesting. Forest farming invites a remarkably different perspective: that a healthy forest can be maintained while growing a wide range of food, medicinal, and other nontimber products. While it may seem to be an obscure practice, the long view indicates that for much of its history, humanity lived and sustained itself from tree-based systems. Only recently have people traded the forest for the field.

The good news is that this is not an either–or scenario; forest farms can be most productive in the places the plow is not: on steep slopes and in shallow soils. Much of the cool temperate forest that remains today is the result of land that was just too hard to plow up, coupled with the massive abandonment of small family farms, which left fields fallow and gave forests the chance to rebound. Further, since forestry has emphasized timber harvesting, much of the forest cover in the modern world is low value, leaving people to wonder how to support their woodlots. The opportunities presented in forest farming are a perfect complement to these scenarios, which is why this book was written.

Forest farming is one of many agroforestry practices, specifically focused on growing crops underneath the canopy of an existing forest. The novelty of this idea and its relative obscurity in the mainstream indicates just how far society has removed itself from the source of much wealth in the form of food, medicine, and wood products. It also indicates how a thirst for timber resources has

overwhelmed forests to the point where little else is seen as valuable. Yet chances are that on any given day the average Westerner has consumed multiple forest-grown products. Many of the daily indulgences taken for granted—coffee, chocolate, and many tropical fruits, for instance—all originate in forest ecosystems. Few know that such abundance is also available in the cool temperate forests of North America.

Definitions

In dynamic fields such as agroforestry and forest farming, there are multiple definitions by different authors of some of the key concepts presented in this book. Each definition contributes something to the conversation, with no single one being entirely right or wrong.

By discussing and defining the key terms repeated in this book, the reader will be able to better understand where the authors are coming from. The most important items to define at the outset of this text are these:

- Agroforestry and six main practices in temperate climates
- "Cool temperate agroforestry" as a geographic region used in this book
- The practice of forest farming, including:
  - Nontimber forest products, including cultivated versus wildcrafted
  - Productive conservation and forest health

Agroforestry

Forest farming is one of six agroforestry practices designated for the temperate climate. Although indigenous peoples have been practicing tree-related agriculture and gathering of forest products for millennia, it wasn't until 1973 that a Canadian forester, John Bene, coined the term "agroforestry."[1] This led, several years later, to the establishment of the International Council for Research in Agroforestry (ICRAF), in Nairobi, Kenya, in 1978. In addition to documenting indigenous agroforestry practices, one of ICRAFs earliest efforts was to explicitly define the field to facilitate communication among investigators, practitioners, and other stakeholders. In 2002, ICRAF was renamed the World Agroforestry Centre (WAC). Since agroforestry has components of agriculture, forestry, and associated social dimensions, it is not surprising that it means different things to different people according to their experience and their goals. WAC in Nairobi, Kenya, defines agroforestry as:

> a collective name for land-use systems and technologies where woody perennials (trees, shrubs, palms, bamboos, etc.) are deliberately used on the same land-management units as agricultural crops and/or animals, in some form of spatial arrangement or temporal sequence. In agroforestry systems there are both ecological and economical interactions between the different components.[2]

This definition of agroforestry is process oriented ("land-use systems and technologies"), whereas a contrasting definition of agroforestry from the National Agroforestry Center in Lincoln, Nebraska, is more outcome oriented:

> Agroforestry is a multidisciplinary approach to agricultural production that achieves diverse, profitable, sustainable land-use by integrating trees with non-timber forest crops.[3]

An important consideration that is missing from both of these definitions is that agroforestry is not simply integration of trees and crops, but also of people and society. A more recent definition from Garrett adds this dimension:

Figure 1.2. An agroforestry planting of staghorn sumac, hazelnut, and hybrid larch in New Brunswick, Canada. The planting is an interesting case study, as dozens of species were planted and not maintained for almost ten years other than occasional mowing. Part of the tour on the 13th North American Agroforestry Conference.

> Agroforestry is a dynamic, ecologically based, natural resource management system that, through the integration of trees on farms and in the agricultural landscape, diversifies and sustains smallholder production for increased social, economic, and environmental benefits.[4]

The appeal of this definition is the recognition that these practices are of great benefit to small landowners and that there are social benefits in addition to economic and environmental ones. Ultimately, people are the variable in agroforestry, as production of 20 pounds of shiitake mushrooms from 100 logs is nothing more than a rotten mess without the mushroom pickers who harvest the mushrooms at the right time and get them to market and bring home the money. The place and time where agroforestry can occur is dependent on, first, consumer interest and then market demands, which make production profitable.

Figure 1.3. Grove of hybrid chestnuts at Badgersett Research Farm in Canton, Minnesota, which primarily does research on agroforestry nut cropping, including hazelnuts, chestnuts, and hickory nuts

While each of these definitions brings something to the table, there is a desire by the authors to simplify and highlight the critical function of forest farming in the postmodern world. Without attempting to further complicate things, an additional definition of agroforestry crafted by the authors is this:

The combination of crops (plants, animals, fungi) and trees in forest-inspired agricultural systems that benefit human communities through greater connection to landscape, improved stewardship of resources, and enhanced economic opportunities.

Ultimately this speaks most directly to the important parts of agroforestry. Regardless of the specific practice, each approach seeks to mimic the forest in its design and implementation. The end goal is to produce agricultural crops—whether for home use or commercial sales. And the outcomes for people are reconnection, better caretaking, and improved livelihood. With all of these definitions, the words are crafted to suit the desires of the authors and reflect the context and thinking of different places and times.

Six Temperate Agroforestry Practices

These practices are defined here to provide a general sense of the focus agroforestry has taken in the temperate climate. Other resources provide further discussion of each practice and define agroforestry in more detail.[5]

1. *Alley cropping* is the most spatially structured agroforestry practice. It involves planting trees in straight rows or along contours, with one or more other crops grown in the alleys (spaces) between the rows of trees. The trees themselves may provide one or more products, such as fruits, nuts, fodder, or even timber when the trees reach sufficient size. A wide range of crops can be intercropped with the trees, such as hay, wheat or other grains, vegetables, woody plants such as blueberries, or semiwoody plants such as blackberries. The trees in an alley cropping (or any) agroforestry system may include so-called fertilizer (nitrogen-fixing) trees, which provide a biological source of nitrogen "fertilizer" from the air.

2. *Riparian and upland buffers* involve the practice of planting trees, shrubs, and grasses between agricultural fields and water bodies (rivers, streams, creeks, lakes, wetlands) and on hillsides around and within agricultural fields to prevent erosion, filter excess nutrients (fertilizers or manure), and capture other undesirable runoff (pesticides, herbicides, etc.). The planted components may or may not provide one or more yields; for example, fruits, nuts, or biomass. Planting riparian buffers usually includes multiple zones of planting that provide a gradient (see figure 1.4) and a focus on species that can adapt to various conditions presented in this edge habitat.

3. *Silvopasture* involves grazing livestock beneath the canopy of a woodlot or the practice of bringing trees into pasture in ways that mutually benefit both the animals and the trees. In essence, this is a three-way symbiosis between trees, the grazing animals, and grasses. For livestock prolonged exposure to hot sun is a source of stress that can reduce weight gain or milk pro-

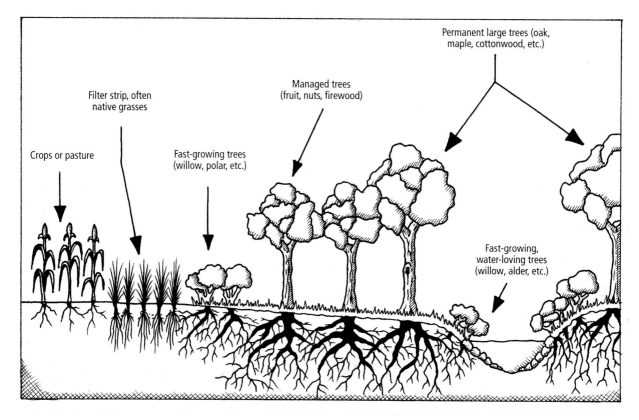

Figure 1.4. Riparian and upland buffers are usually a gradient of vegetation, with each "zone" performing a different function to slow and gradient runoff from neighboring agricultural fields. Even if cropping isn't occurring, creating tree corridors along waterways is a good idea. Illustration by Travis Bettencourt

duction.[6] Access to shade reduces stress, thereby improving animal health. The trees benefit from the manure left behind by the grazing animals as well as the cycling of fertility from the grazing of grasses. In most cases, the forest canopy must be maintained to around 50 percent canopy cover to ensure good sun exposure for the grasses. Thus, a savanna-like ecosystem emerges as a template for silvopasture systems.

4. *Windbreaks* are an agroforestry practice in which complex assemblages of trees and shrubs (multiple rows, multiple heights, and multiple species) are deliberately planted in configurations designed to decrease the velocity of the prevailing wind, thereby reducing livestock stress, soil erosion, and water loss from crops. The trees can, of course, also provide various other products while providing shelter from the elements. Rather than a single row of trees and woody shrubs, the most successful windbreaks are those that employ multiple rows and densities of planting to offer a way to slow and buffer the effects of wind on a site. Some common examples of temperate species used for windbreaks include black locust, honey locust, Siberian pea, willow, alder, and some conifers. Simpler versions can include single rows; for example, deciduous

trees planted in rows specifically to act as a "snow break" in the Plains states.

5. *Forest farming* is commonly defined as the cultivation of crops under a forest canopy that is intentionally modified or maintained to provide shade levels and habitat that favor growth and enhance production levels. Possible yields include medicinal plants, food crops, mushrooms, ornamentals, and a variety of wood products. The practice of forest farming is largely the subject of this book. Since forest farms are working within the forest canopy, the palette of species is much more limited to those with at least some shade tolerance, unless planted along forest edges or in gaps or clearings.[7]

6. *Forest gardening* has origins as far back as any agroforestry practice, although it has not been included in recent agroforestry publications. It has elements of both gardening and edible landscaping in its practice. Dave Jacke and Eric Toensmeier define edible forest gardening as "the art and science of putting plants together in woodland-like patterns that forge mutually beneficial relationships, creating a garden ecosystem that is more than the sum of its parts."[8] It is in this phase of succession that the most layers of trees, shrubs, and herbaceous plants can

Figure 1.5. Silvopasture of Black Angus underneath black locust at Cornell Extension forester and farmer Brett Chedzoy's Angus Glen Farms near Watkins Glen, New York. Photograph courtesy of Brett Chedzoy

Figure 1.6. Red alder is a fast-growing nitrogen-fixing tree being established as a windbreak at Wellspring Forest Farm, Mecklenburg, New York.

Figure 1.7. Forest gardening is a good strategy for urban agroforestry, as demonstrated in this backyard in Holyoke, Massachusetts. Photograph courtesy of Eric Toensmeier

thrive together. In contrast to the other agroforestry practices, forest gardens, because of their smaller scale and increased intensity, are likely to contain many more species within a very small area.

Agroforestry Practices and Succession

One way to draw out some distinctions between different agroforestry practices is to consider succession influences on ecosystems. While this ecological phenomenon is discussed in more detail in chapter three, for now the existing ecology sites can be generally categorized as one of the following: *early succession* (open field ecosystems dominated by grasses and herbaceous plants); *midsuccession* (mixture of grasses, herbaceous plants, shrubs, and trees in patches); and *late succession* (a closed canopy of maturing trees).

From the standpoint of the relationship of the trees to the crops, the six agroforestry practices fall into

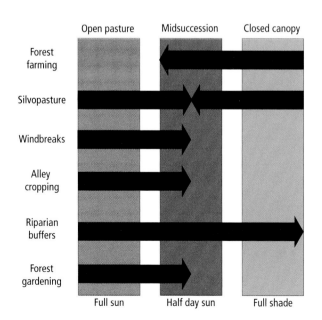

Figure 1.8. Agroforestry practices compared in relationship to stage of ecological succession and light qualities. Arrows indicate the direction in which the systems often head successionally over time.

FOREST FARMING VS. FOREST GARDENING

Forest gardening is not usually considered to be one of the five temperate agroforestry practices. Yet when comparing definitions it appears to the authors that forest gardening, with a focus on "woodland like patterns to create garden ecosystems" fits well with "the intentional mixing of trees and shrubs into crop and animal production systems to create environmental, economic and social benefits," as put forth by the latest USDA Agroforestry Strategic Framework.[9] From this comparison, forest gardening fits well into the overall concept of agroforestry.

Another question then becomes, is forest gardening sufficiently different from any of the other agroforestry practices to warrant its being recognized as a separate temperate practice? At first glance the practices of forest farming and forest gardening can appear to be similar, and they are often confused. A key difference between the two is that most often forest gardening gradually builds a forest "from scratch," whereas forest farming begins with an established forest. Even simpler, as Dave Jacke and Eric Toensmeier have said, forest gardening is "gardening like the forest," whereas forest farming is best described as "gardening or farming *in* the forest."

Species diversity, too, is a key distinguishing characteristic between the two. Most forest farmers dabble with a few crops, adding a bed of ginseng or leeks and a few stacks of mushrooms to their woodlots. The focus is most often on producing larger yields of a fewer number of species. Forest gardens, on the other hand, often focus on designing polycultures, or mixtures of multifunctional plants that occupy different niches in time and space. These polycultures are designed to maximize the potential of light falling on a site for production. Yields per species may be lower when compared to forest farming, though a total yield from such a large mix of plants may be larger.

One additional question: Why does this even matter? Is it any more than a word game or does it have any practical significance to people currently practicing or considering agroforestry? From the authors' vantage point, it is a useful distinction for people considering engaging in agroforestry who have only a little land and no forest. They could easily be discouraged from engaging in agroforestry because of its perceived large scale. Recognizing that a small-scale agroforestry practice that is suitable to their home is possible would encourage many to become involved in agroforestry (see figure 1.7).

Based on these considerations, the authors encourage the North American agroforestry community to consider forest gardening as a type of agroforestry, a sixth version of the practice.

Each aspect of agroforestry offers a slightly different perspective on tree-based agriculture—but in the end all achieve similar results. In short, it might be best to summarize this conversation on agroforestry by highlighting the main goal: to increase the presence and value of trees on farm and forest landscapes to achieve multiple goals that both benefit the environmental character of the site and provide a diversity of yields for its inhabitants.

Further, there is a tendency of academics and others who write, conduct research, or teach agroforestry to think and describe agroforestry practices as discrete and mutually exclusive. On the other hand, most practitioners (farmers and landowners) do not see them this way at all but rather have a tendency to overlap and blur the lines between the practices as they see fit.

groups based on the stage of succession they are most well adapted to, as well as to the direction succession is being driven, as can be seen in figure 1.8.

With the exception of forest farming and in some cases silvopasture, the other agroforestry practices all tend to begin with open field, which can be considered early succession. This perhaps explains why much of the focus of agroforestry education is in working with field crop farmers in the Midwest and encouraging them to bring trees into the landscape. Alley cropping, riparian buffers, windbreaks, and forest gardening systems tend to start out with planting young trees and waiting for them to grow for at least five to ten years before they begin to produce nuts, fruits, and other products.

In contrast, silvopasture and forest farming systems most often start with existing forests and thin them

Figure 1.9. Coauthor Ken Mudge explains ginseng cultivation at the MacDaniels Nut Grove Open House, Fall 2013, when over 150 people interested in forest farming visited for the day.

to varying degrees to provide the right conditions for understory production. It's interesting to note that most of these agroforestry systems, because of the desire to mix trees with other crops, tend toward midsuccession conditions. Most of this pattern has to do with the availability of light to the plants, as early to midsuccession systems tend to offer better growing conditions for the vast majority of species.

Simplistically, one might say that alley cropping, riparian buffers, windbreaks, and forest gardens involve *bringing the trees to the crops*, whereas forest farming and silvopasture involve *bringing the crops to the trees*. Of course, to accept this characterization of silvopasture, one must buy in to the view that cows, goats, or chickens are "crops." Robert Blake, a professor of animal science at the University of

California at Riverside, describes livestock as the "crop that walks."[10]

In writing this book, the authors have had a number of conversations with academics, farmers, Extension associates, students, and others interested in agroforestry. With regard to forest farming, a common question that has arisen is, When can one say that a forest has an "existing" canopy; that is, when does a field become a forest? While technically speaking succession stages are classified by the percentage of canopy cover (see chapter 4, Getting a Yield: Light and the Forest), it becomes tricky when considering a few examples in practice.

When we visited Badgersett Research Farm (see chapter 4), the challenge of putting the common definition of forest farming into practice was made readily apparent; here is a research plot where the vegetation was in some places forty years old. The trees had grown in many places into a closed canopy. So in every sense it was an "existing forest." Yet would the same quality be true if the site had been visited ten years ago? Or twenty? It's hard to say where to draw the line.

The MacDaniels Nut Grove, too, is an existing forest at Cornell University that was once open field. The intention of this planting was to create a nut orchard, where trees would be evenly spaced and managed to maintain some open light. Yet when abandoned, (see chapter 2) the grove reverted to forest for several decades; when it was rediscovered about ten years ago it underwent a third transformation into a forest farming system. Today this forest is over ninety years old and exists as a great example of forest farming. Yet it did not begin as an existing forest. Does this mean that at one point the cultivation was not forest farming, and now it is?

These examples provide a bridge between theory and practice. The various agroforestry practices, which in theory cover a wide range in appearance, species composition, and stage of succession, all use the forest as a model for design. The yields are more than just timber and aesthetics but can also include a long list of fruits, nuts, medicinal plants, aesthetic plants, animals, woodworking materials, wildlife habitat, and soil stabilization. All the practices create systems in which the whole system is more complex, robust, and dynamic than any of the individual parts. And if humans were to leave the picture, a forest of some type would be left in their absence. These factors bring all the practices together, recognizing a common goal as their end.

In a time when extreme environmental, economic, and social stresses plague the planet, the benefits of trees and forests for carbon sequestration, for improving the yields from farms, and for the contributions they can offer to society are what is most important. There is not sufficient time to argue petty details, but rather time needs to be devoted to education and development of all and any agroforestry practices. The correct combination and use of the practices will ultimately be determined by three variables:

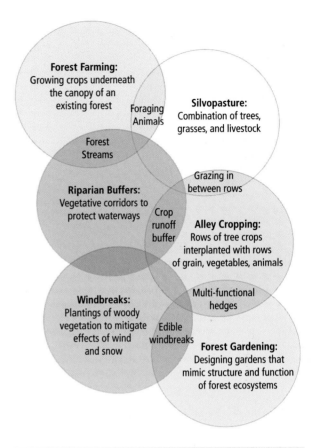

Figure 1.10. This diagram depicts the inherent overlap and relationship in the various temperate agroforestry practices. For example, animals can be grazed in riparian buffers during dry times, and strategies from both systems may be used.

1. The character of the local landscape, including the limiting factors, species, and ecosystem succession specific to a place and time

2. The goals for a project, including consideration of how outside social structures affect management; for instance, if local market demand justifies growing a crop on a commercial scale

3. The willingness of the participants to use agroforestry as a general concept, mixing, matching, and blurring the lines of each together, exploring the relationships between the various practices to determine a unique agroforestry for the place being designed

A final consideration is that many farms and projects already use the term "forest farm" and "forest farmer"

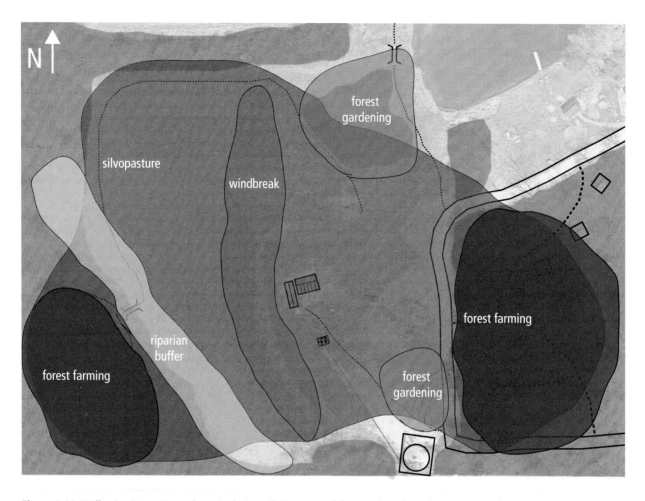

Figure 1.11. Wellspring Forest Farm schematic design utilizing many of the practices of agroforestry. As can be seen in this map, often the practices relate and overlap in practice.

regardless of what agroforestry practices they might be utilizing. This is simply due to the fact that these terms are more familiar and understandable compared to "agroforester" or "alley cropper." A forest farm, then, might be a place where many of these practices are used together and where there is a fundamental difference in the way agriculture is practiced, best summarized as farming in the image of the forest.

As an example, at Wellspring Forest Farm (Steve Gabriel's farm, which he runs with his wife, Liz Falk), there exists the opportunity to utilize each of the agroforestry practices, all on just 10 acres of land. The farm contains 2 acres of existing forest, which is composed almost exclusively of sugar maples ranging from fifty to one hundred years of age. This setting is the perfect

place for mushroom cultivation (shiitake, lion's mane, oyster, and Stropharia), as well as for maple sugaring and ginseng (forest farming). The remainder of the site, about 8 acres, is all open field. Here the main goal is to restore a productive forest, first by planting trees to mitigate some of the effects of the cold and gusty winter winds that affect the site (windbreaks). The lower edge of the property contains a seasonal creek bed, which is being designed with walnut and pawpaw production as the primary crops, mixed with floodplain species and earthworks to capture and filter water and sediment (riparian buffer). Several old hedgerows and portions of the field are being allocated to small, intensive patches mixing trees, shrubs, and herbaceous plants (forest gardening). And finally, across most of the property, animals will be rotated for forage and soil building; in this case, sheep and ducks (silvopasture).

In the spirit of this interdisciplinary approach, this book will stick largely to the topic of forest farming, while allowing for some flexibility and "reach" into related and connected aspects of agroforestry as they present themselves.

FOREST FARMING

It has been only recently that the term "forest farming" was applied specifically to cultivation of nontimber forest products (NTFPs) beneath an established forest canopy. The term "tree crops" was used much earlier by J. Russell Smith in *Tree Crops, a Permanent Agriculture* (1929),[11] to describe large-scale tree planting on hilly land to provide erosion control as well as nuts and other food crops for people. This is sometimes cited as an early example of and inspiration for modern forest farming. Similarly, Douglas and Hart's *Forest Farming* (1985),[12] used the term "forest farming" to describe extensive tree planting in open pasture to produce fruit and mast for cattle.

Hill and Buck's "Forest Farming Practices" (2000)[13] was the first to define forest farming in the "modern" sense that it is used in this book.

> intentional manipulation of forested lands to produce specific products, most specifically food or medicinal products, although other non-timber forest products as well

Since then, other definitions for forest farming have been proposed by the National Agroforestry Center (NAC), the Association for Temperate Agroforestry (AFTA),[14] and by Ken Mudge, one of the authors of this book.[15] They all agree that the cultivation of NTFPs beneath an existing forest canopy is at the core of forest farming, but each goes on to add or exclude peripheral considerations, such as the necessity of management of the forest not only for NTFPs but also for timber (NAC, AFTA). Hill and Buck's and AFTA's definition explicitly exclude wildcrafting, and Mudge adds ecological and economic sustainability to the core definition. What follows is a definition that focuses on

Figure 1.12. Co-author Steve Gabriel stacking freshly inoculated shiitake logs in his yard. Mushroom cultivation is currently one of the most practiced forest farming strategies in the northeastern United States. Photograph courtesy of Jen Gabriel

the essential characteristic of forest farming (preexisting forest), with a number of qualifiers that may or may not apply to any given situation.

> Forest farming is the cultivation of nontimber forest products beneath the canopy of an existing, actively managed forest.

Depending on the site and circumstances, it may include:

- Manipulation of the light environment by pruning or tree removal to meet the needs of specific crops
- Management for timber as well as NTFP production
- Management to restore or maintain a healthy forest ecosystem
- Intentional cultivation, as of crops on hobby and commercial scale
- Gathering of wild NTFPs in addition to cultivation

Forest farming is increasingly being adopted in the cool temperate United States, where most notably there is a drastic increase in the number of mushroom growers in the northeastern states (see chapter 5). Many forest owners who view their forests as recreational for hunting, hiking, and even wildcrafting are increasingly interested in forest farming, whether they call it that or not.

Core to the practice of forest farming are two concepts: the growing of nontimber forest products and the focus on the concept of "productive conservation," where the production of crops is balanced with sound forest management practices.

Nontimber Forest Products

NTFPs include any plants, fungi, animals, and wood products from the forest other than timber. The term "nontimber forest products" and its abbreviation, "NTFP," recur throughout this book and any discussion of agroforestry practices. NTFPs encompass both the deliberately cultivated (i.e., farmed) products of any agroforestry practice including forest farming, as well as similar naturally occurring materials collected from the wild (wildcrafting or foraging). In other words, any living thing of value (objectively or subjectively) taken out of the woods is an NTFP (except trees for timber).

Figure 1.13. Wildcrafted ramp bulb with flower stalk and Dryad's Saddle mushroom, collected near Ithaca, New York in spring 2012. The mushroom is considered a moderately good edible.

Table 1.1: NTFPs Cultivated, Wildcrafted, or Both in Cool Temperate Climate

Cultivated ONLY	Wildcrafted ONLY	Can Be BOTH Cultivated and Wildcrafted
Shiitake mushrooms	Chanterelle mushrooms	Oyster mushrooms
Hostas (nonnative)	Porcini mushrooms	Stropharia mushrooms
Lady's slipper orchids		American ginseng
		Wild leek (ramps)
		Nut trees

Cultivation versus Wildcrafting of NTFPs

Of those academics and growers who like to debate the nuances of such topics, there is an ongoing debate about how to view wildcrafting in the context of forest farming. Some want to specifically exclude any NTFP that is not cultivated, while others are more flexible. In this book we proposed the concept of "assisted wildcrafting." In practice most who are interested in (cultivated) forest farming are also engaged in some wildcrafting. Some NTFPs can only be cultivated (shiitake mushrooms, for example), while others can only be harvested from the wild (chanterelle mushrooms), and a third group can be acquired through both means (wild ramps and ginseng).

Wildcrafting may appear at first glance to be the easiest approach to gaining yields from the woods, where all one has to do is wander the woods in search of sustenance. The reality of it is far from this notion. Because nature works in pulses of abundance, the discovery of an amazing mushroom patch or the harvest of a masting nut crop one year may yield nothing or very little for several years until the conditions are again just right. In addition, because of the large-scale degradation of forests, there simply isn't an abundance of healthy habitats to support the growth and survival of many of the desirable species.

The danger in wildcrafting is, of course, overharvesting. This can happen because a large number of individuals visit a patch and harvest at a rate greater than the population can regenerate, or even if a single individual gathers too much over a season or a longer period of time. Certain species are particularly vulnerable to this, and some of these are even illegal to harvest (ramping is illegal, for example, in Ontario, Canada).

For those wild species we can cultivate, then, it might be best to view the practice of cultivation also as a practice of conservation. Wild leeks are a great example, as a huge interest in local and seasonal foods has led to a boom in wild harvesting for restaurants, festivals, and farmers' markets in recent years. This has also led to some concerns that ramp populations are being decimated, a fear vetted by research by several institutions, indicating that only a very small percentage of any population can be harvested to ensure the population sustains itself (see chapter 4).

Concepts of "wild" or "primeval" ecosystems need to be challenged as well, since evidence suggests that humans have been cultivating ecosystems for a long time, far before the advent of plow agriculture. The forest has long been held in romantic regard, retaining some sense of wildness if not "improved" for agriculture or other development (see chapter 2). Further, some of the crops that are popular foraging species, such as stinging nettle, are adapted to disturbed environments from such activities. Is a plant or mushroom that one wanders into a forest to gather a "wild" specimen if either directly planted or indirectly supported by a past human activity?

The disconnection of people and the natural world is fundamental to a conversation on how best to relate to the forest, whether for gathering or for growing goods for home use or commercial sales. The need to connect and learn from forest landscapes in order to relate is in fact what draws many people to both wildcrafting and forest farming in the first place.

Cultivated crops are the general focus of this book and the authors' definition of forest farming, partially because if everyone went out wildcrafting, populations would be even more threatened. If any reasonable amount of sustenance is desired from forest ecosystems at this point in time, it needs to be cultivated.

Productive Conservation: Farming for Forest Health

So far this chapter has stressed that the outputs and value of forest farming lie in production of NTFPs (medicinal plants, edible crops, and ornamentals). Equally important to the practice of forest farming, however, is attention paid to managing forests to be as healthy and productive as possible. Forest farming is one form of "productive conservation," which is by nature (pun intended) an activity that both produces a yield while conserving or improving the environmental status of a site. This is a compelling argument for the potential power of these systems in a culture where it is often seen that agriculture and conservation of land are at odds.

In addition to providing crops, trees and forests provide a number of critical ecosystem services, including:

- Erosion control, which trees provide by slowing down the movement of water, allowing it to infiltrate rather than run off
- Nutrient cycling, especially in a deciduous forest, where leaf decay helps maintain soil organic matter
- Clean air as a by-product of tree and plant photosynthesis
- Clean water, filtered through the rich soil and root systems
- Shade, which moderates extreme temperatures and is essential to many shade-loving/tolerant NTFPs
- A buffer against drought, because water is captured and cycled within the trees
- Wildlife habitat for a number of birds, insects, reptiles, amphibians, and mammals
- Recreation, sustenance, and spiritual fulfillment for people

By one measure there are three components of forest health that are a bit more specific to forest farming because they emphasize the interdependence among forest ecology, forest management, and forest farming: *resilience*, *diversity*, and *sustainability*.[16]

Figure 1.14. Forest farming means inevitable management of the forest through at least some tree selection and felling. These efforts can be used to benefit production of NTFPs. Photograph courtesy of Jennifer Gabriel

RESILIENCE

Forest *resilience* refers to a forest's ability to regenerate quickly after disturbance. Aldo Leopold, one of the mostly highly regarded conservationists of the twentieth century, and author of *A Sand County Almanac*, said

> [Forest] health is the capacity of the land for self-renewal . . . Any definition of forest health must consider the capacity for forest replacement within the time span of successional processes.[17]

This idea of "replacement" is more accurately regeneration, or the forest's ability to grow another generation of trees from the current stock. Over time,

Figure 1.15. Sugar maple seedlings grow well in full shade and "wait" for a gap opening in the forest, then grow into the canopy. This forest is even-aged, in that most of the trees are the same age. It should be managed to encourage more understory seedling growth.

plants have evolved strategies to make this work, since space is tight in the woods. Seedlings of certain shade-tolerant tree species such as sugar maple will proliferate beneath the canopy of mature trees for decades, grow only several inches tall, and maintain that stature for many years. If a gap in the canopy occurs because of a blowdown or high grading, the suppressed seedlings are "released" in response to higher light and begin to grow rapidly, refilling the gap.

An analog of this same natural phenomenon may occur in forest farming when perennial NTFPs such as woods-cultivated ginseng are harvested at maturity. Harvesting requires that the entire plant (shoot and root system) be removed, so if the ginseng grower hasn't been planting a new crop from seed each year, there will be a long gap between one harvest and the next. There are two different systems for cultivating American ginseng, one that suffers from this limita-tion (woods-cultivated ginseng) and one that does not (wild-simulated ginseng). These will be discussed in detail in chapter 3.

THE IMPORTANCE OF DIVERSITY

Another characteristic of natural forest ecosystems that contributes to forest health and overall forest pro-ductivity is *diversity*. Diversity applies at many levels of composition of a given forest ecosystem, including:

- Species (white ash, black locust, black walnut, for example)
- Genera (*Acer*, maples; *Quercus*, oaks; *Fagus*, beech; and so on),
- Families (Fagaceae, oaks and beech; Juglandaceae, walnuts and hickories; etc.)
- Kingdoms (Plants; animals; fungi; bacteria; etc.)
- Genetics within a species

BIODIVERSITY AND INVASIVE/NONNATIVE SPECIES

Forests often include native (indigenous) as well as nonnative (exotic or invasive) species. The conversation around invasive forest species is a tricky one and ultimately is a policy decision each forest farmer must make for himself or herself. Lumping a bunch of different species together in the "invasive" or "exotic" category further muddles the situation, because in reality some species pose a greater threat than others. And in some cases, the species come with both positive and negative qualities.

An example is the Japanese knotweed, which is often found in heavily disturbed sites along highways and in eroded creek beds. While the rhizomatous root system makes eradication near impossible, the ability of the plant to stabilize erosive soils is an important quality. Further, the early spring shoots are edible and highly nutritious[18] and even medicinal; research is being done to look into its potential effects in fighting the symptoms of Lyme disease.[19] This is not to say that necessarily these plants should be encouraged to stay in ecosystems where they are not native, but perhaps these species fill niches and respond to the environmental conditions of the current time. It might be one's perception that needs some further examination. In this case, when a species is examined from the perspective of the *functional* role it plays as opposed to *where it comes from*, it's not so black and white.

A 2011 article coauthored by nineteen prominent ecologists encouraged science to take on just this type of approach:

We are not suggesting that conservationists abandon their efforts to mitigate serious problems caused by some introduced species, or that governments should stop trying to prevent potentially harmful species from entering their

countries. But we urge conservationists and land managers to organize priorities around whether species are producing benefits or harm to biodiversity, human health, ecological services and economies. Nearly two centuries on from the introduction of the concept of native-ness, it is time for conservationists to focus much more on the functions of species, and much less on where they originated."[20]

This doesn't mean that all recent arrivals should be celebrated. A common plant to show recently in eastern forests is garlic mustard, which grows and spreads rapidly, producing copious amounts of seeds each season. Research indicates that it also exudes a toxin from its root, which suppresses the growth of native fungi in the soil.[21] In this case, there is a better reasoning for removal.

Ultimately, the individual or group managing a particular landscape defines whether or not a species is "invasive" or "problematic" for their goals. Much like the concept of a "weed" species, the beauty (or hatred) of a species is in the eye of the beholder. The plant can be native or nonnative in origin, but if it gets in the way of the goals of those working the land, they will seek to remove it.

For the purposes of this book and the practice of forest farming, both native and nonnative species are presented. The general approach is to favor native species, then encourage use of nonnatives where there is a track record of "naturalization" or use where there is no evidence that the species inflicts harm on native ecosystems. Practitioners should take careful notes and with nonnative species keep a close eye on potential problems that emerge. As with anything in ecosystems, the only constant is change.

A strong principle in forest ecology is that there is a strong positive correlation between biodiversity and overall forest productivity.[22] Exactly how one contributes to the other is not entirely understood, but one factor that is involved is the ability of species to engage in "resource partitioning" with respect to

their ecological niches. While it's always true that species compete for limited resources in a forest, this doesn't mean that each species is at war with the other. Resource partitioning describes the evolution of species to occupy different parts of an area, which gives better access to the resource (nutrients, water, light) and effectively captures the sum total of that resource, so none is "wasted."

The underground sphere is not the only component of structural diversity in a natural forest or a forest farm. A healthy, natural forest (and a well-designed forest farm) is vertically stratified aboveground to partition limited light resources. There are different ways to classify this concept of architecture in forest ecosystems. The Society of American Foresters defines "life zones"[23] of the forest as:

1. Emergent trees that grow above the general canopy;
2. Canopy trees, which act as the "roof" of the forest;
3. Understory small trees and shrubs; and
4. Forest floor vegetation that includes grasses, ferns, flowers, mushrooms (and other fungi), as well as the belowground components of some species, including ginseng, goldenseal, bloodroot, fungi, and so on.

These distinctions make sense from a forestry perspective. In the case of forest gardening, Robert Hart further delineated forest layers[24] because it helped define a more intensive architecture in the understory, which is often a focus of forest garden design:

1. Canopy
2. Low tree
3. Shrub
4. Herbaceous
5. Vertical (vines and climbers)
6. Groundcover
7. Rhizosphere

Both compositional diversity (species, genus, family, kingdom) and the structural diversity of above- and belowground architecture are important factors in healthy forests. Forest farmers can both introduce new species (seeds, transplants, and so forth) and create conditions in the forest (wildlife habitat, gaps, fire management) to increase diversity.

SUSTAINABILITY

A third pillar of forest health is sustainability, which has become one of the most overused words in the

Figure 1.16. American ginseng is an endangered species in many states that can sometimes be found in the wild and also cultivated in a forest farm.

English language, but in the context of forest ecology, it is a critical factor in maintaining forest health. In the case of forest farming and other agroforestry practices, sustainability can be described as the long-term ability of a forest to continue growth and reproduction of the desired species.

An example of questionable sustainability can be found with the wildcrafting of American ginseng. In parts of its range it is in decline because of overharvesting, deer browse, and habitat destruction. Obviously this is not sustainable for its population. Some environmentally conscious ginseng hunters and, more recently, several state regulatory agencies have attempted to offset ginseng's decline by mandating that any ripe seed be sown in the immediate vicinity of the harvested plant. The effectiveness of this strategy is unknown, so only long-term studies would clarify if this approach leads to a form of sustainability.

Measuring sustainability, then, requires looking at indicators of success. One example of this is in the forests managed for hundreds of years by the Menominee tribe in Wisconsin, which reports that even though more than a half billion board feet of lumber has been harvested since the mid-1800s, there

Figure 1.17. Spent shiitake logs composting at Ozark Forest Mushrooms, Salem, Missouri. Logs can be harvested as part of a timber stand improvement that benefits the forest. The logs are then partly decomposed by mushrooms and eventually decompose completely and feed forest soil.

is more standing timber today than there was 150 years ago. With these numbers the tribe can justify saying it is practicing sustainable forest management. As forest farmers of cultivated crops, a question then becomes how to measure and thus know if sustainable management is happening.

YIELDS AS A BY-PRODUCT OF FOREST HEALTH

One option to consider as a starting point for farming the woods is aiming to produce yields that are the by-product of an act to support forest health. For a given crop, then, a parameter (or several) must be established to measure the impact of the activity on forest health. For example, if animals are going to be grazed or allowed to forage in a woodlot, regular soil sampling can provide details on the availability of nutrients and percentage of organic matter in the soil over time.

This idea further ensures that actions are not taken for the sake of production alone and can in fact change the entire relationship of farmer to landscape. For example, while mushroom cultivation guides often cite that logs for production should be taken from straight, disease-free, and healthy-looking trees, these are also often the trees that are the best candidates for healthy, long-term growth, which means good genetics and therefore good seed. And mushrooms, which decompose dead wood, don't need to have the best logs to thrive (though sections that are infected with other fungi should not be used). In the end harvesting the best trees leaves the forest worse off than it originally was.

Instead, a forest farmer can mark trees that are the poor performers, or that have evidence of disease or defect. In this way the main product of the action is an improvement to the forest, while the by-product of this action (downed trees) can be utilized as a substrate for mushroom production (see chapter 5), and likely other forest products, too (see chapter 8).

Defining Cool Temperate Forest Farming

The classification of climate zones is widely varied, depending on the source, and there is not a single commonly referenced version of climate designation. In this case, cool temperate is based on the hardiness zones of 4 through 7, with areas characterized by warm to hot summers and rainfall of 30 inches or greater (see figure 1.18).

The reasoning for this designation is twofold: One is simply that the authors mainly have experience and expertise in forest farming within this region. The other is to limit the possible scope of topics to maintain integrity of the practice as covered in a book this size. Just as the range of any species extends in many directions yet often contains a "sweet spot" where it thrives best, the content for this book is centered around species and strategies that work well in the northeast and midwestern United States up into parts of Canada. Readers in other cool temperate climates should consider comparing differences between their own bioregions and this climate and proceed armed with information as to how a given crop may or may not succeed in another climate zone.

So while the book sets a realistic scope of geography that matches the competence of its authors, the specific species, strategies, and techniques may apply to a much wider geography and cultural context. For instance, shiitake mushrooms can be cultivated in a wide range of temperate climates, from those that are cool to those that get much hotter. The basic process is the same, yet strain selection and management can be quite different. If a reader is not residing in the heart of the geography defined as the main focus of this book, then extra care should be taken to "transfer" the knowledge to another place.

Considerations for Beginners

There are many different reasons that people want to learn about and ultimately practice forest farming

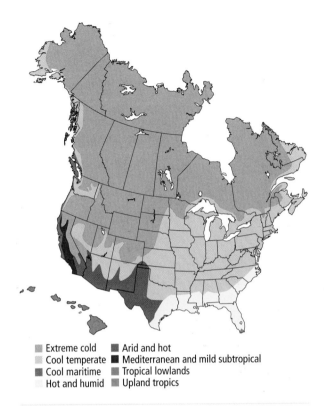

Extreme cold ■ Arid and hot
Cool temperate ■ Mediterranean and mild subtropical
Cool maritime ■ Tropical lowlands
Hot and humid ■ Upland tropics

Figure 1.18. One version of a climate zone map created by Eric Toensmeier for his book *Perennial Vegetables*. Eric cross-indexed USDA hardiness zones, American Horticultural Society heat zones, and Sunset gardening zones for the map. His definition of cold temperate is hardiness zones of 4 through 7, with warm to hot summers and 30-plus inches of rain, which fits the authors' definition of cool temperate for this book. Illustration courtesy of Eric Toensmeier

and/or other agroforestry practices. The practice is inherently interdisciplinary, calling upon a range of skill sets, depending on the scale and scope of the operation.

On the most basic level, forest farming will be difficult without the ability to cut, process, and move trees around the woods. This means that one of the first skills needed is to develop safe habits for the use of chain saws and in some cases use of machinery, such as a tractor. Basic construction skills are also a necessity for being able to build raised beds, deer fences, and other necessary structures. And finally, a familiarity with how to use the array of tools common in gardening and farming will be useful in earth shaping, path construction, and the development of spaces conducive to cultivation.

Forest farmers are often hobbyists, in which case the stakes are much lower, and less time and attention may need to be invested into designing a setup that maximizes the efficiency of the operation. Of course, care should still be taken to minimize impact on the forest. Hobby growers may be more able to preserve the "wild" nature of the forest and may forgo strategies that are more resource intensive. For instance, some hobby mushroom growers choose not to soak their logs, which means eliminating water infrastructure and tanks from the situation. For commercial growers, soaking will be a necessity.

Commercial growers should start small. Indeed, many commercial agroforestry ventures began as a hobby. Scaling up should only occur if those doing the managing find they enjoy the type of work the project demands, as scaling up too fast can turn an enjoyable task into torture. Potential commercial growers should also take care to research potential markets and ensure that sales and distribution will be a reasonable task before going all out. Too many beginners believe they will love growing a crop and are confident that it will sell easily, only to find out they hate the work and the demand isn't as high as they'd thought.

Some examples of common groups of people interested in forest farming are as follows:

BEGINNERS

This group consists of forest owners who have little or no farming or forestry background or those simply intrigued by forests and cultivation for forest crops. Sometimes they are upper middle-class folks who moved to the country to "get away from it all" or who have a second home in the country. The entry into forest farming often begins as a recreational venture, but it may end up becoming a secondary income stream. This group also includes people who may have owned forested land for many years but don't engage in active management. They often indicate that they just want to "do what is best for their woods," seeking advice from more experienced folks in their community.

Figure 1.19. A student in the "Practicum in Forest Farming" class at Cornell works on terrace construction. Forest farming requires a wide range of skills, including use of hand tools to shape and manage cultivation spaces, as machinery is often challenging to use in the woods.

FARMERS

The second type of prospective forest farmers is current farmers who produce cash crops (often organically) on small farms. Many small farms in the eastern United States have a woodlot that may or may not be managed for timber or other purposes. Engaging in forest farming may offer the farmer an opportunity to diversify the kinds of crops produced on the farm. The challenge is that often field crop farmers don't want to cultivate additional acreage of the forest, and many don't have the chain saw and related skills needed for forest farming.

WILDCRAFTERS

The third type of prospective forest farmer is someone who is an experienced "outdoors (wo)man," having spent a considerable amount of time in the forest hunting or collecting NTFPs from the wild. While their experience with forest ecosystems may prove to be a valuable asset, they may lack the skills to manage production-based systems that the farmer likely possesses. They may or may not have forestry skills as well.

FOREST OWNERS

Another common group are those who have owned and managed a woodlot themselves for many years, sometimes to produce their own firewood and sometimes as a way to relax on the weekend. They often posses the forestry and chain saw related skills but may not be as familiar with production-focused work, as they are usually harvesting existing trees. Still, their interest is often related to wanting to spend time in the woods, and often they are curious about other options for things they can do for fun, or sometimes sell as well.

Of course, by no means can a few stereotypes cover all the possible types of people attracted to forest farming. In addition, conservation groups, permaculture practitioners, chefs, food producers, and many others have been known to come to a workshop or inquire about agroforestry. For those interested, this book offers a "wide angle" look at the possibilities, along with a number of practical how-to strategies to get those who are eager started in a number of different projects.

An Invitation

With forest farming defined in the context of the larger practices of agroforestry, it is time to dive into the historical and ecological frameworks that will equip people of all types to successfully implement forest farms that benefit themselves and their environment. As with any good forest, the material in the coming chapters is complex and offers much to think about. A good forest takes time to grow, and readers are encouraged to pace themselves with the information presented in this book.

2 Historical Perspectives on Farming the Woods

It's important to emphasize that the practice of forest farming is nothing new. Around the world traditional and modern cultures have long valued systems that either make productive use of existing forests or grow new ones with a mixture of beneficial tree crops, shrubs, and herbaceous plants. In other words, agroforestry and forest farming are not a new concept but are in many senses the way people grew and gathered food and other materials for much of the time humans have spent on earth. In the eastern forests, for example, much of the assumption is that Native American peoples roamed the woods, mostly foraging from the bounty that primeval forests offered them. In actuality, while the native populations certainly wildcrafted and hunted for some of their needs, there is ample evidence that they also both cleared forests entirely and cultivated a mosaic of woodland areas, orchards, and forest gardens.[1]

As settlers arrived in North America in the fifteenth century and began to dominate the landscape, a new cultural context and attitude began to infiltrate the land, perpetuated largely by the notion that land could be owned, and that to own land one must "improve" the landscape, defined by Europeans largely as clearing trees off the land entirely. This approach, coupled with a general fear of the wild-forested landscape,[2] began a cycle of rapid forest decline and with it the viewpoint that the most valuable land was that which could be tilled or grazed. This further expanded as settlements grew and forests were harvested en masse for building new towns and as a key export to Europe, which had long deforested its landscape. Because this stage of development was largely human and animal powered,

Figure 2.1. Horse logging in Minnesota, circa 1940. The rate and intensity of forest harvesting has fluctuated over time and has been related to the technology available. Photograph courtesy of Department of the Interior, Bureau of Indian Affairs, Minnesota Agency

however, the rate of harvest, while impressive, did not match the thrust into the industrial age, during which machinery became the primary management tool.

Land use in the twentieth century can be correlated with the advent of both world wars, which sent development of technology and the demand for resources into hyperdrive. As discoveries of coal, oil, and gas increased, fuel wood became less and less important, to the point where today firewood is often only harvested as a by-product of timber extraction. Modern forestry now focuses almost entirely on timber production, which continues to be the primary force driving decision making in many cool temperate forests, frequently with disastrous consequences. And as with agriculture as a whole, the negative consequences of global agriculture and a growing population began to surface in relationship to forest management. Along with this came a small faction of agronomists interested in the ideas of tree cropping systems as a solution to the modern world issues of soil erosion, pesticide toxicity, and even global food shortages. While the arguments by these writers and thinkers were (and remain) compelling, the cultural paradigm shift required to adopt agroforestry and forest farming practices has kept the movement small since the ideas were first presented. This chapter provides context to the forest farming approach, which in many ways combines traditional and indigenous cultivation strategies with modern science and research on the values trees provide to the landscape. If we better understand the past and the origins of a movement, we can make better choices for the future.

Historical Forest Use in the Eastern United States

A common perception of Eastern American forests at the time European settlers arrived paints a picture of dense, thick, and seemingly endless forests as far as the eye could see. In fact, the landscape was much more complicated: a patchwork of forest, field, and everything in between.[3] This was from both the natural patterns of forest ecosystems and the management practices that people practiced. In *The Pristine Myth: The Landscape of the Americas in 1492* (1992), W. M. Denevan writes:

> The myth persists that in 1492 the America's [*sic*] were a sparsely populated wilderness, 'a world of barely perceptible human disturbance.' There is substantial evidence, however, that the Native American landscape of the early sixteenth century was a humanized landscape almost everywhere. Populations were large. Forest composition had been modified, grasslands had been created, wildlife disrupted, and erosion was severe in places. Earthworks, roads, fields, and settlements were ubiquitous.[4]

Settlers, who viewed the abundant forests as ripe for conquering, had the most devastating impact on forests, partly due to the fact that America became a massive exporter of wood products to Europe, which had long starved its forest resources. England, in particular, needed wood for its large navy and rapidly expanding trade markets, to the point where by 1770 it was bringing from its American colonies 14,000 tons of timber, 6 million feet of boards, and 5 million staves for barrels, which translated to the felling of over thirty thousand trees.[5]

Little known is the role that wood products and forestry played in the American Revolution. While people tend to think about taxation and tea as the key icons of the American independence from Britain, in fact wood arguably played a much larger role, as it was the main source of wealth and thus leverage for colonists. Britain was so concerned about its access to trees, notably the tall and stately white pine, that it declared several areas of the Americas "Royal Forest" and at one point forbade the cutting of "any white or other sort of pine fit for masts" that was larger than 24 inches in diameter. Enforcement proved difficult, and colonists came to see the felling of trees as a revolutionary act.

When the larger patterns of forest use over the past several hundred years are examined, the state of American forest use can best be described as devastating. Comparing the censuses of 1810 and 1880, it's easy to see a dramatic shift in attitude. The earlier census talked of the almost burdensome nature of the forests, which were viewed as obstructing the ability to cultivate land in traditional fashion, with the plow. By 1880 the tone had changed significantly, as the publication noted that forests in New York, Pennsylvania, Ohio, and Indiana were depleted beyond much marketable value. Another report from the same time frame claimed "the states of Ohio and Indiana . . . so recently a part of the great East American forest, have even now a greater percentage of treeless area than Austria . . . which have been settled and cultivated for upward of one thousand years."[6]

Another study, which looked specifically at land use in Tompkins County, New York, from the years 1790 to 1980 using land survey records, aerial photographs, and field work, noted that "forest cover dropped from almost 100% in 1790 to 19% by 1900 then increased to 28% by 1938 and over 50% by 1980."[7] While the percentage of forest cover has indeed increased across much of the cool temperate United States, due largely to the abandonment of farmland, this isn't to say that a recovering forest has any degree of the same value and integrity of the ancient forests, most of which are long gone.

This pattern of valuing the forest as a timber resource ripe for rapid and extensive exploitation has not significantly changed in the last few hundred years and has left the work of modern loggers to plucking the few remaining large trees scattered around. Sawmills used to accept trees only 24 inches in diameter or greater, but in the last ten years that threshold has shrunk to 12 inches, and in very recent years mills began accepting trees of 10 inches or greater. The impact is so widespread that it is now rare to see a valuable timber species in the woods dying of old age.

The perception that has pervaded each new generation since the arrival of European settlers is that forests appear resilient and can handle the type of harvesting that "takes the best and leaves the rest" (high-grading). Little effort is made on the ground to define and create limits for what a sustainable harvest looks like. Make no mistake: This is a choice, not a necessity of management. As mentioned in chapter 1, the Menominee tribe in Minnesota has been harvesting their forests sustainably for a long time.[8]

In addition to the continued practices of most modern foresters and loggers, several factors, including species loss, the introduction of new pests and diseases, and degraded soils from decades of agricultural use, are all inhibiting the pathway to our ever seeing healthy, mature forests as a common landscape feature again. At this point in time good management may arguably be the critical role of restoring even a fraction of the wealth that was once here. Leaving the forests to their own devices may be a "too little, too late" type of strategy.

EASTERN NATIVE AMERICAN AGROFORESTRY

To contrast the previous story, it is worth examining the historical land use perspectives of the peoples native to the cool temperate forests of North America.

This analysis offers some principles for the ways in which postmodern forestry can be a positive force in the world. It becomes apparent from this story, when contrasted with that of the path settlers chose to take, that the paradigm and beliefs of a culture have much to do with the way it acts on the ground.

Native Americans are believed to have inhabited the temperate regions of North America for thousands of years before settlers arrived; various sources cite upward of ten thousand to twenty thousand years.[9] Their land use ethic and practice was, in many ways, the complete opposite approach from that taken by European settlers. Key factors in this intentional management style were pulses of use and abandon that accompanied seasonal changes, the use of fire as a tool, and perspectives on ownership and property rights.

One of the most striking observations made by settlers who first encountered native peoples in the eastern United States was their disbelief that natives appeared to be so poor while the landscape appeared to be so rich. What settlers misunderstood, and ultimately learned the hard way, was that native cultures had developed an intimate relationship with the natural world that demanded they be mobile and flexible with the changing seasons.

Native American manipulation of the landscape was less permanent: a shifting mosaic of cleared forest, burned woodlands, and wildlands for hunting and fishing that were visited periodically, abandoned, then revisited. Settlers, accustomed to land claims, fences, stone houses, and domestic animals, had little understanding of this seminomadic approach to land use.

Consider that while the past cultural customs of Native American cultures are being discussed here it is not to imply that these tribes are not in existence today; too often the history books imply that indigenous peoples *were* an element from the past, rather than a piece still present in the complex cultural fabric of America today. In addition to offering a perspective that is critical as we deal with the many challenges of the modern world, these cultures have an inherent right to exist and be honored simply because they exist. Furthermore, it is a mistake to think that any culture, whether considered "native" or not, is ever perfect, or that as a modern society we have any ability to turn back the clock and live in this way. Instead, analysis of the complex historical stories help shed light on where things are at today. If anything, strategies that appeared to have worked better in the past have to then be translated into the modern context.

On the Move

A key aspect of traditional Native American land use patterns was the semipermanent nature of dwellings and villages. Populations would move with the seasons, following food sources to the banks of streams for fishing, the edges of forests during hunting season, and the expanses of fields for growing summer crops.

The *pulse* nature of arriving at an area and using it intensively for a relatively short duration before moving on meant that the landscape had time to rest and change. This practice has been called "sequential agroforestry" or "shifting cultivation," in which forested areas would be cleared and set to croplands (corn, squash, etc.) and used for five to ten years before being abandoned to grow back to forested lands over another ten- to twenty-year time frame. This resulted in a "patchwork" landscape that improved the age and species diversity of forested lands.[10]

The key point from this perspective is that the ecosystems of temperate forests appear to have benefited from intensive use followed by a fallow time, when the land could naturally regenerate. This type of thinking has more recently been confirmed as beneficial in many different agricultural systems, including rotational grazing.[11] While returning to some sort of hunter-gather lifestyle is not realistic for modern humanity, the concept of pulsing or "intervention, then rest" are concepts that can be applied to forest farming practices in a number of ways, which will be further expanded upon in the latter parts of this book. The challenge as forest farmers, then, is to be able to see and work with shifting patterns in both space and time, and to cultivate the ability to think in the longer term.

Fire as a Management Tool

Lacking metal tools and domestic animals, the main management tool native cultures employed to

Figure 2.2. This 1908 painting by Frederic Remington depicts Native Americans of the Plains using fire to ward off enemies, though fire was primarily used as the main management tool of many different tribes to manage land, so much so that it shaped ecosystems across the continent.[12] Illustration by Frederic Remington, Wikimedia Commons

manipulate the landscape was fire. While historical documents largely cite the reason for fire management to be maintaining good hunting land and for ease of travel, there is also evidence to support the use of fire to promote berry, fruit, and nut tree growth; though initially these may have been unintended consequences.[13]

Periodic, low-intensity fires (contrasted with high-temperature and high-intensity fires, such as those that devastate West Coast forests) tend to have several beneficial effects on ecosystems. They increase the rate of nutrient recycling into the forest, which improves growth for grasses, shrubs, and nonwoody plants. They also tended to thin forest canopies, which dried out forest soils and brought more light to the forest floor. Burning also destroys populations of plant diseases and pests. Nut trees also flourished under fire conditions. Oak, hickory, and chestnut trees have a relatively thick bark that can withstand the heat from low-intensity fires. They have an additional ability to resprout from the stump, and the decay resistance of oak and chestnut increases resistance to basal wounds, which are an inevitable result of burning.[14]

Nut trees were further promoted by the reduced presence of trees with thinner bark, including the more fire-susceptible birches, maples, beech, and hemlock. This opening of the canopy would likely result in

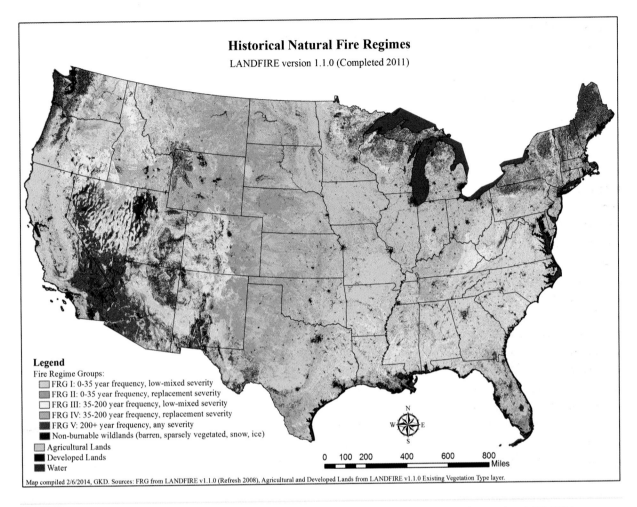

Figure 2.3. Historical fire regime types based on Kuchler's Potential Natural Vegetation types. Source data are from LANDFIRE (www .landfire.gov/fireregime.php) and created by the USDA Forest Service, Fire Modeling Institute (Greg Dillon)

improved nut production. Burning also makes available significant quantities of potassium, which is an important nutrient in the nutrition of nut trees. For the native populations this practice not only contributed to the improvement and diversity of grasses, herbs, berries, and nut trees but also improved their hunting. Forest animals important to the native diet, including bear, moose, deer, and turkey, are all heavy mast feeders when it is available. In addition, the opening of the canopy and understory meant a greater diversity of foods for foraging.

The use of small, low-intensity fires as a modern management tool needs more exploration. Modern conservation groups (as well as federal, state, and local agencies) such as the Shawangunk Ridge Biodiversity Partnership (SRBP) have already been experimenting with this practice, commonly called prescribed burning. SRBP describes its work as ". . . the intentional setting of fire under a rigid set of specifications to restore fire adapted forests and/or reduce large accumulations of flammable material on the forest floor."[15]

While fires carry a degree of risk, if done correctly they can be safe and effective at reducing future wildfires. Controlled burns are typically low to the ground and burn at a much lower temperature than the typical wildfire. Between the years of 1996 and 2001, federal government agencies conducted 31,200 prescribed burns in the United States, of which only 0.5 percent burned outside the specified boundaries.[16] That said, however, this is the work of trained professionals, not necessarily the random landowner.

In addition, with the emerging and increasingly intense pest and disease problems, an epidemic outbreak of tick populations and Lyme disease, and a prevalence of invasive species, fire may prove to be both effective and economic as a management tool. Its ability to address so many problems at once make it worth further exploration if we want to see forests improve in health and vigor.

Concepts of Ownership

In addition to the effects of aforementioned pulse management and the use of fire, another key factor that greatly influences land use is concepts of ownership and property rights. In fact, it's actually difficult to discuss this in the construct of Native American cultures, because some of the concepts didn't really exist in the way modern people might think of them. The concepts are further muddled because ownership is relative; that is, claims of ownership are only "real" insofar as they are recognized by an authority. In the case of much of the Americas, the sovereignty of native tribes was often overlooked by European governments that staked a claim to land and then granted it to settlers. This led the way to much of the horrible death and destruction of Native cultures, who often signed treaties with settlers, only to find that the settlers did not uphold them.

With native cultures the most common pattern of ownership had to do with individual goods, as people owned that which they made with their own hands. Items such as pottery, clothing, and structures were only retained insofar as they proved useful, and there was not as much accumulation because when something became useless, it was easy to get rid of it, because it was made of all biodegradable components. This approach was likely part from necessity, as tribes were often on the move. Once you settle in one place, it's easy to accumulate and store more stuff.

When looking at land rights, the concept was generally that indigenous people had a right to "use" a space for a season or two, yet they only "owned" the products of their labor, rather than the space itself. Since this sense of ownership was only for use of space, native peoples could not (and did not see the need to) ban others from trespassing or gathering other goods from a land.[17]

While there are many more layers to the story, this concept, sometimes referred to as *usufruct*, is a framework worth consideration to apply in the modern world. The word has its roots in civil law and literally means the right to use (*usus*) the fruits (*fructus*) of the property. It promotes the idea that people can have the right to certain uses of land but don't necessarily need to own the entirety of the land outright.[18]

Modern land use is an entirely new paradigm; today there is a resurgence in farming, agriculture, and land use. It is characterized by young, enthusiastic individuals who want to get started but often lack the finances and ability to purchase land and pay taxes, all while trying to start a farm. In the same breath there are many who own tracts of land but lack the ability, skills, or time to manage it properly. A usufruct system, where a landowner leases the right of a particular use to a tenant, might prove to be a critical strategy to get more people back on the land. For example, in a forest farming system, there could be usage rights doled out to a number of different individuals. Perhaps one person would have rights to the canopy, to manage the forest for long-term timber, fuel wood, and nut production. Another could retain the rights to the cultivated understory, growing mushrooms and medicinals for market. And yet another could focus on wild foraging and the cataloging of existing vegetation and wildlife species.

This type of land use system has multiple benefits. It's a real challenge to keep up with all the needs and potential yields of a landscape, and in addition, it's an expense to own the various pieces of equipment needed to harvest all the potential of a forest system. Yet in this model common costs could be shared, while individuals could focus on specializing in specific areas, thereby making the system more efficient. The natural overlap and connections between systems would mean that ultimately all the individuals would need to communicate, and collaborate, to ensure success.

Early Twentieth Century Proponents of Tree Crop Agriculture

As America reached the twentieth century, alongside the rapid and expansive growth of industrial agriculture were a small group of academics and agronomists who proposed strong arguments for the role tree crops could play in a more sustainable food system. Unfortunately, attention was really only paid to high-value commodity tropical crops such as coffee and cocoa, and little was paid to nut trees, legumes, and the potential of agroforestry in colder climates, where the basic limitation is that trees grow slowly. Regardless of how successful adaptation was, the general concepts and principles presented by these authors still hold some fundamental truths to be central to the practice of forest farming.

A large emphasis by all the individuals mentioned here is that these systems are suitable for marginal lands and thus would not compete or interfere with field cropping production. Steep slopes and land with poor soils are particularly appropriate for agroforestry systems. The other major commonality is that agroforestry is the long-term form of agriculture needed to change the ultimately destructive path of tillage agriculture.

Examining these arguments from the early twentieth century (almost a century ago) makes the "technology" of tree crops feel like an old idea that the modern culture has never been able to come to terms with. The problems and future concerns are so similar to those of today that these books suggest that modern humanity hasn't come very far—and indeed still has a long way to go—if modern agroforestry is to become a serious pattern of land use. As the effects and consequences of rising food costs, the degradation of soils, and the changing climate are felt, perhaps more will pay attention to the long-term, regenerative, and stable form that is forest agriculture.

J. Russell Smith: *Tree Crops: A Permanent Agriculture* (1929)

Sometimes when we are scanning the shelves of an institutional library, the search feels much like a treasure hunt; perhaps a long-lost text will be discovered

that will be the answer to all the world's problems. Discovering *Tree Crops* by J. Russell Smith[19] for the first time is a good example of this. In it Smith, who was born in Virginia in 1874, argues that agriculture must be "adapted to physical conditions" and that "farming should fit the land." He aptly observed in 1929 that there was a worldwide catastrophe brewing of hill agriculture, whose cycle he described as "forest — field — plow — desert." However, rather than take on conventional plow agriculture and argue for a complete overhaul of American agriculture, Smith chose instead to begin his strategy with two clever caveats:

1. Focus not on arable land but on marginal slopes and soils for tree agriculture, land that was already considered unproductive to the farmer
2. Rather than claim that foods from tree crops could replace all the needs in a human diet, focus rather on replacing the grain inputs for livestock with tree crops on forage-based systems

With this approach, Smith avoided both of the major arguments against tree crops—that "replacing" field crops with tree crops would be impossible to compare in terms of the yields and that scaling up production would be difficult because the machinery and technology was not (and still isn't) widely available. By utilizing land that could not (or should not) be in tillage agriculture, and by focusing on utilizing animals to harvest the yields, Smith provided a window for agroforestry that is still a good starting point for implementation today.

Upon Smith's death, his estate made the book available online for free[20], and it continues to be a timeless resource. While numerous species are catalogued in the text, most notable in the context of this book are the following species and the characteristics that Smith describes to lend argument to why these tree crops are so critical for successful agroforestry systems.

Stock Food Trees: Carob, Mesquite, and Honey Locust

Smith was fascinated with the potential of leguminous trees to both fix nitrogen and produce high-caloric

Figure 2.4. "Stock" trees as defined by Russell Smith, clockwise from top: Chilean mesquite (*Prosopis chilensis*), Carob (*Ceratonia siliqua*), honey locust (*Gleditsia triacanthos*). Note that all are pod-producing legumes for different climate zones. Wikipedia Commons

pods for both animal and human consumption. He details the profile of the "best of the best" for each major climate area around the world: the algaroba or kiawe (*Prosopis juliflora*) for the tropics, the mesquite for arid regions, the carob for warmer temperate zones, and the honey locust for cold temperate areas. Since this book is focused on cooler temperate zones, the honey locust will be explored a bit more in detail, but readers are encouraged to review the original text for specifics on the other species.

On all the above "stock food" species, Smith notes,

All of these bean-bearers have very ingeniously bedded their seeds in a sugary pod which is greedily eaten by many ruminants. The seed itself no beast can bite, bruise, or digest. It passes with the excreta, dropped on every square rod of pastureland and bedded down in fertilizer to help it start its new life. Nature is indeed ingenious! All of these beans and their pods are much alike in food service and in food analysis. In nutritive value, both protein and carbohydrate, they are much like wheat bran—that standard nutrient of the dairy cow. Therefore, it seems fair to call these bean trees "bran trees" because some are already used as bran substitutes and others may be made to afford a commercial substitute for bran. This gives the possibility of their being major crops of American agriculture.

For the honey locust (*Gleditsia triacanthos*) in particular, Smith notes several promising characteristics of the species, which serve as good food for thought as we think about species selection for our forest farming activities.

1. Honey locust is a good timber tree with strong and beautiful wood.
2. It grows quickly. Annual growth can be 2 feet in height and ½ inch in diameter under good conditions.[21]
3. It fixes nitrogen.
4. It has an open compound leaf structure that allows a decent amount of secondary light through to the lower layers of the forest.

5. It's productive and a regular bearer and produces large-size pods.

6. It's very easy to propagate from seed and suckers.

It's important to qualify some of these traits. For example, reports from the University of Virginia indicate that there is considerable variability in yields depending on where the trees are growing. According to the publication, "Annual yields of 180 kg (396 pounds) from individual mature trees have been reported from South Africa, New Zealand, and the US. Ten-year-old trees grown in the southern US yielded 43 kg {96 pounds} per tree. Honey Locust have produced considerably lower yields in the shorter growing seasons of the middle and northern USA."[22]

Considering the implications of this for the temperate United States, while we may not be able to get the yields that are achievable in other parts of the world, the possibility of growing highly nutritious animal feed on-site, in any quality, is appealing. While the pods can be ground and fed to cows, pigs, and even poultry, it seems that sheep are the best candidates, as they can digest the pods whole. Researchers in France found that the pods could not only meet the maintenance needs of sheep but also keep up with production growth.[23]

Another ongoing question is the ability of the tree to fix nitrogen, a process in which special nodules on the root structure of plants host bacteria that can pull nitrogen from the atmosphere and make it available in the soil to the plant. The honey locust is part of the legume family, but not all leguminous plants fix nitrogen. Originally, since scientists observed no nodules on the roots of the tree, it was assumed that the tree was not able to fix nitrogen. More recent analysis from Yale University indicates, however, that the tree is able to fix nitrogen, despite lacking the nodules.[24]

The honey locust is but one of the species that Smith talks enthusiastically about in his book. While there are parts of his work supported by data, other sections are based mostly on good storytelling. It is these parts of the book that bring to the forefront that at heart (and in his career) Smith was an "economic geologist" and not a scientist. He was, and he makes this quite clear in his writing, proposing a vision that he wanted agronomy experts to carry out. Indeed, this book proved to be pivotal in inspiring generations of agroforestry researchers and practitioners. It was the subtitle of this book (*A Permanent Agriculture*) that, in fact, inspired two scientists, Bill Mollison and David Holmgren, to coin the phrase "permaculture" (meaning permanent agriculture) in the 1970s. The permaculture movement is closely tied to agroforestry and has achieved many successes in promoting the concept of perennial and tree-crop agriculture.[25]

Reading Smith's monumental text leaves the reader with two main feelings. One is the sense of awe and potential at the abilities of multiple tree species to present so many benefits in the form of shade, food production, habitat, and erosion control for the farm. It is difficult to not become rather excited at the prospects for tree crops to "save the world," as it were.

The other feeling is one of confusion, and frustration, at the fact that Smith's concepts have been rarely adapted or even researched further. Why is this? Perhaps it was because larger governmental agencies such as the USDA and US Forest Service didn't heed his call and promote the ideas. Maybe it was simply a concept that was ahead of its time. In reality, both of these points are only partially true. The USDA and other agencies have supported and developed research into agroforestry development, though widespread awareness and adoption is still low. And while Smith's writing came at a time when industrial forms of agriculture were in the beginning stages and the full range of consequences were not yet realized, it in many ways would have been much easier to adopt the practice early in the industrialization of the food system. Today the system is so invested in crop agriculture that tree crops seem like a drastic departure, and with each crop there are many mountains to climb.

ROBERT J. HART: FOREST FARMING: A SOLUTION TO WORLD HUNGER (1976)

Another marginalized yet important figure in the advancement of forest farming was Robert J. Hart, who wrote the book *Forest Farming: Towards a Solution to Problems of World Hunger and*

Conservation in 1976. Hart also maintained a smallholding named Highwood Hill farm in Shropshire, England. His original intention was to provide a healthy place for him and his brother Lacon, who was born with severe learning disabilities. Hart started out with large annual beds, livestock, and orchards before eventually abandoning all but the perennial vegetables, herbs, and tree crops that he found tended to look after themselves with little or no intervention.

Hart's journey started, as he admits, with reading *Tree Crops*, which he comments just "made sense." Further, he saw that "the answer" had been there in front of him for some time: that "agriculture is for the plains, while silviculture [forest management] is for the hills and mountains." As Hart traveled the world he saw great evidence of this wisdom and also notably discovered what he called the great "generosity" of nature:

All my life has been a journey of discovery of the generosity of nature. I started out thinking that we had to do everything ourselves, and of course we couldn't. But then I discovered that everything will be done for us, provided only that we realize our "nothingness" and thereupon start to search for a way fitting-in with the great processes of nature, and making the best of them, for our purposes.

The book reads more like a manifesto than a technical manual, taking readers through conversations on how tree crops can revitalize rural landscapes, about the historical relationship of trees to man, and the basics of forest ecology, before getting into more detail of some practicalities of planting and maintaining forest farms. Hart's own forest garden is thought to be the first temperate forest garden in the world, one that, as noted by Dave Jake and Eric Toensmeier when they

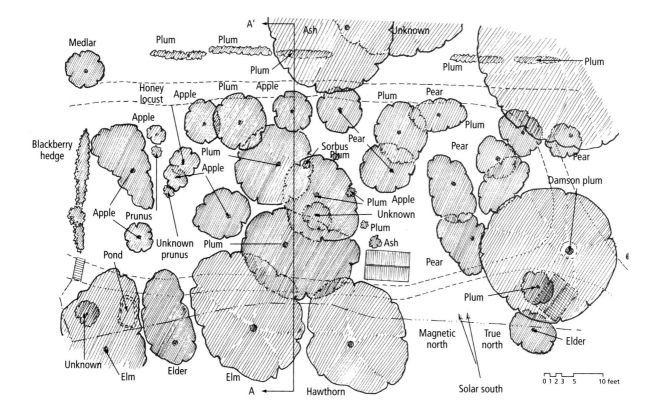

Figure 2.5. Map of the overstory species of Robert Hart's original forest garden in Shropshire, England. Illustration courtesy of Dave Jacke

visited and mapped the garden in 1997, "It felt like a forest, but it also felt like a garden. This combination truly holds a special magic."[26]

While his writing includes a much larger vision for forestry and land use, Hart's attention ultimately went to forest gardening, a form of agroforestry that, as mentioned in chapter 1, is often smaller in scale and more diverse in species composition. Hart wrote a second book, published in 1991 and simply called *Forest Gardening*, in which he discusses the concept of a "mini-forest" and many of the same concepts as the first book, albeit in garden form. This appropriation of the larger-scale agroforestry concept to the backyard is perhaps the greatest concept Hart gave to the larger movement. In his words,

> Obviously, few of us are in a position to restore the forests . . . But tens of millions of us have gardens, or access to open spaces such as industrial wastelands, where trees can be planted . . . and if full advantage can be taken of the potentialities that are available even in heavily built up areas, new "city forests" can arise.

In *Forest Farming*, Hart opened up the forest farming concept to the modern world of the 1970s, when awareness of environmental destruction from agriculture and forestry was just gaining widespread awareness. He seemed to refine this vision to his garden, which he quietly went about his way tending until his death in 2001. During this time he adopted a vegan 90 percent raw diet and harvested almost all his food in the form of fruit, nuts, and leafy green vegetables from his forest garden.

PROFESSOR LAWRENCE MACDANIELS

After graduating from Oberlin College in 1912, Lawrence MacDaniels enrolled in graduate studies at Cornell and was appointed instructor in botany two years later. He was subsequently appointed assistant professor of pomology in 1919 and four years later achieved the status of professor. At that time he taught and conducted research on basic aspects of pollination of apples, tree wounds and bracing, and anatomical aspects of pollination of flowers and fruits fallen from their plants. In 1940 Cornell appointed him head of the Department of Floriculture and Ornamental Horticulture, a position he held until his retirement in 1956, at which time he received emeritus status.

The now-mature trees at the MacDaniels Nut Grove were planted about ninety years ago by Dr. Mac, as he was known. MacDaniels also planted hundreds of nut trees in the Ithaca area, including walnuts, hickories, filberts, chestnuts, and pecans. The most numerous and greatest concentration of nut trees (more than one hundred) is on the 5-acre site along Cascadilla Creek, then known in historical records simply as "the wood-lot" and named the MacDaniels Nut Grove upon its rediscovery in 2002.

Figure 2.6. An incompatible graft of a shagbark hickory cultivar onto native pignut hickory rootstock. This was one of the signs that led to the rediscovery of the MacDaniels Nut Grove in 2002.

Figure 2.7. Students sort through the nut harvest, searching for the biggest nuts, which were saved for sampling and planting. The nut crop of 2013 was a particularly excellent one, the best witnessed since the grove was rediscovered.

The site was cleared and terraced in 1923 to facilitate planting of and provide sufficient lighting for growth of seedlings that he would later graft onto with improved selections. MacDaniels planted nut tree seedlings and grafted promising varieties onto the rootstock in the '20s and '30s. The site was largely abandoned several decades ago and gradually reverted to unmanaged secondary forest, choked with a dense undergrowth of honeysuckle. While Dr. Mac left more than 39 cubic feet of paper's worth of documentation in the Kroch Library archives, surprisingly little sheds any light on the history and work done on the MacDaniels Nut Grove. So far, no map or plan associating individual nut trees on the site with specific variety names has been uncovered. But it is clear that most of the hickories and walnuts were deliberately planted because of their obvious

graft unions and the orderly rows of trees on parts of the site. The identity of six individual trees is not in question. Remarkably, Dr. Mac's original metal identification tags, bearing the name and accession number of the cultivar, are still attached. It's likely that all of the original nut trees were similarly labeled, but the labels were engulfed by bark as the trees grew.

The species MacDaniels planted include these:

- Shagbark hickory (*Carya ovata*)
- Shellbark hickory (*Carya laciniosa*)
- Shagbark x shellbark hybrid hickory (*Carya*)
- Mockernut hickory (*Carya tomentosa*)
- Pecan (*Carya illinoiensis*)
- Black walnut (*Juglans nigra* sp.)
- Japanese walnut (*Juglans ailantifolia*)

- Filbert (*Corylus* sp.)
- Chinese chestnut (*Castanea mollissima*)

Scores of hickories and walnuts remain on the site, but there is only one filbert and only three or four struggling chestnuts (most perished in the deep freeze of 1933–1934, when temperatures dropped to −35°F; see chapter 4). In addition to the nut tree overstory, Dr. Mac planted edible mid-story fruit trees, including pawpaw (*Asimina triloba*) and persimmon (*Diospyros* sp.), some of which still survive on the site. An old notebook suggests that he planted blueberries as well, but none remain.

Although agroforestry did not exist as a recognized discipline during Dr. Mac's lifetime, his experiences with nut and other trees culminated in his advocacy of a very similar concept called "tree crops agriculture." In 1979 he and Professor Art Lieberman published a paper in *BioScience*, in which they described tree crops agriculture as:

> … the growing of perennial crops in such a way that the soils are at virtually no time exposed to erosive forces, as contrasted with mechanized orchard culture. In its broadest sense, although primarily trees are concerned, the concept includes shrubs and perennial herbaceous plants …[27]

He also expressed his philosophy well in a 1981 Extension bulletin, *Nut Growing in the Northeast:*[28]

> Planting nut trees is particularly appropriate because of the loss in recent years of the American elm to the Dutch elm disease and the decline of the white ash and hard maple in some areas. Fence rows and other areas now growing up to weeds and brush if planted to appropriate nut trees would contribute substantially to future food supply, erosion control, wildlife refuges, and in the case of black walnut, to a valuable timber resource . . . planting of nut trees for noncommercial purposes should be encouraged. . . . Whenever a shade tree is planted it might as well be a nut tree of one of the better varieties.

While MacDaniels didn't provide a great record of his work, the papers, publications, and plantings speak of a man who was deeply interested in the subject, even if the early experiments were largely low in their success rate. The multiple plantings around campus were done at a time when agriculture funding was abundant and experimentation plantings widely encouraged. And today the nut grove serves as a place for students to learn about an expanded vision of forest farming, as discussed more in later chapters of this book.

PERMACULTURE: AN EXPANDED VIEW OF TREE CROPS

Permaculture originated in the 1970s, when David Holmgren and Bill Mollison coined the word, proposing that agricultural systems needed to mimic the patterns found in nature if they were ever to become sustainable. Much of the writing and early thinking about permaculture was influenced by agroforestry books, most notably Smith's *Tree Crops: A Permanent Agriculture*. In *Permaculture One*, written by Mollison, it is noted that permaculture means "permanent agriculture."[29] (Some sources indicate the name was directly influenced by Smith's book.) Indeed, permaculture emphasizes using perennial and tree crop–based systems as the bulk of its cultivation strategies.

Zone Planning

The biggest difference between agroforestry and permaculture is that the latter goes much further than just tree crop or forest-based systems, also designing for integration of gardens, animal systems, and even the home system (heating, cooling, etc.), so that the entire system promotes sustainability, efficiency, and resilience. Permaculture systems can exist to support individual families (i.e., a homestead) or be developed for commercial or community use (i.e., farms and community gardens).

This difference can be demonstrated in the Zone Planning tool of permaculture, which designates six zones on a landscape based on each zone's type of designated land use:

Table 2.1. Comparison of Permaculture and Agroforestry Systems by Zone

Zone	Types of Permaculture Systems	Appropriate Agroforestry Practices
0—The home	Container gardens, window boxes, food preservation and storage, energy generation	Windbreaks to protect the home and outbuildings
1—Intensive gardens	Mixed vegetable gardens, berries, medicinal plants	Forest gardens, windbreaks
2—Broadacre crops	Beans, grain, garlic, potatoes, squash, poultry, nursery production	Alley cropping, windbreaks
3—Orchards and pasture	Fruit trees, nut trees, larger grazing animals	Silvopasture, windbreaks
4—Managed woodlot	Mushrooms, woodland medicinals, shade nursery production	Forest farming, silvopasture, riparian buffers
5—Wild conservation land	None—for observation and recreational use	Riparian buffers, wildcrafting

Zone 0: The home, or central structure

Zone 1: Intensive gardens of annuals and perennials that feed the household

Zone 2: Broadacre (large scale) crops and small animals (poultry)

Zone 3: Orchards and pasture for larger animals

Zone 4: Forest areas managed for timber, fuel, mushrooms, and other products

Zone 5: Areas to remain "wild" and uncultivated; conservation land

By this distinction forest farming is but one of many strategies in a whole systems landscape. In permaculture each zone definition goes into great depth as to the types of systems, technologies, and strategies employed. For instance, in zone 0 homes are designed for passive solar gain, high thermal mass, and multiple heating systems that could include radiant floor heating, masonry stoves, or rocket mass heaters. Each of these concepts is a book (or several) on its own. Permaculture approaches land use as a whole system, and it becomes complicated quickly. In the end, is it realistic to separate home heating needs (zone 0) from the management choices of the forest farm (zone 4)?

The boundaries of the zones are not rigid, nor are they uniform in their composition. Instead, the zone tool arranges various elements based on their intensity and need for human interaction in relationship to zone 0, which is where humans will spend the most time. This type of format can also shed some light on the role the various agroforestry practices can play in a whole

landscape. Table 2.1 describes some of the possible combinations of systems.

Permaculture Design

The other important aspect of permaculture is the element of design. Before any intervention is made, the site goes through a complete assessment, goal setting, and design process that helps sort out site characteristics, such as soil, water, access and circulation, microclimate, and site aesthetics. The goal with design is to match cultivation of crops and animals with the unique character of each site, along with the expressed goals of the people who will manage it. It is here that permaculture offers a direct set of tools for forest farming. The design process and its application to forest farming is discussed at length in chapter 10.

More recently, in 2005 the phrase "edible forest gardening" became popularized with the publication of Jacke and Toensmeier's *Edible Forest Gardens*, a treasure trove of well-researched information on ecology and design, of home-scale gardens that mimic forests. As mentioned in chapter 1 there are many companion concepts and strategies to forest farming presented in this highly recommended book.

In short, permaculture principles and design strategies offer good tools for organizing the thinking around forest farming systems, not only in isolation, but in connection to the other needs of life. The approach acts to strengthen the ability to succeed in developing ecological and economical ways of production. In addition, permaculture takes the concept of sustainable living not just into the forest but also

Figure 2.8. A student at the Finger Lakes Permaculture Institute works on a design project for a forested homesite during a summer permaculture design course. This 72-plus-hour class is offered around the world and provides an overview of sustainable land use.

into the garden, home, pasture, and everything in between. More principles of permaculture, and the full designing process, will be discussed in depth in chapter 10.

Cool Temperate Forest Farming: The Future

This chapter cites the multiple influences that serve to influence the scope and definition of forest farming in the postmodern world. With an extensive perspective here that encompasses a vast history of agroforestry and forest farming, from the Native American stewardship of temperate woodlands for thousands of years to the wholesale destruction beginning with the arrival of European settlers and the revival of eco-conscious concepts throughout the twentieth century, the next question naturally is, "Where to next?"

The content of the following chapters is intentionally written to build upon the knowledge base of historical land use patterns, the writing of early and late twentieth century advocates, and the research and experience of academics and practitioners who are on the ground, building upon the knowledge base. In many of the works mentioned above, the authors all came to a similar conclusion; that the pieces of information are there to justify the idea that agroforestry (and forest farming) practices are a beneficial component of agriculture. The reasons are well laid out, as are many of the "how to" strategies for implementation and management. The work left in many senses is to connect the dots and present forest farming in a whole-systems context.

It is the authors' vision for forest farming that it be seen as a practical, reasonable, and accessible approach toward a diversified management of healthy forest ecosystems. Those who want to work with it for home production as well as those looking to make a profit can benefit from integrating a greater diversity of plants and animals into their woods, all the while improving the forest, increasing yields, promoting conservation of endangered species, and increasing our ability to provide the things needed to survive and thrive in a changing climate.

For this to happen, aspiring forest farmers need the tools of forest ecology, plant biology, propagation techniques, and design and management skills, to which the rest of this book is devoted.

CASE STUDY: TROPICAL HOMEGARDENS

As we imagine what agroforestry-based approaches to land use can look like in the cool temperate zone, it is useful to consider other places in the world. In tropical regions in South and East Asia and in the Western Hemisphere, there is a practice called tropical homegardening that is in many respects quite similar to temperate forest farming. Tropical homegardens, an ancient form of agroforestry, typically involves cultivating a diverse multistrata array of NTFPs beneath an overstory of tropical trees. Robert Hart (see page 31) was inspired by tropical homegardens, which he called "forest farms" and which contributed to his envisioning temperate forest farming. Tropical homegardens typically involve cultivating NTFPs beneath an overstory of tropical trees. This overstory serves not only to create shade for the understory plants and domestic animals but also is a source of nontimber forest products such as coconut and a dizzying array of fruits, medicinals, and other useful species. The larger trees in the overstory serve as a long-term timber bank for future harvest. A key function in tropical ecosystems is the provision of shade; indeed, the rapidly growing and thick density of tropical vegetation serves to modify temperature and humidity and is a critical pattern in the success of plants in the system.

A homegarden is a traditional family farming system that may have been passed down in some families for generations. Indeed, some forest gardens have been around for centuries, attesting to their sustainability. For example, in the 1500s the Spanish Catholic Bishop Diego de Landa described intensively managed, multistrata homegardening in the north-west corner of the Yucatán Peninsula, which provided goods for trade, sale, and personal as well as family consumption, for many centuries.[30]

P. K. Nair, who has studied homegarden systems in various parts of the world, has described homegardens as "glorious examples of species diversity in cultivated and managed plant communities," in contrast to single-species stands of crops, which he calls "biological deserts." For example, a homegarden of $\frac{1}{10}$ acre in Central America may contain as many as twenty-five species of trees and food crops. This is something for temperate forest gardens to aspire to, and while extreme, it may not be unattainable.

On the other hand, species diversity in some tropical homegardens goes far beyond what could be expected for a temperate forest farming site. For example, on the island of Java, Indonesia, where homegardens cover about 20 percent of the arable land, no fewer than 191 species were observed at a single homegarden. These included 37 species of fruit trees, 11 species of food-producing plants, 12 medicinal species, 21 herb species, 18 vegetable species, 45 species of decorative plants, and 47 species of plants used for fuel wood and for construction — all growing at a single site. This is an extreme example, but it is clear that plant biodiversity can be considerable in traditional tropical homegardens.

Homegardens are found in many tropical regions of the world, including Asian Pacific islands, India, Africa, and Central America. Take the Chagga people, who live at the base of Mt. Kilimanjaro, Tanzania, in Africa, for example. In one of the early

Table 2.2. Species Diversity in Tropical Homegardens at Various Locations[31]

Location	Number of Species	
	Max. No. Species in Any One Garden	Total for Geographic Location
Cianjur, West Java	72	—
Northeastern Thailand	60	230
Cilangkap, Java	58	—
Santa Rosa, Peruvian Amazon	74	168
Kerala, India (woody sp.)	25	127
Belize	30	164

modern reports of tropical homegardens, Fernandez reported that the average-size homegarden was 0.68 hectares.[32] Banana is one of the most important crops, yielding an average of 275 bunches of bananas on that acreage. Such an impressive yield of this staple crop was not due to the introduction of improved varieties, improved methods of disease or pest control, or the use of mineral fertilizers out of a bag but rather a stable sustainable agro-ecological system that has persisted virtually unchanged over at least several hundreds of years. In that respect the forest gardens of the Chagga people would qualify as "primitive." Certainly, to the casual observer they would appear random, even chaotic—crops not planted in rows or grouped together by species. The irony (or enigma) of advances in modern agriculture, unlike traditional homegardens, is that modern improvements that resulted in increased productivity have been the shift from intercropping to monocultures and the consequent loss of biodiversity.

A somewhat different example of tropical homegardens occurs on the island of Sri Lanka, off the southern tip of India, where approximately 12 percent of the total land is under homegarden agrofor-

estry. Pushpakumara estimates that there are about 1.4 million homegardens in this country of 21 million people.[33] In the wetter area, where rice is the staple crop, the overstory of the homegardens is dominated by coconut palm. Other overstory trees include areca nut (betel nut), jackfruit, and mango. Beneath the overstory are multiple vegetation layers that include smaller fruit and spice trees, many of which are sold along the roadside. These include black pepper and vanilla vines, banana, durian, jackfruit, rambutan, cinnamon, clove, cardamom, tea, coffee, mango, sour sop, ginger, castor bean, cassava, allspice, rubber, black pepper and passion fruit vines, papaya, cocoa, and cashew cloves. Gliricidia, a small woody legume, is also grown in the midstory as green manure or fodder for goats and other animals. Understory crops include sweet potato, taro, pineapple, and passion fruit.

Tropical homegardens occur both in moist and dry zones, where coconut typically is the most important canopy tree, followed by areca nut (betel nut), jackfruit, and mango. Pushpakumara estimates there are 1.42 million homegardens in this country of 20.3 million people, occupying 12 percent of total land area.

— Ken

Figure 2.9. A typical tropical homegarden in Puerto Rico, where bananas, coffee, two palm species, and at least twenty others are grown, many of which are situated close to the house and others on the steep hillsides nearby.

Figure 2.10. A typical tropical homegarden in Sri Lanka. The layered character of this garden is readily apparent, with coconut (trunk on far right), mango, and jackfruit in the making up of the high canopy. Black pepper vine climbs up the trunk of the jackfruit. Its watermelon-size fruit is near the top of the picture. Midstory trees, coffee, and banana can be seen down the path, and taro and pineapple are toward the front, in the far left and right, respectively. Many other species are growing here as well.

3 Mimicking the Eastern Forest in a Changing Climate

Looking at the forested landscape anywhere in the world, only one thing is common: change. While some of this change is due to human activity—a tendency for societies to devalue woodlands, choosing instead to cut them down, sell off the timber, and grow annual cash crops—there is also the slower and often more subtle change of species composition that comes at the hands of wind, water, natural disaster, and the animals and birds. These smaller and slower changes tend to be a positive thing in the mid- to long-term outlook. As the situation changes, so does the species composition, soil dynamics, and other characteristics that define ecosystems at a point in time. And while we can make generalizations about forest types based on the mix of factors, in the end each and every woodland is a unique place and should be designed with this in mind.

As the term would suggest, being a good forest farmer means knowing the ecology and appropriate management strategies of the forest, along with the production and marketing skills of farmers. Another way of saying this is that a balance must be found between the goals of long-term management and the relatively short-term harvest of yields for hobby or commercial use. And as mentioned previously, if our legacy is to leave a forest in our footsteps, then it is necessary to begin by understanding the multiple factors in forest growth and health.

In the end it is also about how people relate to the forest. For all the stories of clear cutting and slash-and-burn agriculture (both of which, on a small and appropriate scale, can be good things), which industri-

alize the forest as a commodity, there are an equal and perhaps more compelling set of stories that tell of a different human relationship to forests—a relationship in which humans gain sustenance and livelihood, while the forest also benefits. This is our fundamental goal: to understand the forest ecosystem and walk hand in hand as we marry our agricultural needs with the support of tree-based systems.

This chapter discusses the elements that help define a particular forest, and it looks at some key principles of forest ecology that aid our understanding as forest farmers. Further, a look at some specific forest types in temperate climates offers templates for mimicking as future forest farms are designed. The chapter ends with forest farming strategies in response to the reality that significant climate change will be the norm as work is done to develop forest farms over the next several generations.

Forest Ecology 101

One way to consider the ecology of the forest is to draw an analogy to the narrative of a story. Like a good plotline, the forest contains characters (the species), which are constrained in the limits of a setting (the environmental factors). These parts are then taken through the plot, which is a series of conflicts and resolutions, analogous to the concept of forest succession over time, which is the progression the whole system follows while experiencing many disturbances small and large. Within each of these parts, and the big-picture pattern, there are many variables. This is why

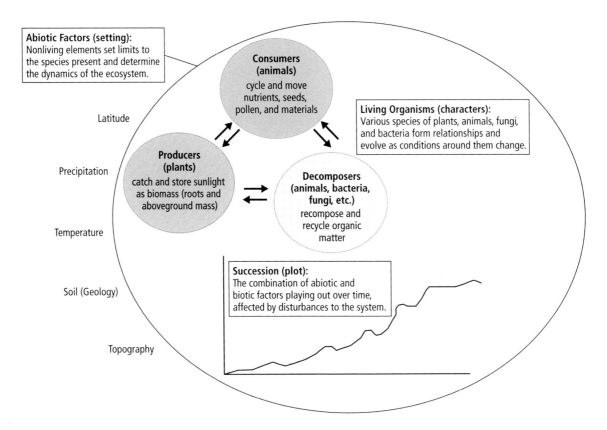

Figure 3.1. Diagram of the entire "big picture" of ecosystems, with the analogy of being similar to a good storyline. The entire system is limited by abiotic (nonliving) factors, which, like the setting of a story, provide the context. Biotic (living) organisms are either producers (plants), consumers (animals), or decomposers (animals, bacteria, fungi, etc.), each of which has a unique role to play, like characters in a story. The combination of these factors, along with unexpected twists and turns along the way (disturbance), is called succession.

each forest is a truly unique combination of factors that are specific to one place and point in time. Recognizing that all forests are unique, however, doesn't mean that some generalized patterns and management strategies cannot be utilized.

The Setting: Abiotic Factors

The word "abiotic" means nonliving, and abiotic factors constitute the elements that a forest farmer should first recognize when assessing his or her forests. Since nonliving elements are not easy to change, these parts are often referred to in ecology as limiting factors—defined as elements that control a process.[1] By identifying the limiting factors of a particular place, a forest farmer can design to work with those realities and in the process save time, energy, and resources.

The major abiotic factors affecting the forest farm are latitude, landform, precipitation, temperature, and soil. The combination of these effects in a forest farm site determines to a great deal the species and strategies that are employed. More examples of assessing and designing for these elements will be covered in more depth in chapter 10. Here, abiotic factors are arranged in order from those that limit the system most to those that are less (though still significantly) influential.

LATITUDE

The amount of sunlight available to any place on earth is directly related to its latitude. The farther north the latitude, the more dynamic the amount of available sunlight throughout the seasons. The dynamic nature of the available light has to do with the fact that over

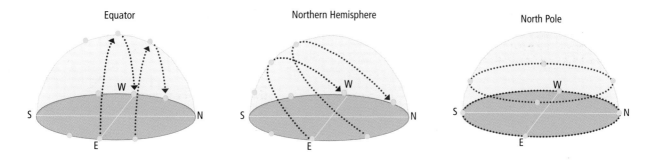

Figure 3.2. Patterns of the sun throughout the seasons; the winter solstice (December 21) is green, the equinoxes (March 21 and September 21) are blue, and the summer solstice (June 21) is red. The differences between the three become more extreme as latitude moves north or south from the equator.

the course of the seasons, the sun will change both in the total amount of daylight and in the vertical angle of the sun in the sky. For example, in central New York State the shortest day is December 21 with 9:03 hours of daylight; the longest day is June 20 with 15:17 hours of daylight. On December 21 the sun peaks at 24 degrees from the horizon, whereas on June 21 the sun is 71 degrees from the horizon. The equinoxes, occurring on March 21 and September 21, are the middle point, when there are around 12 hours of sunlight and the sun is at 44 degrees.

The dynamic change of the sun over the seasons in a place like central New York State is partially what defines cool temperate climate zones. Since forest productivity is determined by trees and other photosynthesizing plants, light affects much of the character of a forest. The change in daylight and corresponding temperatures may mean that fruits, nuts, and seeds will not be able to ripen in some more northern climates. Latitude can also relate to landform (see below), in that a south-facing slope, for example, becomes more significant as a microclimate the farther north a forest farm is. In general, a wider range of tree and other crops can be grown the closer to the equator the site is.

LANDFORM

The various elements of landform, including elevation (in relation to sea level), slope (how steep a hill is), and aspect (the direction the slope faces) are all extremely critical in understanding forest ecology, especially in

temperate systems. For instance, sugar maple (*Acer saccharum*) is often found on steeper north-facing slopes, while oaks (*Quercus* spp.) prefer the warmer south- and southwest-facing hillsides, versus black walnut (*Juglans nigra*), found on warmer, wetter bottomlands as a general rule. Of course, the steepness of a slope also has a larger determination on how easy the site is to use and access. Steeper slopes also have more challenging issues with water, and ultimately, the patterns of water are the patterns of landform. Finally, elevation has effects on other abiotic aspects of the site, including the flux in temperatures from day to night and the potential exposure to wind.

PRECIPITATION

The common adage "water is life" couldn't be more true. Whether sites have a relatively even distribution of rainfall or pulses of wet and dry, the amount and persistence of available water has significant implications for the forest ecosystem. Precipitation variables can be broken down into two parts; the overall precipitation a site receives, as well as the distribution of that precipitation over the course of the year. In some cases the proportion of precipitation that falls as rain versus ice and snow is also an important consideration. For the northeastern United States most sites are lucky to have a rather even distribution of precipitation throughout the seasons, which makes agriculture a much easier proposition. On the edges of the cool temperate map (see figure 1.18 in chapter 1), distribution of precipitation may be more variable. Of course, all bets

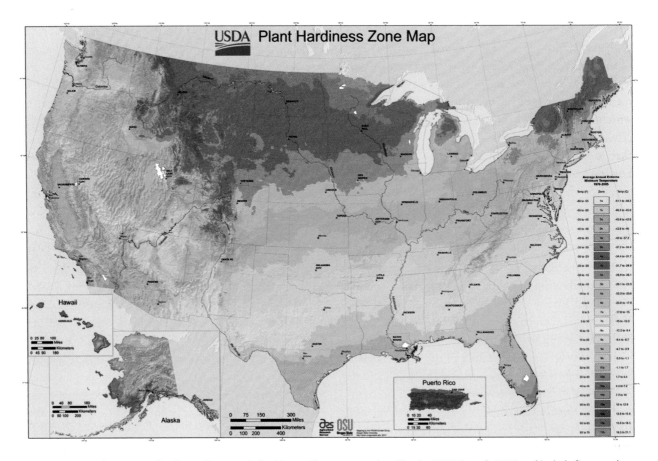

Figure 3.3. Hardiness zones for the northeastern United States. These were updated by the USDA in early 2012 and include finer resolution of winter low temperature zones. Hardiness zones are useful in describing extreme low temperatures for a place but do not tell the whole story of the wide range of abiotic factors that contribute to effects of temperature, including extreme high temps and microclimate effects. The duration that temperatures remain in a high or low range also has a great effect. Illustration courtesy of USDA

are off with the onset of climate change, as both overall precipitation and the seasonal distribution is predicted to change, for some areas quite drastically (see The Big Change: Climate Chaos, page 61).

TEMPERATURE

In its most basic sense temperature is the result of the combination of latitude, landform, and weather. The extremes of temperature are what can be most challenging for planning purposes, especially on the low end. The concept of "hardiness zones" (see figure 3.3) defines geographic areas based on the lowest possible temperature that could occur. This is a useful system, then, to classify plants as hardy to a particular zone, at least in terms of survivability. Just because a plant

won't die in a particular place doesn't mean it will thrive there. Care should still be taken to consider the combination of needs when sighting elements in a forest farm.

Of course, extreme high temperatures will also have effects on all the other living elements as well. Combinations of low precipitation and high temperatures can be severely crippling, if not deadly, to many crops, even within a moderated microclimate such as a forest. Depending on the species, too, temperature has more minute effects. Take mushrooms, for example, which upon observation appear to be extremely sensitive to changes in temperature and will wait days and even weeks, not growing until the temperatures are just right.

SOIL

Only part of the soil can be considered abiotic, as much of the upper layers of soil (humus) are a living part of it. However, the structure and texture of soil is fundamentally determined by its parent material (bedrock), which forms the physical parts of soil: sand, silt, and clay. The parent material also determines the nutrients that may be abundant (and deficient) in a given soil. For instance, limestone in soils as a parent material tends to offer high rates of calcium to plants growing in it. Since sugar maple trees are calcium accumulators, this may be one reason (though likely not the only) that the species thrives on limestone soils.

So in one sense soil is fixed, as no amount of effort could change the layers of bedrock beneath or the resulting structure. Heavy clay soils will largely remain as such, as will sandy soils. What can be done on the part of the forest farmer is to stimulate the living layer of soil, mostly through incorporating organic matter. Compared to the other abiotic factors above, soil has the most options in terms of management.

The above mentioned are just some of the potential abiotic factors to consider in a forest ecology. Of course, in practice none of these elements exists alone, and the importance is in recognizing the relationship of one to another. At the MacDaniels Nut Grove we can characterize our site in terms of the following abiotic factors:

- The site is located at around 41 degrees latitude.
- The site is in hardiness zone 5a; the extreme temperatures can go as low as −10°F
- Annual precipitation is around 35 inches and is evenly distributed over the seasons.
- Steep slopes fill most of the site and include aspects of south and southwest.
- The lower portion of the site is a seasonal creek where cold air readily settles.

The combination of these bits of information makes for a site that is relatively warm and dry, with a high erosive effect and low fertility in the soil. The lower portion has a slight frost pocket where sensitive plants could be vulnerable, especially in the springtime. Matching these qualities suggests that the site may be favorable for nut trees (which like warm, dry slopes) and pawpaw, while it would more likely be unfavorable for ginseng, ramps, and sugar maple (which like wet, cooler microclimates).

Taking the time to research and learn the unique combination of abiotic factors for a particular forest allows a greater understanding of the more dynamic cast of characters that make up a story of the forest: the living plants, animals, fungi, and bacteria that evolved to adapt to the particulars of place. Recognizing these factors also preemptively avoids the headaches that come with selecting species that don't fit the character of a site, then dealing with those consequences. This aspect of ecology gets at the concept of permaculture mentioned in the previous chapter; instead of the forest farmer's deciding on a crop, then molding the site to fit its needs, there is far less energy used in matching the right crop to the right site.

The Characters: Producers, Consumers, Decomposers

Within the living components of a forest ecosystem, groups exist that help define the roles and relationships of different organisms to the whole: *Producers* are plants, which photosynthesize to convert sunlight into biologically available energy; *consumers* (animals) eat plants or other consumers; *decomposers* break down the waste debris and make it available again as matter, nutrients, minerals, and so forth.

More important than simply identifying these elements is understanding the unique role that each group, and even members within each group, performs. By offering a "job title" for each category, we can better appreciate the complex and critical roles that come with each group.

PRODUCERS: CATCH AND STORE ENERGY

The amazing and unique quality of plants lies in the distinction that they are able to create energy from sunlight, then store and distribute that energy to others. This process happens as photosynthesis, where carbon dioxide (CO_2) and water are transformed into

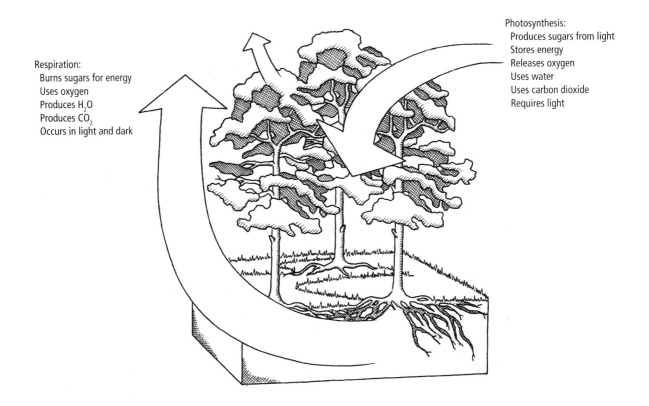

Respiration:
  Burns sugars for energy
  Uses oxygen
  Produces H_2O
  Produces CO_2
  Occurs in light and dark

Photosynthesis:
  Produces sugars from light
  Stores energy
  Releases oxygen
  Uses water
  Uses carbon dioxide
  Requires light

Figure 3.4. Photosynthesis rates depend on many things, including the temperature and weather dynamics on a site and available nitrogen in the plant mass. As plants open stomata to draw in carbon dioxide, they inevitably release water into the air, which pulls water up from the base of the tree, bringing in more water through the root system. Illustration by Travis Bettencourt

glucose (sugars), with by-products of water and oxygen emitted from the plant.

Producers are also incredible at capturing, storing, and cycling water out of necessity. Compared to animals, plants need a much larger amount of water, but why? The answer is that plants not only use water to transport nutrients throughout their tissues but also in the exchange of gases necessary for photosynthesis. Air containing carbon dioxide enters the plant through stomata, tiny holes that open and close depending on light, temperatures, and humidity. The release of water through the stomata brings more water up through the plant through the natural cohesive property of water, which "pulls" water up, seemingly defying gravity. This loss of water is a necessary "cost" of photosynthesis, which in the big picture means that plants, and trees especially, are constantly cycling water, which moderates the temperature and humidity in a forest.

Considering that all plants undergo this fundamental process, starting with the same inputs (sunlight), it's rather amazing to consider the incredible diversity of shoots, leaves, root structures, flowers, seeds, shapes, and sizes that plants come in. This diversity is the result of evolution, in which various plants engage in a "choose your own adventure" experiment of how to go about growth and reproduction. And this transformation, in the end, is what makes all other life possible on earth.

The net collection of all the roots, vegetation, seeds, fruits, and so on is called biomass, and the ability of an ecosystem to produce a certain amount of biomass is called "net primary productivity."[2] Compared with more conventional forms of agriculture, temperate forests produce almost twice as much biomass in a given time frame, for two important reasons. The first is that forest systems are based on perennial plants, which as a whole tend to leaf out earlier in the spring and stay in leaf cover later into the fall, maximizing the season's gift of sunlight and banking as much as possible until dormancy arrives. The second reason is surface area. The amount of plants stacked in the forest means that very little sunlight is wasted. Canopy trees catch sunlight, but in most cases allow some amount of diffuse light into understory and ground vegetation. Compared to the single vertical layer of a cornfield, the productivity is quite astounding.

Forest farmers can leverage plants in their systems by making the best use of space in their designs. By introducing species and managing relationships, a balance can be struck between productivity and yields for home use or commercial sales. In the end much of the management becomes making choices about light characteristics in the forest. The practical side of this will be further discussed at the beginning of chapter four, since food crops need extra attention when it comes to light quality and quantity.

CONSUMERS: CYCLE MATERIALS AND NUTRIENTS

This group, otherwise known as animals, are as vast and varied as the plant producers—the millions of species of birds, insects, mammals, and amphibians one can encounter in the temperate forest is amazing; yet as people grow and cultivate forest farms these elements might be regarded as an afterthought. After all, what role do forest creatures offer other than as pests? This is, unfortunately, the general perspective of much of the agricultural community, that animals are either a nuisance if wild or fit for a single-product role if domesticated. One of the truly sad realities of humanity is how much we have missed the potential benefits, and perhaps necessities, of engaging positively with animals in agricultural systems.

What unique roles do animals play? One word: movement. Animals are the great distributors of seeds, fertility, and materials in forests. And in the case of domesticated species they can complete tasks that forest farmers consider work with far more ease and skill. The patterns of animals as they feed, reproduce, migrate, run from predators, and defecate are sometimes straightforward but often rather eloquent.

One of the most important groups of animals that should be attracted into forest farms are insects, specifically those that are predators and parasites for the pests that might show up in cropping systems (beneficial insects) and those that pollinate, including bees, wasps, yellow jackets, and ants. For these species to show up, suitable homes and food sources are important to have. Beneficials are easy to attract by planting members of the sunflower family (Asteraceae), onion family (Amaryllidaceae) and the mint family (Lamiaceae).[3] Inevitably some of these plants will grow best along forest edges and in gaps created through management.

Pollinators love the sheltered yet partially sun-exposed environment offered by hedgerows; "edge" environments, in other words. Of course, honeybees can provide pollination and another yield for the forest farm; several tree species provide excellent forages, including American basswood, both honey and black locust, and American redbud. In addition to honeybees, which are not a native species but imports from Europe, native pollinators such as bumblebees, sweat bees, ground bees, wasps, and yellow jackets are also important,[4] as each species pollinates different flowering species, as well as offering a buffer for crops traditionally pollinated by honeybees.[5] Of course, there is a place and a time for removing these species if their choice of home sites becomes problematic, but in most cases simply avoiding nests will allow for a peaceful and productive coexistence.

In the forest farm there are many reasons to welcome and encourage animals—to a point, of course. This is in fact critical because unlike more conventional farming methods, forest farming is a practice that seeks to integrate with the normal functionality of a forest, even if this means some negotiation with certain species during the shorter term.

Domestic animals, such as turkeys, chickens, ducks, goats, and even pigs, may possibly have a place in the mix, too. Many variables are at play, and extreme care should be exercised when considering introducing livestock

SUPPORTING WILDLIFE IN THE WOODS

An important strategy in forest management is to leave plenty of standing dead trees (also called snags) in the forest for habitat purposes. When trees are taken down, it's also important to leave substantial material on the forest floor (called coarse woody debris in scientific terms), as this is food and shelter for wildlife of all sizes.

Figure 3.5. This black cherry is slowly dying, but along that route it will serve as a home for many animals. Evidence near the base indicated that ants burrowed into the heartwood long ago. The opening in the photo was likely begun with a boring insect, then opened wider by woodpeckers searching for food. Once large enough the opening may become a home or storage location for rodents. Photograph courtesy of Jen Gabriel

Figure 3.6. Piles of debris and brush are useful as "habitat piles" for their creation of homes and hiding places for a range of animals, including squirrels, chipmunks, snakes, salamanders, and insects. Forest farmers can make these from treetops after felling and processing trees for other purposes (see also Hugelkulture at the end of this chapter, page 67). Photograph courtesy of Jen Gabriel

to the woods, as damage can be rapid and irreversible. Chapter 9 discusses this in more depth and more specificity regarding the considerations for various species.

Of course, all this isn't to imply that pest animals are not going to show up in the forest farm. As mentioned in the anecdote Putting It All Together on page 50, rodents enjoy mushrooms and their mycelium, which could in some cases cause a problem. Growing an acre of ginseng may cause a pest outbreak. Ultimately forest farmers must approach the relationship with animals cautiously and consider the pros and cons of their participation in the forest. Too much of anything usually becomes problematic for the overall health of the system.

Figure 3.7. Mycorrhizal fungi (white threadlike parts) colonized on the roots of eastern white pine (*Pinus strobus*). The symbiotic relationship greatly expands the root system of the tree, allowing for greater absorption of water and nutrients. The fungi benefit from sugars that the tree synthesizes through photosynthesis. Photograph courtesy of Dr. David Johnson, University of Aberdeen

DECOMPOSERS: THE RECOMPOSERS

This group of animals, bacteria, and fungi have one "simple" job: take complex compounds and break them down to essential nutrients, which are then made available to both the producers and the consumers. Without these organisms the forest would be buried in waste. Perhaps the most enchanting processes and relationships exist in this group, as these creatures have evolved rather creatively in their pursuit of decomposition.

Fungi, including yeasts, molds, and mushrooms, often take an initial crack at the debris. White rot fungi (including shiitake and other cultivated fungi discussed within this book) are unique in that they are the only microorganisms able to break the tight bonds of lignin in wood, making the food then available to all the other organisms in the soil. Fungi also have the unique ability to colonize vast areas throughout the soil and unlike bacteria don't need a lens of water in the soil to survive. Fungi have a wide range of roles in the forest. Some are decomposers (saprophytes) that break down organic materials. Mycorrhizal (meaning fungus-root) species form symbiotic relationships with trees and plants,

Figure 3.8. The bacteria *Frankia alni* forms large masses on the roots of common alder (*Alnus glutinosa*). Alder is one of the better nitrogen-fixing trees for temperate climates. Photograph by Cwmhiraeth, Wikipedia Commons

wrapping their vegetative hyphae around the surface of the root and between its cortical cells (ectomycorrhizae) or penetrating cortical cell walls (arbuscular mycorrhizae), making it hard to distinguish where one organism ends and the other begins. The fungi trade nutrients and water for sugars from the plants, and as a result both the plants and the fungi benefit.[6]

Bacteria, the earliest form of life on earth, are important decomposers of soil, second only to fungi. They can grow and reproduce in massive quantities and are chemists of the soil, key players in converting nutrients into useful forms for plants. Plant-associated bacteria live around the root zones of plants feeding on root exudates. Their presence is key to nutrient retention in the soil and in the case of nitrogen-fixing bacteria, bringing this atmospheric nitrogen into the system through fixation from the atmosphere, where nitrogen is abundant.

Other members of the soil community include archaea, protozoa, nematodes, earthworms, and larger reptiles, mammals, and birds. This community of creatures orchestrates a back and forth that takes dead plant and animal material and turns a waste into a resource, recycling nutrients and matter over and over again. In fact, for the most part plants are heavily reliant on the organisms for access to limited nutrients.[7]

In the process of digesting organic matter of all shapes and sizes, what the decomposers do is create soil: a rich, healthy, living medium that the plants and ultimately all creatures depend on for survival. In the end any natural system is only as productive as the health it contains in the soil. With this in mind it may not be surprising to learn that the vast amount of biomass (over two-thirds that is produced in a forest ecosystem) goes directly to decomposers.[8] The implication of this is that to produce good soil a lot of biomass must be produced solely for decomposition. Forest farmers should take note of this and consider how to meet this end, whether by cover cropping, mulching, composting, or other techniques.

Living Organisms: The Big Picture

Considering all these factors, the forest can be a rather dynamic place to farm (and live). The role of the forest farmer is to recognize who is in the forest and what implications may occur when other living organisms are introduced. In this way forest farming is a form of orchestration for the farmer, as the living forms are really the ones doing the work. Observation, interaction, and feedback before making decisions are the real skills of good farming. The plants, animals, fungi, bacteria, and everything else will do the work, if only they are supported to do so.

Putting It All Together: An Example from the Pacific Northwest

As compelling as it is to ponder each group, a full appreciation of the complexity and coevolution of the relationships between living organisms is really the key to understanding ecology. As much as science has offered in our understanding of this, there are relatively few well-documented instances of the incredible symbiosis that brings together the abiotic factors, producers, consumers, and decomposers in an incredible theater of activity that is unique in each time and place. One example, well documented in the temperate Pacific Northwest region of the United States by scientist Chris Maser,[9] offers a glimpse at this impressive combination of forces.

The story begins in coastal forest dominated by the Douglas fir. The climate is temperate, with considerable precipitation. Living high in the tops of these trees, which can sometimes reach hundreds of feet, is a most curious creature, the northern flying squirrel. This rodent, and many other squirrels, mice, chipmunks, and so on, are often seen as useless to the overall forest and are generally called a pest species. Yet those who make this distinction are missing the subtle and critical functions these small animals play.

Little known is that the favorite food of the flying squirrel, and of many of the smaller forest-dwelling rodents, is mushrooms, and sometimes the mycelia (the actual body of the organism of the fungus). In this case truffles and false truffles are the delicacy of choice for the squirrels. Saying "truffle" is much like saying "apple"—it describes a larger group of mushrooms that form their fruiting bodies underground. Thus the squirrel must sniff out and dig to get at the highly nutritious foods, which offer a high-protein, nutrient-rich food source.

Figure 3.9. A depiction of the complex web of relationships in the forest. This example highlights relationships between producers (Douglas fir), consumers (flying squirrel and owl), and decomposers (fungi, bacteria, and more!). If one of these relationships is fragmented or if one of the "characters" is removed from the system, everything changes. Illustration by Carl Whittaker

The fascinating part is that as the squirrel consumes the mushrooms it also inevitably picks up bacteria and minerals through contact with the soil. The result is a "pill of symbiosis"; in other words, the feces is a package of nitrogen-fixing bacteria, the spores of the mushrooms, and the complete mix of nutrients to support their establishment of these microorganisms in the forest. These organisms, fungal and bacterial, have evolved to survive their trek through the digestive tract of the squirrel.

It is remarkable that, through the process of feeding and defecating, the squirrels are in effect inoculating the forest with these essential, life-supporting organisms. The fungal masses offer many benefits, well described by Maser:

> For its part, the fungal symbiont mediates the plant's uptake of nitrogen, phosphorus, other minerals, and water and translocates them into the host. In addition, the mycorrhizal association promotes the development of fine roots; produces

antibiotics, hormones, and vitamins useful to the host plant; protects the plant's roots from pathogens and environmental extremes; moderates the effects of heavy metal toxins; and promotes and maintains soil structure and the forest food web . . . Further, nitrogen-fixing bacteria occur on and in the ectomycorrhizal mushroom's mycelium, where they convert atmospheric nitrogen into a form that is usable by both fungus and tree.[10]

To close the web, the final character is the primary predator of the northern flying squirrel, the famous northern spotted owl, an icon for activism around deforestation in the northwestern states that is considered an indicator species of healthy, old-growth forests. The owl exists because there are ample food sources, such as the flying squirrel. The owls, of course, keep squirrel populations in check, which means the mushrooms and bacteria are not overharvested.

LESSONS FROM NATURE

Taking a step back to view the big picture, it can be seen that in ecosystems the setting (abiotic factors) sets limits on how the characters (producers/consumers/decomposers) facilitate a series of transitions facilitated by the interaction between each of its organisms.

This example is one of the more documented ones. Case studies and evaluations of these complex interspecies connections are rare, partly because it's challenging on the part of the scientist to document the relationship thoroughly. Thankfully, though, this type of work has been done, as a number of lessons can be gleaned from the observations of this interaction.

First, it encourages valuing the role of rodents in a new light. Mike DeMunn, a local forester in the Finger Lakes Region, has said, "The squirrels are the tree planters." He leaves trees standing that other foresters would often cut, such as the large hardwoods that lose a branch and develop a large cavity, known to foresters as a "den tree." Squirrels in all temperate forests bury caches of nuts at a much larger rate than they consume them, partially because most nut trees fruit in a cycle of "mast years" (see Masting of Nuts in chapter 4, pg 102): one year the tree puts out hundreds of nuts, then might only produce a limited few for the next several before again producing a mast—likely an evolutionary development to limit both the energy expenditure for the tree and the overpopulating of a forest with nut-loving mammals. Even if the reason a species exists isn't readily clear, this is a sign to take a step back and consider.

Another take-home from this example is that the bacterial and fungal kingdoms are the web that connect all things; they are critical to the restoration and sustaining of forest ecosystems. The adage "build it, and they will come" is key to supporting soil health. Further, it's quite clear that all the organisms in this system have come to not only connect to one another but actually depend on the others for their continued existence. Remove any one element of the picture—the trees, the squirrel, the fungus, the owl—and the entire community suffers.

While forest farmers might not be able to develop systems to such a high degree of symbiosis, this example provides inspiration for the types of ecological relationships to strive for and highlights the fact that forest farmers are not just ginseng or mushroom growers but stewards of an entire networked ecosystem.

BRINGING ECOSYSTEM STRUCTURE TO FOREST FARMING

This type of ecosystem framing can be applied to cultivated systems, though they are not likely to have the same level of symbiosis and interdependence, at least not at the outset of system development. It's important to remain humble in claiming to design ecosystems and relationships and to recognize that all evolutions in nature are the result of many decades at the very least and more likely thousands of years of work on the part of the organisms. The best that can be done is to set up the opportunity for organisms to interact and adjust the system as more is learned from it.

At author Steve Gabriel's Wellspring Forest Farm, for example, there has been a focused effort to look

Figure 3.10. At Wellspring Forest Farm, ducks have been used experimentally to help control slug damage to mushroom production. The key is to catch the slugs long before they get to the mushrooms, for then it is too late. Photograph courtesy of Jen Gabriel

at the relationship among the forest (where shiitake mushrooms are raised), ducks that inhabit the forest floor, and the slugs that feed on, among other things, the mushrooms being grown. This can be considered a four-way relationship that has been intentionally cultivated to work together (or, in the case of the slugs, to reduce or eliminate the population).

The woodlot is a 1-acre forest almost entirely composed of sugar maples. About one hundred of these trees are tapped each winter for maple syrup production, which averages around 20 gallons of finished syrup when all is said and done. Necessary to the continued health of the sugar maples is periodic thinning of diseased, crowded, and structurally unsound trees. This thinning process produces a decent yield of logs in the range of 4 to 10 inches in diameter, which can be inoculated with shiitake mushrooms as another yield. There is an inherent mutual interdependence in this relationship from a management perspective, since the fact that mushrooms are a high-value cash crop means the farmer has an incentive to thin the woods, which ultimately makes the forest a healthier system.

Any mushroom grower has had plenty of interface with the slug world; it's sort of a given if you are going to grow mushrooms in the woods. Taking a nod from permaculture cofounder Bill Mollison's quote, "You don't have a slug problem, but a duck deficiency," the farm brought ducks into the mushroom yard starting in 2010.

Ducks like it cool and moist, so the forest becomes a perfect refuge for them, especially in the summer months. Putting the animals on a rotation is key to reducing negative impact on the forest floor, however. The ducks don't "hunt and peck" slugs off the logs but instead reduce the slug population before they can get to the logs.

While after two seasons the ducks have proven to be useful as a slug deterrent, they are not a perfect solution. As with any farming system, the weather, the timing of when ducks are moving into the mushroom yard (and moved out), and the ability of the farmers to observe and make decisions to change this or that as the system moves through the seasons are all part of the evolution. (See the insert, Can Ducks in the Woods Provide Slug

Figure 3.11. An attempt to mimic the parts of an ecosystem in design of systems at Wellspring Forest Farm. Sugar maple (the primary producer) provides shade, habitat, and sap. Mushrooms (the decomposer) break down wood and provide food for consumers (humans and ducks) in the form of mushrooms and slugs, respectively. *Illustration by Carl Whittaker*

Control for Shiitakes? in chapter 9 (page 301) for more details on this system and the results of the research.)

In this example, there are abiotic factors that influence the living organisms in the system. The maple trees are the major primary producers that feed the system from the ground up, thanks to photosynthesis. Mushrooms act as primary decomposers for the wood, which ultimately also feeds all the other elements of the soil's biological community, then ultimately the root systems of the trees. The consumers interact to balance populations (ducks eating slugs and humans eating ducks or duck eggs). As the ducks go about their business, they fertilize the trees with a high-phosphorus manure. It isn't a full, intact ecosystem, but it's a start.

Designing farming systems to be more like natural ecosystems is at the very least headed in the right direction, toward increased interaction and symbiosis. In the process time, labor, and resource inputs into the system are reduced. At Wellspring Forest Farm, the ducks don't spend all their time in the woods but have become critical weed and pest control agents on the entire farm, for the bettering of all the systems that are being implemented. At this point, with the setting (abiotic factors) and characters (living organisms) defined, an ecosystem is beginning to take shape. Of course, a few years of trials are no match for the realities of succession, to which the conversation now turns.

The Plot: Forest Succession and Disturbance

As if the abiotic factors and biological organisms were not enough of a dynamic to work with, it's critical to remember that each of these elements is changing in relationship to the others and in the context of events taking place over time. Forest ecosystems in particular are especially long in their cycles of growth, death, and decay. The challenge with this element of time is that the variability of outcomes greatly increases, and the exact timing and magnitude of change are unpredictable. With each change the species composition adapts in both short- and long-term responses. As forest farmers our best-case scenario is to understand the basic patterns and design systems that are flexible and adaptable. Those looking for the signs of change and who are ready for them are most likely to succeed in the long term.

The change of vegetation architecture and species composition in an ecosystem over time is what truly defines its character, and the path of this progression is called succession. The concept is that ecosystems naturally evolve from scattered, competitive, and disorganized groupings of plants and animals that eventually find balance (if only for a brief moment in time) before another disturbance enacts a different chain of related events. The succession of an ecosystem is relatively short term, compared to the concept of evolution in species and environments, which is the long-term track.

Readers should note that the following summary of succession concepts is extremely simplified, with the goal of covering the basics and setting a tone for discussion as specific crops and systems are discussed throughout this book. By simplifying these concepts, general patterns useful to forest farming are highlighted. It is highly recommended to review the content in volume 1 of *Edible Forest Gardens*, in which Jacke and Toensmeier offer a thorough review of the science and the intricacies of this topic.[11]

PIONEER OR EARLY SUCCESSION

This stage is characterized by fast-growing, often annual and opportunistic species that quickly move in to capture the free flow of water, nutrients, and sunlight made available by a recent disturbance. Species that thrive in this stage are sun loving and often tolerant of a wide range of limiting factors, such as drought or excessive water. One can find many nitrogen-fixing species in this group. Within this stage a further distinction between primary and secondary succession is made. Primary succession occurs when an ecosystem is starting from scratch, usually after an extreme disturbance. Secondary succession is a more minor disturbance, such as a large, old tree falling to the forest floor. The distinction is that, whereas primary succession leaves no trace of the previous system, secondary succession offers some "legacies" of the previous ecosystem. Therefore, as forest

Table 3.1. Stages of Succession

Stage	Pioneer/Early	Midsuccession	Late/Climax
Species types	Grasses, legumes; sun-loving, fast growing; annuals dominate	Mix of groundcover, herbaceous woody species	Long-lived tall trees, limited understory
Relationships between organisms	Individual, competitive	Cooperative, emerging	Coevolutionary, interdependent
Example description	A gap in the forest left by a falling tree	Thickets of shrubs and trees in abandoned farmland	Dense, open forests with trees dying of old age

farmers we will most often work with secondary succession, which can take many forms.

MIDSUCCESSION

The best word to describe this phase of succession is "diversity." Here in midsuccession, a wide mix of groundcovers, herbaceous plants, shrubs, and small trees prevail. Light density is often 40 to 50 percent and is sometimes called open woodland. Picture an orchard as a template, with many species packed into every available space. Cool-season grasses thrive in this environment.

LATE SUCCESSION/CLIMAX

In this phase the forest canopy is close to, if not entirely, filled out, nearing 100 percent cover of the forest floor. The potential species that can live in this environment are diminished. Nutrients, water, and sunlight are "locked up," as each plant carves out a rather specific and limited niche for itself and secures the necessary elements for survival.

The Dynamics of Succession

While the different stages classified have been more or less agreed upon by ecologists, the understanding of the patterns or how these stages play out over time has changed rather dramatically in the last few decades.

LINEAR MODEL OF SUCCESSION

The first notion of succession is simple: Ecosystems move through one stage to the next, from pioneer to midsuccession to eventual climax, which in early thinking was a straight line—the climax was seen as the end point, or apex, of ecosystem succession. Once an ecosystem reached the climax, there was some variation, but mostly the system remained in a more or less "steady" state indefinitely.

PULSING

The problem with the above characterization was that scientists simply couldn't find *any* examples of ecosystems that had existed in the climax steady state, at least not for very long. Each time an ecosystem was getting close, a disturbance would come along and change the

dynamic, sometimes dramatically. Thus, it appeared that the model of linear succession, which suggested reaching a destination, the climax, was flawed. Indeed, in nature each ending is a new beginning, and scientists began to consider that disturbances to this upward trend—fires, floods, large weather events, and so forth—were essential and important elements in the equation. And rather then seeing them as setbacks to the development of the system, they began to be seen more as key components in characterizing a given ecosystem. Rather than interruptions to a desired end, they were merely milestones on a pathway that is long and meandering and has no fixed destination.

DISTURBANCE AS A POSITIVE FORCE

As we examine the different concepts of succession, the key point to remember is that disturbances happen, and while often damaging or even catastrophic in the short term, they tend to propel an ecosystem forward in some positive way. For example, floods deposit nutrients into floodplain ecosystems, and fire often assists in the germination of many dormant seeds in the forest, while also reducing pest populations and breaking down forest litter, making it more available to the soil.

Like it or not, disturbance happens. The more this can be accepted and taken into account as forest farms are designed, the better off the system will be in both the short and the long term. While the specific timing, magnitude, and breadth of a disturbance cannot be predicted with much foresight, we can do some "worst-case scenario" thinking and plan ahead, to the best of our ability. We can also design disturbances to work to our advantage.

With the onset of more rapid climate change, resilience to disturbances, both environmental and social, will be critical. Forest farmers are fortunate in that forests are already more buffered against disturbances, yet they are not immune.

SHIFTING MOSAIC

More recently, it has been recognized that the linear form of succession, even with the addition of disturbance as an inevitable force, was not a useful way to think about the process of succession. A new model of succession, named the "Shifting Mosaic Steady State," acknowledges the idea that, rather than ecosystems climbing the ladder to the top of the mountain (climax), ecosystems of all types simply go though shifts from one type to the next. This model represents the forces at work as a cycle while still recognizing the stages of succession and the disturbances that transition the ecosystem from one phase to another. No stage is necessarily better than the other. What is more important is the ability to identify the stage within which a specific ecosystem resides and consider where it might be heading. A system naturally wants to move toward the climax phase, but disturbances of all types change that course again and again.

This concept is complicated by the factors of time and scale. A forest may remain in one stage of succession for centuries, while a field may turn into forest in only a few decades. The scale can range from a catastrophic event such as an ice storm or hurricane to a gap created in the forest when just one tree falls from the canopy, opening the forest floor to an influx of light. It's important, then, to take the overall pattern into consideration but recognize that each site will be a unique iteration of the process.

Keep in mind that succession theories are just that: a concept that tries to summarize a very complex subject material. Scientists will continue to try to best describe this phenomenon in more detail, but the takeaway is that context matters. Recognizing the past successional patterns of a forest farm and preparing for future possibilities is the best forest farmers can do.

Applying Succession to Forest Farming

It's a well-known axiom in forestry circles that the consequences of the actions taken by one individual forester won't be seen in his or her lifetime but will instead be passed to the next generation. This is a challenging aspect of forest stewardship and forest farming, yet as practitioners we need to take the implications of this into account. While the conversation on forest succession is interesting, it's easy to be unsure as to how these concepts apply to the short- and midterm planning of a forest farming operation.

Prepare for Disturbance

If one accepts that disturbances are not a matter of "if" but "when," then it's worth spending some time preparing. In permaculture this is often referred to as "design for catastrophe," and there is plenty of evidence that communities and individuals that take time to prepare ahead fare better when the disturbance happens.[12]

Planning for Disaster

The following chart provides a starting point for developing a plan for dealing with disturbance. Possible disturbances depend on the site specifics, of course, but could include fire, flood, heavy rain events, drought, cyclones, storms, hurricanes, earthquakes, and volcanoes.

It's important to keep in mind that this planning tool can also be used for "microdisturbances," which are the small things that happen around the farm or homestead that can interrupt the basic functioning

Table 3.2. Questions for Disaster Planning

Disaster	Questions
Cause(s)	Natural or human-made or both?
Frequency	How often could it potentially occur? Is it consistent or sporadic?
Duration	How long could it potentially last?
Speed of Onset	What's the warning time? How fast does it move?
Scope of Impact	Is the impact concentrated or spread out over larger areas? Is the damage short or long term?
Destructive Potential	How severe are possible impacts? What is the population density? How is the environment likely to respond?
Predictability	Does it follow a pattern? What patterns can we rely on? What are unpredictable factors?
Controllability	What human responses are there? What is the time frame to restore order?

Adapted from Morrow, 2006[13]

Figure 3.12. Gaps in the forest benefit sun-loving canopy trees growing on the edge, such as white oak (front left), and also provide space for regeneration of tree seedlings and herbaceous material in the center. Photograph courtesy of Jen Gabriel

of the site; for example, the breaking of a pump that serves to get water to soaking tanks for mushrooms, or the predator that consumes your flock of ducks overnight. The same questions can be useful to consider. Again, the more planning ahead, the less devastating the consequences.

Intervene with Minor Disturbances, as Necessary

Disturbances can also be seen as a tool for management. For example, creating periodic gaps in your forest "resets" the succession and can lead to the introduction of new species or the sprouting of seeds of trees that need sunlight to germinate (such as tulip poplar, black locust, and black cherry). In general, a gap is a space that is one to two times the height of the canopy, while a clearing is defined as two to four times the height.[14] For example, in a woods where the canopy was 40 feet tall, a gap would be an area with a diameter of roughly 40 to 80 feet, while a clearing would be 80 to 120 feet wide.

Other possibilities of induced disturbance include strategies discussed later in the book, including coppice management, the use of animals, and strategies in forest management by felling and girdling trees, including timber stand improvement and crop tree management. Keep in mind the scale and intensity of managed disturbances. A forest that is thinned out too quickly can, with some species, result in epicormic branching in trees, in which trees send out dozens of otherwise dormant side shoots in response to the increase in available light.[15] A good rule of thumb to avoid this is to never thin more than one-third of a stand of trees in one season, even if more trees should be removed in the longer term.

Consider Changes in Forest Farm over Time

Take some time to consider the changes the forest farm will undertake in five, ten, twenty, and even fifty-year increments. Consider what the forest will look like at these intervals, and also how it might be expected to change if various disturbances occurred.

For example, at Wellspring Forest Farm, as mentioned, the main mushroom/maple sugaring operation takes place in a 1-acre woodlot that is almost 100 percent sugar maple trees, of which most are twenty to thirty years of age. In fifty years these trees would be around eighty years of age, reaching the peak of their growth. The forest would have a thicker canopy than the current arrangement. Since the population is almost entirely maple, and likely seeded from only one or two "mother" trees, the stand is vulnerable to any number of diseases and pest outbreaks. A creek that runs adjacent to the woods could potentially overflow and flood the forest. And certainly in fifty years, with a shortage of timber resources, the trees would be under the threat of removal for timber. Recognizing all these factors, known and unknown, allows for planning that stretches beyond the next week or next year and into the next lifetime. Further, taking possible disturbances into consideration, the scenario could look very different. This is critical for the long-term success of forest farms.

Forest Types as a Method to Mimic

Equipped with the broad patterns of ecosystem relationships and succession that are applicable to the wide range of systems found around the world, it's time to bring the conversation more specifically to the context of the cold climate temperate regions of the northeast United States, southeastern Canada, and many of the midwestern states. Much of this conversation can also apply to other regions around the world, with some minor tweaks to compensate for the differences, especially in the abiotic factors mentioned at the beginning of this chapter.

The concept of examining the natural forest types that exist for a particular region offers a starting point for thinking about the patterns of site management that are likely to succeed in the long term. Farmer

Figure 3.13. One constant reminder of how forests change over time are the streams and waterways that run through the woods, carving new pathways and depositing sediment. The past, and sometimes the future, can be read by observing the water and its effects on the landscape. Photograph courtesy of Jen Gabriel

Mark Shepard, who manages a 106-acre commercial farm in southeastern Wisconsin, offers three main steps toward developing what he calls "restoration agriculture," which seeks to grow a suite of perennial food crops that can be produced at the farm scale:

1. Identify your biome
2. Find the "key" economic species
3. Imitate the system

For Mark this means mimicking the structure and progression of the oak savanna, which is the natural late-successional environment of the place where

Figure 3.14. New Forest Farm in Viola, Wisconsin, is designed to mimic a savanna ecosystem, the climax ecology of the region. Woody trees and shrubs such as hybrid hazelnut, hybrid chestnut, poplar, black locust, oak, and others are planted in this section. Grass is grazed on rotation by cows, pigs, and poultry.

his farm is located. As he identified and researched this ecosystem, the same species kept showing up as common in the cast of characters. They included oak, chestnut, beech (all Fagaceae family); hazelnuts (*Corylus*); apple (*Malus*); cherry, plum, peach (*Prunus*); raspberries and blackberries (*Rubus*); gooseberries and currants (*Ribes*); and grapes (*Vitis*), as well as a number of pasture grasses.[16]

Filtering these options for economic prospects, the system now focuses largely on chestnut and hazelnut, with the fruits integrated in various patterns throughout the farm. Different combinations of these plants are scattered throughout his landscape, seeking to find

the balance between successful cropping systems and this idea of ecosystem mimicry. Mark's system will be further discussed in the case study at the end of this chapter. The take-home message is that every forest farmer should start by learning the forest types localized to her area. The material in this book is a good reference, but ultimately conversations with local foresters, Extension agents, and those knowledgeable about the local forest cover will be able to offer the customized information, as each variation in slope, latitude, soil type, and so forth offers a slightly different take on any generalized patterns offered here. Naturally, this is the best place to begin planning a forest farm venture.

Table 3.3. Common Types of Eastern Forests and Associated Species

Forest Type	Associated Species	Subspecies
Maple–beech–birch	Hemlock, elm, basswood, white pine	Bluebead lily, Canada mayflower, starflower striped maple, various ferns, mayapple, trillium, wild leek, cohosh, American ginseng
Oak–pine	Red and white pine, oaks, gums, hickories, yellow poplar	Sassafras, mountain laurel, flowering dogwood, foxglove, black huckleberry, Juneberry, low- and highbush blueberry, wintergreen, wild indigo
Oak–hickory	Yellow poplar, elms, maples, black walnut	Flowering dogwood, blueberry, mountain laurel, hawthorn
Spruce–fir	White cedar, tamarack, birch, maple, hemlock, yellow birch, American mountain ash, pin cherry	Rhododendron, thornless blackberry, mountain cranberry, wood fern, northern lady fern, clubmoss; wildflowers include mountain wood, sorrel
Loblolly–shortleaf pine	Oaks, hickories, gums	Bayberry, inkberry, mapleleaf viburnum, arrowwood, green brier, blackberry, Virginia creeper, lowbush blueberry, wild grape, witch hazel, and sumac
Oak–gum–cypress	Cottonwoods, willows, ashes, elms, hackberry, maples	Buttonbush, Virginia sweetspire, cyrilla, buckwheat, dogwood, leucothoe, yaupon, southern bayberry, possumhaw, swamp rose, poison sumac, greenbrier, supplejack, decumaria, cross-vine, pepper-vine, and poison ivy
Aspen–birch	Red maple, balsam fir, paper birch, big-toothed aspen, quaking aspen	Shadbush (at low cover), black huckleberry, lowbush blueberry, bracken fern, Canada mayflower, sheep fescue, dicranum moss, large hair-cap moss, reindeer lichen

COMMON TYPES OF EASTERN FORESTS

Eastern hardwood forests are also some of the most diverse in terms of species composition. While many forests around the world are dominated by only a few tree species, it isn't uncommon to walk into a forest in Virginia or New York and immediately count dozens of different trees in just a few acres.

The table and map below suggest various forest types for the eastern United States. Keep in mind the name of the type does not declare the *only* species in the forest, but the common and dominant species. It is also rather broad in scope, and it's worthwhile to do some further exploration for local resources that more specifically characterize forest types. Keep in mind that in the eastern forest, multiple types can grow in proximity to each other, and occasionally a "transitional" forest blends two or more types.

Consider a given forest type as a starting point for design. For example, an upland, south-facing woodlot

that shows a preference for oak and hickory trees provides one direction for a forest farm, while the cool and wet north-facing slopes often support more of the maples, birches, and beech trees. In the first example, a nut production system might be the best model for a forest farm, versus a maple and birch sugaring and ginseng or goldenseal operation for the latter forest.

The situation gets more complex, however, as we bring in aspects of climate change. Until this point the discussion of forest ecology has mainly come from the historical perspective and with it the assumption that forest succession will look more or less the same over time, with only slow and relatively small-scale disturbances being the norm. Yet with climate change forest farms will be faced with more frequent, more intense, and more unpredictable circumstances. While forests are some of the more resilient systems to be cultivating in such circumstances, it is nevertheless important to take a look at the environmental

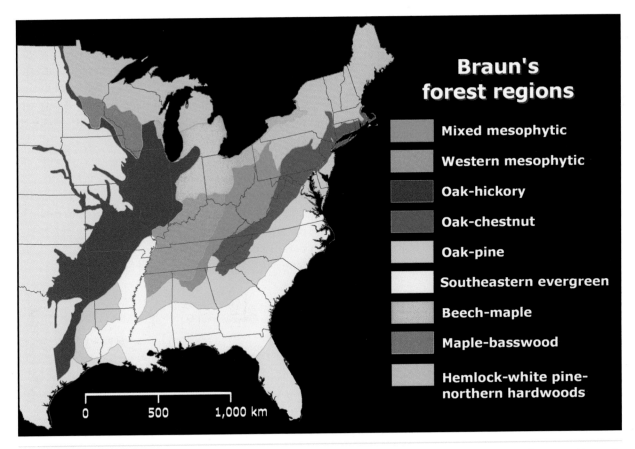

Figure 3.15. Forest types are generally classified differently depending on the source and offer a starting point for researching and mimicking in forest farming design. This map is based on the work of ecologist Dr. E. Lucy Braun from *Deciduous Forests of Eastern North America*. The mesophytic forests are diverse mixtures of trees and occur across the Appalachian region, considered by many to be the epicenter of highest development of the eastern deciduous forest. Illustration by http://www.ohio.edu/people/dyer /forest_regions.html

challenge of this lifetime, and the potentially most critical factor in the success of forest farming, if not agriculture as a whole.

The Big Change: Climate Chaos

These modern times are a truly unique and unprecedented period of change, spurred by human consumption of fossil fuels that are warming the planet and causing more flooding, droughts, and "super storms" all around the globe. A 2013 report by the Intergovernmental Panel on Climate Change noted that "a large fraction of anthropogenic climate change resulting from CO_2 emissions is irreversible on a multi-century to millennial time scale, except in the case of a large net removal of CO_2 from the atmosphere over a sustained period."[17] In other words, climate change is here to stay.

Forest-based agricultural systems offer some important advantages in the wake of these changes. For one, forests are key in their ability to buffer against extremes; anyone who has walked through a field on a really hot day in the summer, then slipped into the cooler, moister-feeling forest has felt this. Forests can also remediate the effects of high winds and flood events, though the degree of remediation is variable. The reality is that of all the land-use strategies available, those that include trees might be some of our best bets.[18]

Climate change brings to the table all the natural biological and succession-driven elements that we have

seen on Planet Earth, yet brings them forward less pre-
dictably, more frequently, and more intensely. Forest
farming has to be prepared for this reality, and we must
also be flexible in the way farms are designed.

While models provided by scientists offer a possible
glimpse of our future, the reality is that no one knows
what it will look like. A key indicator of this is that
many of the models that predict a rise in sea level or
melting glaciers have been off—and if anything, have
underestimated the rate of change.[19] What we do
know is that agroforestry and forest farming strategies
are going to be increasingly important in the face of
uncertainty. The forest and trees are resilient biological forms.

THE MANY FACES OF THE DEBATE

Climate change policy makers around the world make
many claims about the various aspects of climate
change and often focus on one item as the silver bullet
for fixing the problem. The reality is that a multifaceted approach must commence on a number of levels
if any significant headway is to be made. While pressure needs to be put on officials to make sweeping
regulatory changes, much also needs to be done in local
communities to prepare for inevitable changes and to
reduce their total impact. Others before us have made
the claim that "this is the time to act" on the issue of
climate change. And the reality is that in some ways
that tipping point was passed a long time ago.

Indeed, there are plenty of changes already occurring. The question now is how much the impacts
can be slowed down while still maintaining a decent
quality of life. The unfortunate truth that masks this
entire discussion is that there is still a significant portion of the population, at least in the United States,
that denies the existence of global climate change.
This unwillingness to examine the facts only slows
down any productive dialogue on taking steps in the
right direction.

The important take-home message, the one we
need to be telling more often, is not that we are
"fighting climate change" but that we are addressing
the challenge *while* also improving the quality of life
for people. This is where we can find some common

ground and get others to listen. As award-winning
scientist and environmentalist David Suzuki notes,
"Doing all we can to combat climate change comes
with numerous benefits, from reducing pollution
and associated healthcare costs, to strengthening and
diversifying the economy by shifting to renewable
energy, among other measures."[20]

The solutions to the climate change challenge can
be broadly separated into two categories: those that
focus on reducing current emissions and those that
focus on sequestering carbon that currently exists in
the atmosphere. Forest farming as a practice addresses
both simultaneously, while at the same time providing
numerous other benefits for the farmer.

STRATEGY ONE:
REDUCTION OF EMISSIONS

In May of 2013, a research facility atop the Mauna Loa
volcano on the island of Hawaii detected the first time
that carbon dioxide levels rose above 400 parts per million (ppm) for 24 hours– the highest concentration that
has existed since humans began walking the planet.[21]
This is but one indicator of the problem: a measure of
the most common greenhouse gas that is being added
to the atmosphere. The CO_2 level remained at around
300 ppm for almost eight thousand years. A sharp
increase in levels correlates with fossil fuel discovery
and increased use during the industrial revolution,
which has continued to grow dramatically.

Climate scientists generally regard 350 ppm as the
safe level for humanity.[22] James Hansen of America's
National Aeronautics and Space Administration
(NASA), the first scientist to warn about global warming more than two decades ago, wrote: "If humanity
wishes to preserve a planet similar to that on which
civilization developed and to which life on Earth is
adapted, paleoclimate evidence and ongoing climate
change suggest that CO_2 will need to be reduced . . . to
at most 350 ppm."

Since the planet continues to increase beyond this
stated capacity, the first step is to decrease the rate
at which new CO_2 is being added. This translates
to recognizing the parts of our farms and woodlots
that contribute emissions and to reduce or eliminate

Figure 3.16. Forest farming practices can replace emissions-heavy methods of food production, including more conventional forms of indoor mushroom cultivation, which requires 24/7 energy for temperature and humidity control. Log-grown mushrooms, after the initial energy to produce the spawn in a laboratory (see Case Study in chapter 5), are a very low-emissions food crop.

those sources as much as possible. Forest farming and agroforestry practices achieve emissions reduction in multiple ways, covered in the following sections.

Less Energy Used to Produce Foods

Overall, forest farming crops are produced in low-input systems. Take the comparison of shiitake mushrooms grown on logs in forest farming systems versus the more conventional method of indoor cultivation on compressed blocks of grain or sawdust. The former method requires no energy other than the electricity to drill holes in the logs and to heat the wax during inoculation and the added human power to soak and harvest mushrooms (and pick the occasional slug). Indoor cultivation requires a 24/7/365 input of energy to regulate temperature and humidity for successful cultivation.

Since the forest is a self-regulating system, a stable and consistent temperature is maintained. There is less need to heat, cool, and pump water for irrigation since soil moisture is maintained. Forest farms can be set up

with a low energy budget and maintained with one that is virtually nothing. The degree of energy use, of course, depends on the site and the systems implemented.

No-Till = Decreased Use of Fossil Fuels

The fact that forest farming systems are inherently no-till means that machinery is a low contributor to the emissions footprint of an operation. The chain saw is perhaps the most utilized tool, and some forest farmers use a tractor, a gator, or an ATV to navigate their woodlot and move materials. However, this impact is far less than that of the field crop system, which requires constant use of a tractor to plow, cultivate, weed, and fertilize.

Less Infrastructure

Because forest cultivation systems are, for the most part, low tech, the cultivation of forest-farmed crops doesn't require a lot of tractor implements (steel), buildings (concrete and wood), and other infrastructure associated with farming, such as water lines, power transmission, and so on. Again, this is not to say that none of these items will appear in a particular forest farm but rather to highlight that the need is severely reduced when compared to other forms of farming and forestry.

Although these factors are true, and forest farmers should work to reduce emissions wherever possible, the reality is that, in comparison with other sectors, agriculture and forestry simply aren't big emissions contributors in a national or global context. Forest farms have the most to offer in terms of the next important group of strategies that aims to recapture carbon already released.

STRATEGY TWO: CARBON SEQUESTRATION

The other set of strategies for addressing climate change seeks to pull existing CO_2 from the atmosphere and sequester it, or store it in both living soil and in plant matter. When the word "sequestration" is used, it might sound as though carbon is somehow permanently locked away, and indeed some forms of "geological" sequestration aim to do just that, though the jury is still out on the effectiveness of these techniques.[23] Forest farming is interested in what is referred to as biological sequestration, in which trees, plants, and soil are utilized to cycle and store carbon for various lengths of time.

In some ways forest farming and agroforestry are uniquely situated for this type of "carbon" farming, which, unless there are government subsidies for sequestering carbon, will have to remain a secondary product of crop production. It's important to remember that everything in a living system is in flux. The issue related to successful carbon sequestration has to do not only with the parts (trees, soil) but also the process, or how the system plays out over time.[24] Carbon cannot be locked away forever. What can be done, however, is to capture as much as possible, delay its release back into the atmosphere (from a few years to hundreds, depending on the technique), and recycle as much carbon back into the system as possible.

Carbon storage in agroforestry and forest systems comprises three main "pools"—aboveground plant biomass (stems, trunks, leaves, etc.), underground bio-

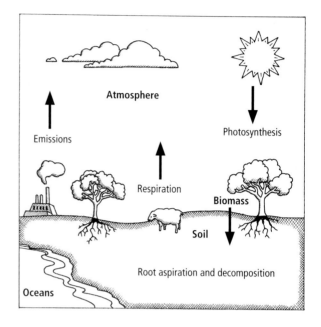

Figure 3.17. Carbon is always in flux, moving from one place to another. The bolded labels are areas where carbon is stored. By increasing storage in biomass (plants) and soils, forest farms can help sequester carbon from the atmosphere. Illustration by Travis Bettencourt

mass (roots, soil organisms, and organic matter), and durable products (i.e., wood harvested for structures and materials). And because agroforestry systems tend to be more diverse in the composition of vegetation, productivity in terms of biomass production is likely to be higher.[25] Maximizing the productivity of the forest as well as the number of nontimber forest products is the key to building forest farms that produce yields and contribute to combating the effects of climate change.

Aboveground Storage

The capacity of aboveground storages to benefit carbon sequestration depends on a wide variety of factors. For example, the climate effects of growing a black locust tree and cutting it down to burn for firewood will have less benefit than harvesting a post that is used for fencing or building materials. In simplest terms, aboveground biomass is an expression of growth and productivity, and its sequestration potential relates to two factors: (1) how well it can allocate carbon into the soil as it grows, and (2) the end use of the product. Of course, with the example above, the forest farm is likely going to need both heating fuel (short-term carbon storage) and fence posts (long-term storage). In the end it's about taking the wide view of the whole system.

Perennial vegetation, and notably trees, is more efficient than annual vegetation when it comes to the percentage of carbon that can be stored in stable forms.[26] Part of the reason for this is the ability of perennial plants to extend the growing season and thus "make more hay when the sun shines." This carbon is stored in the living and dead aboveground tissues until the vegetation dies or is harvested. Therefore, while carbon is ultimately produced aboveground, the long-term potential storage is actually in soils or in harvested wood products. Thinking in long-term cycles, a sustainably managed agroforestry system could theoretically retain and cycle carbon for centuries.[27]

One study out of Germany gets at more details of the relationship of plant growth and management when a short-rotation coppice system of black locust was studied (see chapter 10 for more on coppicing). In this system, a fraction of the carbon is stored in the shoots, which are harvested for use. Another portion is allocated to the stump, or the leftover mass after cutting. A third fraction is in the root systems. Since two of three major storages remain intact and continue to develop as multiple successions of wood are growing and harvested, over their lifetime they form a considerable carbon sink. The total rate of sequestration depends, of course, on the overall productivity of the planting. Black locust is noted because of its ability to grow well on even marginal sites, partially because it can fix its own nitrogen. Therefore, this species has a high capacity for sequestering carbon in its biomass and in the soil, even on poor sites.[28]

Belowground Storage

Research done with modeling in European studies suggests that forest soils are responsible for 30 to 50 percent of all carbon sinks in the forest.[29] When leaves drop to the forest floor in autumn, or if an

Figure 3.18. Black locust regrowth after coppicing, or cutting the tree to the stump at or near ground level. Research on the potential carbon sequestation of locust and other trees in short-rotation systems such as this indicates potential benefits to the climate. At New Forest Farm, Viola, Wisconsin.

entire tree falls down, soil organisms work at remarkable speeds to process and pull materials into the soil, most notably carbon. Of course, some of the carbon is off-gassed back into the atmosphere during this process, so in the end the potential of soils as a pool for carbon depends on the chemical properties of the carbon compounds, the site conditions, and the soil properties, including clay content, moisture, pH, and nutrient status.[30]

Stabilization of carbon can then be differentiated from mere accumulation. One deciding factor is whether soil has an abundance of clay minerals and oxides, which bond with carbon and offer long-term sequestration.[31] Ideally, in such soils increases in organic inputs should increase the amount of stabilized carbon. Soils that don't have the physical properties mentioned above need to have consistent inputs to retain any pooling benefits; in other words, the cycling of carbon will occur at a more rapid pace.

Climate Actions for Forest Farmers

With all the concepts surrounding carbon cycling and sequestration, it's hard to know where to start. The following suggestions are arranged from the most accessible options to those that might take more effort and planning. The good news is that these options are often ones that are most likely going to be done anyway. In the end, forest farmers are going to take the steps necessary to support production of various nontimber products. Carbon sequestration simply adds a side benefit, a layer to the equation that further promotes the virtues of agroforestry as a multifaceted solution to many issues.

Manage Younger Forests for Maximum Growth

There are already several good reasons a forest should be managed for maximum growth and health. Carbon sequestration is just one of them. Engaging with a local forester and thinning appropriate stands with the appropriate methods is the first and easiest thing to do in any woods. Chapter 10 details these steps more completely. Stands should be managed for high productivity with methods that minimize soil disturbance.

Young recovering forests are often the best places to start, as they are riddled with trees that are diseased, dying, and densely spaced. Since a given acre of forest can only capture X amount of sunlight, it might be natural to assume that more trees equals more capture. But since all things are not equal, there are inevitably some trees that are growing faster then others, and thus pulling more carbon. So in many cases fewer trees, growing well, are some of the best forests for carbon sequestration.

Preserve Mature and Old-Growth Forests

Whether on the same land or nearby in the greater community, a widespread focus on preserving mature forests is critical. While younger forests in a stage of rapid growth are rapid sequesters of CO_2, the older stands serve as long-term banks. As with younger forests, there are myriad benefits to preserving older forests, including the protection of wildlife habitat, the preservation of tree genetics, and the benefits to air and water quality. Since so much of the forested land in the temperate United States and Canada is marginal, young forest, it makes sense to start with these with the most intensive, on-the-ground management. Forest farming as a practice naturally promotes this, because healthy forests become desirable, and alternative economic crops relieve pressure from timber harvesting. Preserving old forests is often a task fought with the pen and voice rather than the chain saw. Land trusts and conservation easements ensure that land will be protected well beyond the lifetimes of people active in forest preservation.

Plant Trees That Maximize Carbon Sequestration

A study from Syracuse University in 2006[32] looked at a mix of species that would maximize carbon sequestration while also helping to maintain good air quality. To determine the most effective trees the study used carbon sequestration data and the leaf biomass for each tree species. Some of the best species are hawthorn, staghorn sumac, willow, hornbeam, elm, and honey locust. A common factor is that these species are fast growing and thus can more rapidly pull carbon from the air.

Another set of species to consider are those that have dense and fibrous root systems in addition to rapid growth, since carbon can be stored in the living root tissue and integrated by soil organisms and stored in the ground. Examples of such species are poplar, willow, alder, mulberry, apples (and all *Prunus* species), and birches.

Of course, it's recommended to combine these qualities with management goals; for example, planting poplars as a windbreak and also harvesting on a ten-year rotation for totem logs for growing oyster mushrooms, rather than just planting trees for the sake of carbon sequestration. If the forest farmer can

Figure 3.19. Willow used here on the edge of a farm field on Prince Edward Island in Canada. This planting is utilized to capture water and nutrient runoff from the nearby potato field and is also sequestering carbon with a dense and fibrous root system.

get a yield from his efforts, he is much more likely to manage and maintain plantings.

BUILD HEALTHY, CARBON-RICH SOILS

Forests build healthy soils and are assisted, in large part, by the influx of organic matter. While nature is doing a decent job, there are a multitude of things forest farmers can do to enhance, expand, and increase the soil health in our forest farms, while in the process improving the conditions for the plants and animals we are trying to cultivate.

Hugelkulture

Hugelkulture is a centuries-old practice that is roughly translated from the German as "mound culture." The idea is that woody biomass, in the form of sticks, twigs, and branches can be put to more productive use when buried in soil, essentially turning the material into a carbon and water sponge. Perhaps the most appealing feature of this system is making use of what is often seen as a waste product: the brush that inevitably piles up as undesirable species are removed and trees are cut down.

In practice, hugelkulture mounds are built by laying down woody materials, then covering them with soil. This is a quick and low-labor task if there happens to be a backhoe on the farm, or one can be rented for a few days. The task requires a bit more time and some willing friends if done by hand.

There is no specific recipe for mounds, but a basic pattern is this:

1. Dig a trench in a desired area, or look for a low spot that might benefit from being filled in.
2. Fill the space with the largest logs and debris possible, mounding smaller-diameter material as you build up.
3. As an optional step, you can incorporate high-nitrogen material to help break down the carbon-rich matter faster. Partially composted manures work best. Part of the reasoning for this is that such a high carbon ratio will "rob" nitrogen from soil layers and make growing plants harder in the initial years.
4. Add soil, or compost, on top, ideally as much as you can get your hands on. An alternative option is to cover the bed with straw or chips and let that break down.

With a backhoe and a brush pile, it's easy in a matter of minutes to excavate a hole, bury the brush, and cover with the soil recently excavated. This is a simple, satisfying, and effective process. Proponents of hugelkulture report that irrigation is unnecessary after the first year and that the growing season can often be extended as the material below is decomposing and releasing a small amount of heat. This is an easy way to get raised beds without a lot of soil material, too. By burying the carbon we are also encouraging it to stay in the soil and be cycled longer, contributing to our efforts to sequester it.

Biochar

In terms of the amount of energy and time required, biochar is a jump up from the practice of hugelkulture. Biochar also presents a compelling potential for use of waste biomass, producing a product that will sequester carbon for many years while also providing a boost to soil health.

Biochar is, essentially, wood biomass that has been heated to a high temperature (400–700° F) in a low-oxygen environment. This creates charcoal, which can be used as a soil amendment that is reported to last potentially thousands of years. The material can be further processed, and even fuel can be harvested—to be used for cooking or to run vehicles.

While the burning and natural decomposition of biomass, and in particular agricultural waste, add large amounts of CO_2 to the atmosphere, biochar that is

Figure 3.20. Students constructing a hugelkulture mound at the MacDaniels Nut Grove. A small hole was excavated about 12 inches deep (not pictured). Large logs were placed in the hole (left), then smaller and smaller vegetation, which was stomped down as it was layered (center). The woody debris was then covered with the soil from the initial excavation (right). Hugelkulture is a good use of unsightly piles of brush that are inevitably part of forest farms.

stable, fixed, and what is called a "recalcitrant" carbon can store large amounts of greenhouse gases in the ground for centuries, potentially reducing or stalling the growth in atmospheric concentrations of CO_2. The practice was first discovered in the areas inhabited by pre-Columbian Amazonian tribes, where incredibly rich, dark soil was discovered and ultimately determined to be cultivated by human hands. The natives would bury biomass in trenches or pits and let it smolder into biochar. The practice, and controversies surrounding biochar production, are covered in chapter 8.

External Waste Streams

While the above two practices make use of on-site resources, it's worth mentioning the ample opportunities to collect and recompose wastes that are available in a throwaway culture that does not value organic material. Specifically, leaves from urban areas, wood chips, and animal manures are often readily available and either free or low cost. The rapid influx of organic matter to a system can dramatically improve soil health, and thus its carbon sequestering ability. It's important to recognize that along with the benefits

GAUGING SUCCESSION: TESTING SOIL FOR ORGANIC MATTER

Intentions are all well and good, but how do we know if efforts are really making a difference? While many variables are not easy to measure at home, any soil test conducted through a local Extension office should provide as one of its results the percentage of organic matter (% OM). If this test is done annually in the same season (because organic matter will fluctuate depending on the time of year), you can track results from one year to the next, as management changes.

Make sure you learn the proper methods for taking an accurate soil sample to ensure good results. The basic procedure for sampling is as follows:

1. Determine sampling area. This should be a portion of the woods where the species and site characteristics are all similar. For example, keep a sloped forest of hickories separate from a bottomland creek bed of walnut trees.
2. Take ten representative samples, randomly selected, within the defined sampling area. Use tools that are clean and free of rust. Stainless steel probes or augers are best, but a spade that is clean and rust-free also works. Avoid sampling under extremely wet soil conditions. Sample by digging down 8 to 10 inches into the soil. Ensure that the samples you dig or probe are as uniform as possible and that you are getting equal parts of the profile (imagine a 2 inch by 2 inch by 2 inch piece of cake).
3. Collect samples in a clean plastic bucket or plastic bag. Mix the ten samples thoroughly in the bucket; break up clods, remove large stones, twigs, vegetation, and so on. When the samples are completely mixed, take a sample to fill a bag or box (two to three handfuls). Save several handfuls of the mixed sample in a ziplock bag, in case you want to do other tests with the same medium (see chapter 10).
4. Spread samples on a clean sheet of waxed paper and dry at room temperature. Avoid collecting or shipping wet samples in paper bags or boxes. Wet samples can leach boron out of the paper and contaminate the sample. Air-dry samples prior to shipping, or ship in a plastic bag. Do not use heat, but a fan is acceptable to assist with drying.
5. Submit to your local cooperative Extension or soil testing lab.

The bottom line is, if the % organic matter number increases each season, you are on the right track! The report will also provide valuable information on soil pH and nutrient content as you consider the crops to plant in your forest farm.

Of course, there are other, more complex methods to get detailed results around soil organic matter, but none is very practical or inexpensive for most people. Soil scientists do field tests that use soil color, root presence, and bulk density as measurements, outlined in soil health assessment manuals such as the one from Cornell.[33] Farmer groups could also be taught these methods and take routine measurements on their farm, a practice some organizations are advocating for.[34]

of importing organic matter from other places come some risks of spreading unwanted seeds, disease, or contaminants. Look into the details of the source materials before making the commitment to take it off someone's hands. Another wise management practice is to hot compost all materials before integrating them into planted areas.

Consider first the resource that can be gathered as part of your normal routes of travel: going way out of your way to drive around and get materials may offset any carbon-saving measures that are the intentions behind your actions. But on your way home from work why not stop at the town wood chip pile and fill up?

As an example there is an abundance of wood chips brought on-site at Wellspring Forest Farm for all sorts of uses—from establishing pathways to cultivating Stropharia mushrooms. Bagged leaves are also collected and composted before being added to cultivation beds and used to cover soils in areas that have been trampled by ducks.

In the End: Being Producers rather than Consumers

In the last century there has been a wholesale shift in how people interact with their environment. Especially in industrialized countries, a departure from work that is in production to that which is focused on consumption has taken place. Simply put, more people used to grow their own food and provide their own needs locally, and now a small fraction do, while the rest mainly consume resources.

One way to think of the big picture shift that is necessary to slow climate change refers back to the building blocks of ecology at the beginning of the chapter. More people must engage in acts of production and in the "recomposition" of wastes at a rate greater than that of acts of consumption. Technically, people cannot be producers since they can't photosynthesize, so in the end this really means planting a lot of trees and supporting forest ecosystems.

Forest farms can also be producers by harvesting energy on-site and thereby reducing the need for larger, centralized sources. Both sunlight and water, which are coming to farmscapes in renewable ways, can be captured and stored. Forest farms offer the promise of needing less energy from nonrenewable sources to produce crops, but some energy will always be required. Using low-tech, solar, wind, or hydropowered electrical systems can address electric needs. More time and research needs to be allocated to examining the vast potential of the forest and its ability literally to provide us with food, water, shelter, building materials, and so much more in a less impactful way. Certainly, the effort in this book is to continue to develop and refine this discussion.

Adapting to Change

While forest farmers can do their part to reduce emissions and sequester carbon, the stark reality is that, regardless of how successful humanity is at combating the problem, the planet is changing and will continue to do so into the foreseeable future. It is then important to consider how systems can adapt and even leverage climate change as an advantage.

Taking a brief look at the best predictions and models for what the climate will look like is akin to taking a look into a crystal ball— no one can know exactly what things will look like. That said, however, there are many scientists working hard to provide as accurate a prediction as possible, and taking a look at these guesstimates gives at the very least a starting point when designing resilient forest farms.

What Will the Temperate Forests Look Like?

While again it's hard to say exactly what changes to expect, many scientists have worked with data and modeling to make some best estimates. As trends indicate, some of these patterns are already becoming a reality, and the predictions in some cases have fallen short of the educated guess. The following is taken from the most vetted information sources the authors could locate. Forest farmers should take the following into consideration and plan accordingly, without taking any drastic actions in one direction or another. The key approach is to diversify, be redundant, and be flexible.

Changes in Precipitation

The first large consideration regards the amount of rainfall and snowfall in a given year. How much total precipitation can we anticipate? This question, while looking at total accumulation or quantity, can also be broken down into other factors, including changes in the quantity, duration, and intensity of precipitation events.

There are three key projections for the United States, according to a 2009 report:[35]

- Northern areas are projected to become wetter, especially in the winter and spring. Southern areas, especially in the West, are projected to become drier.
- Heavy precipitation events will likely be more frequent. Heavy downpours that currently occur about once every twenty years are projected to occur about every four to fifteen years by 2100, depending on location (see figure 3.22).
- More precipitation is expected to fall as rain rather than snow, particularly in some northern areas.

In addition to changes in weather patterns, projections suggest that when precipitation does come it will fall as heavier events (see figure 3.22). This has many implications, as three inches of rain in an hour is very different from three inches over three days. Runoff, flooding, and muddy paths and roads are all common issues in the forest that show up along with extreme rain events.

The lack of rain, or drought, is also likely to happen more often. Areas of the temperate climate that may have been used to a relatively even distribution of rainfall are likely to see more extremes: where one month or season there is too much water and then the next too little. Though the forest can often handle short-term droughts, longer-term water shortages can diminish productivity. Crop systems that can get away with no irrigation under normal precipitation conditions may need to have a backup plan for watering plants and hydrating mushroom logs if rainfall becomes scarce.

The implications of these precipitation factors are hard to predict, since they are quite variable. Forests

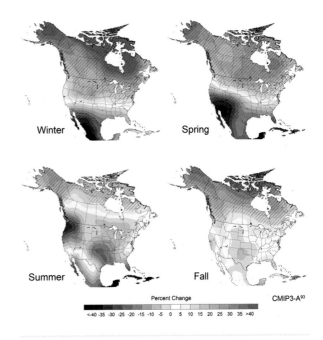

Figure 3.21. Projected changes in precipitation distribution with a high-emissions scenario. For the Northeast, while overall precipitation is likely to increase up to 20 percent, the map shows that that this will mainly be an increase in winter and spring precipitation. Areas with hash marks have a higher confidence than those without. Illustration courtesy of United States Global Change Research Program

and forest farms are able to buffer quite a bit from both extreme moisture and extreme drought. Yet at the same time more frequent flooding may affect especially those forests in floodplains and lowland areas. Droughts will mean that some tree species may be less successful than others in regeneration. Ultimately, the individual site will determine the implications, but planning for the likelihood of more dramatic swings in water resources is critical.

TEMPERATURE

In some ways temperature is a much easier variable to define. It is already warmer than it used to be and will continue to get warmer still. This means overall average temperatures are rising, as well as the length of growing seasons and the frequency of extreme highs and lows.

There is also considerable debate on how much temperatures will change and how fast. This largely depends on how humans respond to climate change.

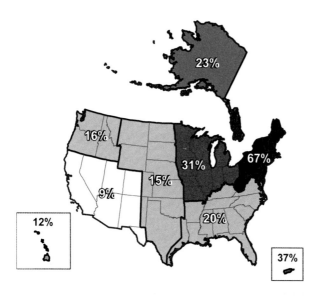

Figure 3.22. The map shows the percentage increase in very heavy precipitation (heaviest 1% of all events) by region (1958–2007).While changes in total rainfall may increase or decrease depending on location, the total rainfall coming in heavy events, defined as the top 1 percent of all rain events, will increase dramatically in some parts of the country, most notably the Midwest and Northeast. More rain, faster, equals more flood events. Illustration courtesy of United States Global Change Research Program

If we do little or nothing, the consequences will likely be more dire than if we make significant efforts to solve the problem. Again, implications of this change are variable. For example, it has effects on the seasonal timing of bird migration. The pollination of flowers by insects, which has evolved as a coevolution of precision timing, is changing drastically. And some tree species, which have adapted to need a certain number of days below freezing for their seeds to germinate, may have a hard time growing the next generation of trees. And warmer temperatures for longer periods mean increases in pest and disease transmission, since winter is the main mechanism in many temperature climates to stop invasions.

SPECIES COMPOSITION

To a large degree the combined effects of precipitation changes and warming temperatures will ultimately have the biggest impact on the species that a given area will be able to support in the long term. It's predicted that there will be drastic change in the ability of some species to survive and reproduce, as well as some significant shifts of overall forest types.

Individual Species: Winners and Losers

The US Forest Service has compiled a "Climate Change Tree Atlas"[36] which provides invaluable information about a wide variety of species under both low and high emissions scenarios. From this resource Cornell Extension educator Kristi Sullivan summarized species composition for common trees in New York State (see table 3.4). The predictions are startling to anyone familiar with the common trees and forest types in New York, as many of the key species are projected to do less well as the climate continues to shift.

An important detail to the results of the atlas is that the maps were intended to show where the future climate would be considered most desirable for a given species, not where species will exist or not exist. Many (including the authors at one point) viewed these maps as suggesting wholesale removal of key species from the climates of the eastern United States. The "losers" category doesn't mean that the trees will disappear overnight. What is implied is that the environmental conditions for these species will be considerably compromised over the medium to long term. This might be from seeds being unable to regenerate because of a lack of colder weather or die-off from an outbreak of insect or disease pressure. Other factors, including land management decisions, will of course play a role in determining the future.

Another fact to consider is that, especially with the keystone species starred in the list, a tree species that is compromised inevitably means that associated species and habitats will also be in trouble. For example, hemlock is a critical species around creek beds. Its year-round shade helps keep streams cooler, which keeps dissolved oxygen levels lower, which in turn supports aquatic life. The unique structure of the tree supports a wide range of bird species, including warblers, thrushes, and flycatchers. Depending on the rate of climate change, the species, which is the most abundant conifer in New York State, is projected to decrease by

Table 3.4. "Winner" and "Loser" Tree Species in New York

"Losers"	"Winners"
Red Maple	Northern Red Oak*
Sugar Maple*	White Oak
White Ash*	Black Oak
Yellow Birch	Chestnut Oak
Eastern Hemlock*	Yellow Poplar
Eastern White Pine*	Sassafras
American Beech*	Eastern Red Cedar
Black Cherry	Hackberry
Quaking Aspen	Honey Locust
Black Ash	Red Mulberry
Balsam Fir	Persimmon
Red Spruce	Elm

*Keystone species; i.e., associated species and habitats will be negatively affected

This chart was taken from a presentation given by Kristi Sullivan of the Department of Natural Resources at Cornell,[37] who based it on data from the US Forest Service Climate Change Atlas data, which highlights that the climate of New York State will be less desirable for the "losers" and more desirable for the "winners."[38]

25 to 50 percent. This could be further exacerbated by the presence of the hemlock woolly adelgid, a pest that rapidly consumes and kills hemlock and has a life cycle that is greatly improved with warmer winter temperatures.[39] Indeed, in the cold temperate forest the low temperatures in winter serve to kill off and shorten life cycles of many diseases and pests. The example of the hemlock reflects the complexity of each of the tree species' struggles in the forest. As was discussed in the earlier ecology section, a loss or decline of a species is the loss or decline of a habitat and, ultimately, the loss or decline of an ecosystem.

Change in Overall Forest Type

Another indicator of change is the bigger pattern of forest types discussed earlier in the chapter. These patterns help us define the abiotic factors and species compositions that make up the overall character of the ecosystem. For much of the eastern temperate United States, there will be a significant change, one that in many cases "simplifies" the present complexity of forest types. Based again on climate modeling, the trend is toward a forest that will largely be based on the oak–hickory type. This forest type, which is currently dominant in the central Appalachian Mountains, will likely compose most of the eastern forests.

While this general trend may be true, based on the dynamic forces of ecosystems it's quite possible that what "oak–hickory" forest means today will not be the same as it will tomorrow. Mike Farrell, director of the Cornell Maple Program, did part of his dissertation looking at trends in oak–hickory and maple–beech–birch forest types. A common pattern was that the dominant understory species were not as expected; sugar and red maple seedlings have been prolific in oak–hickory forests, while beech appears to be dominating much of the maple–beech–birch.[40] The actuality of these findings indicates that the likely scenario for the future is (of course) not black and white but rather that our forests will be a mixture of types, or something new altogether.

This reality is one that is hard to comprehend, and yet some level of acceptance is important, perhaps even critical to humans' survival as a species. Coming to terms with these projections affects the trees that are planted, the management strategies employed, and the types of flora and fauna that should be focused on.

Strategies toward Designing for Change

While the exact timing, scope, and breadth of climate change is unknown and will vary among locations in the cool temperate forest, some general patterns and templates can be a useful way to orient general thinking and planning for a successful forest farm.

ECOSYSTEM ANALOGS

The idea of ecosystem analogs is to examine and learn from ecosystems and climate regions that offer a potential glimpse into the future. This means both traveling to other regions of the world that rest on the same latitude and also traveling south, to warmer regions where, especially in the eastern United States, species will be migrating northward as climate warms.

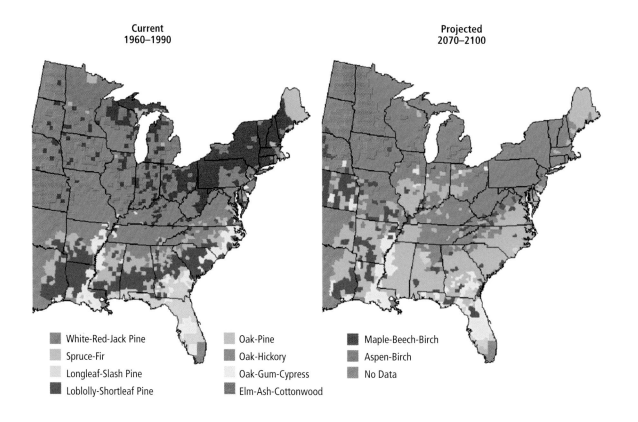

Current
1960–1990

Projected
2070–2100

White-Red-Jack Pine
Spruce-Fir
Longleaf-Slash Pine
Loblolly-Shortleaf Pine

Oak-Pine
Oak-Hickory
Oak-Gum-Cypress
Elm-Ash-Cottonwood

Maple-Beech-Birch
Aspen-Birch
No Data

Figure 3.23. Predicted changes in forest type, based on a midrange warming scenario. Note the considerable shift in the northeastern United States and the southern coastal areas. Illustration courtesy of United States Global Change Research Program

Since New York's climate is predicted to look somewhat like that of Maryland or Pennsylvania under a lower emissions forecast and more like South Carolina or Georgia under the high emissions forecast, a road trip to explore the forests in this range would be prudent for any forest farmer.

In addition, networks of growers that cultivate NTFPs can assist in the sharing of how management and decisions change with changing climate conditions. For example, mushroom growers in the southeastern United States often rely more on warm weather strains of shiitake, which can fruit well under hot weather conditions. Growers in the Northeast may need to incorporate more of these strains as the years become warmer and warmer.

If we think that much of the East Coast is going to be influenced to some degree by the oak–hickory forest type, then we had best learn as much as we can

about it. The current range of this forest type includes the former range of the oak–chestnut forest, which died off from a fungal blight in the early twentieth century. The key species in this forest type are hickories (at least 30 percent) and oaks, with tulip poplar as a common overstory tree. As shown in table 3.5, the species palate of the oak–hickory forest provides some design and management ideas for the future. Likely, the best strategy is to combine species from this forest type along with those of the current forest type a given forest farm has.

When looking at these characteristics in comparison to those of the maple–beech–birch forest type that has been dominating much of the eastern woods, several dramatic differences emerge. For one, the maple–beech–birch has a tendency to become very thick and shaded over time, versus the more sun-loving species of the oak–hickory type. Thus, the

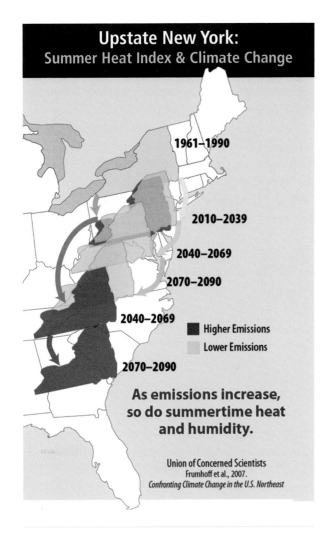

Upstate New York:
Summer Heat Index & Climate Change

1961–1990

2010–2039

2040–2069

2070–2090

2040–2069

2070–2090

Higher Emissions
Lower Emissions

**As emissions increase,
so do summertime heat
and humidity.**

Union of Concerned Scientists
Frumhoff et al., 2007.
Confronting Climate Change in the U.S. Northeast

Figure 3.24. Under low emissions scenarios New York's climate will generally be more like Maryland by the end of the century, but under higher emissions scenarios it will be more like Georgia. Illustration courtesy of Union of Concerned Scientists

Table 3.5. Species Associated with Oak–Hickory Forests

Overstory trees	Red oak, black oak, scarlet oak, white oak, chestnut oak, pignut hickory, bitternut hickory, shagbark hickory
Understory trees	Flowering dogwood, Juneberry, mountain laurel, sassafras, hawthorn, witch hazel, striped maple
Shrubs	Blueberry, viburnum, mountain laurel, rhododendron
Herbaceous	Solomon's seal, mayapple, trillium, black snakeroot, wild ginger, bellwort, aster, cinquefoil, and goldenrod

Recognizing this forest type (or the one likely coming to your area) gives some clear directives for management:

1. Manage existing forests to include and support more of the species of the projected forest type.
2. Consider the practice of "assisted migration," where varieties, plants, and cultivars adapted to warmer climates are brought into a climate zone and grown for adaptability.

The importance of the second action item is that while trees have indeed migrated long distances in the past, they take their sweet time doing it. Migration rates following the last ice age have been estimated to occur at a pace of around 100 meters per year. To keep pace with a shifting climate and regenerate under suitable habitat conditions, trees will need to migrate at a rate upward of 10 kilometers per year, a rate that is a hundred times faster than any previously documented.[41] The entire concept around assisted migration is fascinating, must-read material for forest farmers.[42]

PREFERENCE FOR ADAPTABLE SPECIES

In addition to paying attention to the types of species that will likely dominate the forests of the future, forest farmers can also seek out species that are both highly adaptable and multifunctional. While the list should start with native species, it's important

oak–hickory forest type needs frequent disturbance to open up spaces to new light and thus allow for regeneration of new species. In its more natural mechanism the disturbance takes form in small, low-intensity fires that promote competitive seedlings while decreasing competition. So maintaining a more open canopy (or considering fire management) is critical. The production of high-calorie foods is also dramatically increased with the hickory–oak forest. This has positive implications for humans but also for wildlife.

to also include plants that, while not having origin to a place, provide significant features that may be critical to continued survival in a changing climate. In playing with species that haven't been in the ecosystem for very long, it's important to observe and be vigilant, to ensure that things don't get out of hand. Examples of some species that are highly adaptable include black locust, poplar, willow, and red alder. These species will all be profiled in later portions of the book, as interest in them stems from both their willingness to adapt to changing climate conditions and their usefulness as a food, material, medicine, or other benefit.

DESIGN FOR EXTREMES

Asking the question, "What if there is prolonged drought or flood on my farm?" is a crucial one for forest farmers. In addition to analyzing sites and systems for the types of catastrophes that may become more normal down the road, it should be expected that the "new normal" will be an increase of years with excessive precipitation and also years when drought is present. Designing for this reality ahead of time is important, to save time, energy, and frustration.

As a rule of thumb for drought, consider the water needs on the farm and plan for what you will do in a year when your landscape receives only a third of its normal precipitation. Backup water supplies and strategies should be within reach in a matter of days, should problems occur.

For flooding, a careful analysis of water flows on your farm should be conducted, with the potential of installing swales, ditches, and even small ponds to divert, capture, and direct excessive water away from precious crops, buildings, and farm infrastructure. Design for water catchment will be discussed more in chapter 10.

OTHER STRATEGIES

While several of the key items forest farmers should focus on have been highlighted here, the list is by no means complete. It's recommended that temperate forest farmers take time up front to observe and analyze their particular situation in detail, before jumping into production systems that might fail under a threat of extreme weather or other climate changes. It's not the most fun part of forest farming, but there is no question that some forethought and planning will save one considerable time and money dealing with future challenges.

As with many disciplines, context is everything. Now that readers are equipped with the definitions presented in chapter 1, along with a historical narrative and some basics in ecology, forest types, and effects of climate change, it's time to get a bit deeper into the specific systems of cultivation for the wide range of nontimber forest products that forest farming has to offer.

Although it's fun and compelling to get into the nitty-gritty of how to grow this medicinal or that mushroom, readers are encouraged to constantly reframe their thinking in the big picture presented in the last three chapters. If forest farming is seen only as one thing, then all will be lost. Forest farmers will not persist if they are only in it to make money, or only in it for the fun of spending time in the woods, or only in it for a sane response to climate change. It takes all of these motivations, and more, to build a complete forest farm.

The needs of the people must be balanced with the health of the forest. The short-term gains and yields need to be balanced with the long-term products. Forest farming starts, and ends, in the forest. And the best lesson the forest can teach us is that good, healthy ecosystems take time. Indeed, if the full effects of work are seen in one lifetime, the forest farmer may simply not be thinking long enough.

CASE STUDY: MIMICKING THE OAK SAVANNA FOR RESTORATION AGRICULTURE
NEW FOREST FARM
VIOLA, WISCONSIN

Arriving at the farm on a very hot September day, I was a bit disappointed that I'd failed in successfully scheduling a time to cross paths with Mark Shepard and tour the farm he maintains in southeastern Wisconsin with his wife Jen. Mark was in Michigan giving a talk, probably in relation to his recently released book, *Restoration Agriculture*, which he defines as "the intentional design of productive agriculture ecosystems that are patterned after natural ecosystems." Even though I wouldn't get to talk to the person behind the system, I had come this far, and I figured I would learn something from wandering around the place. Walking around the landscape alone was an incredible experience, as I was forced to discover and observe the system, rather than simply having someone fill in the details.

Since I have a large midwestern family from Illinois and Wisconsin, I figured that as I drove to the farm I would see a landscape like much of the Midwest, one that was flat, open, and largely covered in corn and soybeans, with the occasional farmhouse or grain silo off in the distance. Surprisingly, though, the land in this part of the country is instead rather hilly, carved from glaciers much like my home in the Finger Lakes of New York. For permaculture farmers like Shepard, slope is seen as a good thing. It means being able to work with gravity—and more specifically water—in the landscape. Flat land is certainly good for tractors but not for multifaceted, tree-crop farms.

While the boss may have been out of town, the farm was anything but vacant. A decent-size crew of twenty-somethings were harvesting hazelnuts, as I'd picked the peak harvest season for the timing of my visit. The hybrid hazelnut varieties Mark had begun planting almost twenty years ago had matured far faster than the technology (see the Badgersett case study in chapter 4, page 139), and there is not currently the machinery to harvest at a larger scale. Instead, Mark relies on past students of his permaculture classes, idealistic young traveler types, and local Amish kids to harvest his nuts, which were abundantly weighing down the branches of 10-foot-tall hazel shrubs.

Mark Shepard approaches farming with a mechanical engineering and ecology background as perspective. His approach to farming the 106-acre landscape is simple, at least in theory. Taking nature as a model, Shepard has designed the farm based on the species found in the oak–savanna ecosystem, one of the most prolific and common types in the Midwest. From the species in this system he has plucked the most economical ones (including hazelnut, chestnut, currants, and raspberries), planting them in different patterns around the farmscape. Rather than lay trees out in straight rows, the farm was plowed with a Keyline plow to decompact soil and assist in water infiltration. The resulting pattern works with the contoured patterns of the land while allowing for rows and access by machinery.

Readers should note that the intention of this farm is *not* to become a forest, as this isn't the climax ecosystem of this part of the world. As Mark says:

This land *wasn't* a forest 300 years ago . . . It was a savanna . . . Lightly wooded with an open canopy so that grasses and other prairie plants could flourish. We do *not* intend to mimic and restore a forest, but to mimic and restore a savanna. What we're doing could be used as the bridge to economically get to a closed-canopy forest and *should* be done where closed canopy forest was the dominant vegetation type, but we're planning on "setting back succession" every time the canopy closes too much to grow good grass."

The other point Shepard makes in his book and presentations is that savanna ecosystems are potentially the most productive in terms of photosynthetic surface area of the combined grasses, shrubs, and trees, widely spaced, though much of the data comparing savanna with forest ecosystems says otherwise (see figure 3.14 on page 59); for example, savannah is 3150 k/cal/sq m/year vs. 5850 for temperate forest. Shepard notes that not all the data agrees and that his perspective is based on State University of New York College of Environmental Science and Forestry (SUNY–ESF) data on short-rotation woody cropping systems (SRWC), which are more analogous to what he is doing. Clarifying this point, Mark wrote to me in an e-mail that "natural savannas won't have the hyper-teenage growth of a regularly coppiced and regularly mowed one . . . [and] will be a lot more random in their grazing, browsing and fire (coppice). SRWC will be doing this much more regularly."

Having planted thousands of trees on the farm, Mark has adopted a management strategy that he calls the "STUN method," or sheer, total, utter neglect. The idea is that without extra care, the plants will sort themselves out and those with an advantage will survive. This is another aspect of how New Forest Farm mimics nature (natural selection), while also dealing with the realities of busy farmers. As Mark puts it,

> When it comes to tree planting most people think that they're going to take excellent care of their trees and they want to take immaculate care of them, but somehow life gets in the way and the trees get ignored to a certain degree. Since this is what happens anyway, we might as well dispose of all the guilt and plan to ignore our trees. . . . If some trees of mine want to die, I say, "Good riddance!"

While this method may be appealing to some, it's one that requires a long time for nature to sort out those superior specimens. To this end Mark employed a transitional strategy to the farm that allowed for the economics to make sense. Since the beginning eighteen years ago, Mark has grown and sold wholesale organic vegetables cultivated in the alleys between rows of young tree crops. This provided an immediate yield for the farm, one Mark has personally transitioned from needing as teaching, consulting, and the beginning yields from his tree crops filled the void. Two of the young men I met on the farm were the benefactors of this transition; one grows the vegetables each summer, while another manages the animal rotations. Through this strategy the farm now "incubates" young aspiring farmers, helping them get started.

Figure 3.25. Black locust interplanted with hybrid chestnut. The locust is coppiced and the woody material used as a nitrogen-rich mulch for the nut trees, the longer-term canopy species.

Indeed, this type of knowledge transfer is how I was first exposed to Mark's work. Farmer friends of mine, Melissa Madden and Garrett Miller, spent a season in Wisconsin at the farm, taking on the vegetable production role and learning firsthand from Mark's system. They brought this template back to New York, where they started the Good Life Farm, which combines Mark's system with a number of other strategies that met the particulars of their site and their personal goals.

They too have found a niche in annual vegetable production as a transition while they wait for their apples (mainly for cider), peaches, and asparagus to begin yielding. The farm offers a spring CSA, which fills the gap when most other farms in the area aren't producing. They graze their orchards and asparagus rows, which are arranged in alley cropping form, with turkeys, a lucrative harvest for the holiday market. After the initial establishment of the plantings in 2008, this past season was the first year that all three

Figure 3.26. Asparagus alley cropping between rows of chestnut, hazelnut, and other tree species. Annual and perennial vegetables provided New Forest Farm with initial income streams before the trees started to bear.

Figure 3.27. Successful polyculture at Good Life Farm in Interlaken, New York, of apples, asparagus, and turkeys. Inspiration for this system was gained by time the farmers spent at New Forest Farm.

layers of this polyculture (apple/peach, asparagus, and turkeys) yielded for them.

Walking around New Forest Farm that day, I was struck by the mosaic of different combinations of trees, shrubs, and herbaceous vegetation that I discovered as I explored the property. Here was an example of how to farm the woods if you don't have woods to start with. Evidence of pulses of activity was clear, from the piles of woody brush lying near a coppiced tree to the clear gradient of intensity as I got farther from the main barn, where the vegetable production was focused. I was pleasantly surprised as I rounded a corner or came down a hill and kept stumbling across different animals; first the pigs wallowing and sleeping in the shade of some chestnut trees; then guineas and chickens near the house, foraging underneath a mulberry; and finally the cows grazing down in the bottomlands of the property.

The whole system struck me as semiwild. Here was a farm where trees, other plants, and animals had been set in motion, then largely left to do their thing. I imagined one day the abundance this system could produce, even if neglected by the management of human hands. This farm was growing a forest, or more accurately a savanna, returning the land to an image of its former self, one that balanced the complexity of nature with the need for producing human food. It was a place where an ecosystem was left in the farmer's footsteps, rather than a dry, depleted, and desolate field.

In the context of agroforestry Mark's farm employs a wide swath of strategies. His tree crops, interplanted with rows of vegetables, are straight-up alley cropping. The animals rotating throughout the site are engaging in silvopasture. One portion of the farm that faces the wind is planted in a multifunctional windbreak. And some of the chestnuts that haven't fared so well are cut and used for mushroom cultivation, or forest farming. In most cases, this mixture of practices is the approach farms should take, rather than settling on any one as "the best." In fact, each offers an opportunity to strengthen the others over time.

Mark's book is an inspiring read that is part theory, part practice, and part rant against everything wrong with the current industrial model. The anecdotes and concepts he describes offer a new generation of farmers a lot to consider. His work embodies much of what we could consider a "pioneer species" in a forest; fast growing, experimental, pushing the edges, questioning the paradigm. As I drove away from the landscape where trees, plants, animals, and people all had a place together, the return to the scenery of monoculture corn and soy fields was all the more dramatic, if not frankly boring. For all the ingenuity of humans, I thought, we can do better than cornfields. We can do much better. And New Forest Farm was a clear example, proof that it could be done.

— Steve

4 Food from the Forest: Fruits, Nuts, and More

The Forest Feast

At the Cornell campus in Ithaca, New York, there is a woodlot called the MacDaniels Nut Grove, where students and the public come to learn about forest farming. In the fall of 2006, a memorable culinary occasion was held there called the forest feast, in conjunction with an annual fall course called Practicum in Forest Farming. The feast consisted of an array of edible nontimber forest products. Each of the twenty or so students and their guests brought along one edible to share from the nut grove or a nearby forest. The main course was roast goat. This wasn't just any goat. This goat had been raised at Cornell's Arnot Forest as part of a research project called "Goats in the Woods," to see if goats could be used as part of a forest management plan to control undesirable ("weedy") small-diameter American beech and striped maple (see chapter 9). From spring to late summer the yearling goats fed on the bark of these trees (plus or minus supplemental feed) and were then sold as meat animals. For the forest feast one of these goats was roasted all afternoon on a spit of green ash over a bed of oak and hickory coals.

The MacDaniels Nut Grove has an abundance of hickory and black walnut trees, with a few white oak trees and some Chinese chestnuts, which played a role in the feast. Nyla, who had recently spent her summer internship shepherding the goats in the woods, was smashing hickory nuts, shells and all, in a large wooden mortar and pestle. These were thrown into a pot of boiling water to separate the rich oil, called pawcohiccora (hickory milk), from the nut meat/shell mash (for more information on this see quote by William Bartram on page 114). After everyone at the forest feast got a taste of the hickory milk, the rest was used to fry acorn ash cakes. First, the acorns were leached several times in water to remove the bitter tannins, then ground into a flour to make the ash cakes ("pancakes") that were cooked on a hot stone. Several other students were cooking a stir-fry of pickled ramps (wild leek), forest-cultivated shiitake (*Lentinula edodes*) mushrooms, and wild lion's mane (*Hericium* spp.) mushrooms, as well as wildcrafted black trumpets (*Craterellus cornucopioides*) and porcini (*Boletus edulis*). Jim made a fragrant porridge from the inner bark of slippery elm (*Ulmus rubra*). Forest fruits were in abundance. Abdoul came up with "Cornus mas sauce," made from the fruit of Cornelian cherry dogwood (*Cornus mas*). Just a little tart, it went well with the goat.

There was a soup made from nettle and lamb's-quarter, and applesauce made from wild crab apples. Beverages included tea made from hemlock and pine, and dandelion wine (from Isaac's lawn). For dessert there were brownies made with hickory nuts and walnuts collected from the site. Marguerite made a pawpaw mousse, and Sefra contributed a wild berry torte made from blackberries, raspberries, elderberries, and blueberries!

Although this description of the forest feast may seem like the introduction to a forest cookbook (not a bad idea), the real point is to emphasize the considerable diversity of edible forest products that were sourced from local temperate deciduous forests. Some of the nontimber forest products were deliberately grown (forest farmed) at the MacDaniels Nut Grove (pawpaws, shiitake mushrooms, oyster mushrooms, black raspberries, elderberries, hickory nuts, and walnuts), and others, such as the goat, were from a nearby research forest site. Other NTFPs were wildcrafted, including the blueberries, blackcaps (wild black raspberries), ramps, black trumpet and porcini mushrooms, white oak acorns, and elm bark.

Just as the feast at the MacDaniels Nut Grove combined cultivated with wild NTFPs, any managed forest farm may produce both. As mentioned in chapter 1, forest farming involves deliberate management of cultivated nontimber forest products, including bed preparation, planting, mulching, weeding, and so on, whereas wildcrafting does not involve such cultivation in the strict sense. Much of contemporary wildcrafting is focused on "gourmet" foods, such as wild mushrooms and ramps, and medicinals, such as ginseng, that are intended for sale to consumers. In addition to gathering wild foods, there are many that can successfully cultivate a smaller but significant cast of characters that offer amazing flavors, nutrition, and unique niche crops with commercial potential.

Getting a Yield: Light and the Forest

Besides mushrooms, which will be covered in depth in chapter 5, when we consider the majority of food crops in the forest farm, most are trees and woody shrubs that produce fruit and nuts. Anyone looking into the potential of these will notice that many plant profiles claim the plants will "tolerate" shade, but rarely do you see a mention of such crops thriving in shade. This is simply because production of fruits and nuts, from a biological perspective, is energy intensive. So while many plants will accumulate trunk wood, grow, and photosynthesize in surprisingly low light conditions, most need access to direct sunlight to produce a decent yield.

Literature often emphasizes the need for sunlight in fruiting plants, yet there are examples of high-yielding currants (*Ribes*), blackberries, pawpaws, and other fruits within the forest. As cultivars are bred and released, emphasis has not been on selection for any shade tolerance; in fact, in many cases, plants are selected for performance in open sun. This is another justification for forest farmers to plant seedling stock

Figure 4.1. Raspberries fruiting well in almost full shade. At Edible Acres, Trumansburg, New York.

in a variety of locations, so that selection can be done to maximize efficiency in more forested settings.

That said, there are multiple directions in which to take the concept of "light regime"; that is, the amount of available light and how it will affect both plant growth and fruiting. The good news is that healthy forests need to have a range of light regimes, and as forest farmers it is our job to help cycle them through the various stages.

DEFINING LIGHT REGIMES

Defining the difference in light regimes will assist in understanding their relationship to crop production. In the most basic sense classifications are made on the basis of the percentage of canopy cover; that is, the percentage of shade cast onto the forest floor.

It's important to remember that not all closed canopies are created equal. Even if dealing with an 80 to 100 percent canopy cover, both the type of tree and the height of the canopy can have implications for light coming through to lower layers of the forest. For example, at the MacDaniels Nut Grove, the area called Walnut Island is named for a small zone

Table 4.1. Light Regimes of Different Forest Canopies

Type	Description	% Shade of Canopy	Trees/Acre
Forest	Closed canopy forest	100%	> 25
Woodland	Widely spaced trees and shrubs with a canopy	40–99%	12–24
Thicket or shrubland	Vegetation dominated by shrubs and small trees	< 40%	< 11
Savanna	Open pasture with scattered trees	25–40%	1–11
Grassland	Mostly pasture with an occasional tree	< 25%	< 1

Note: There are no hard delineations between each type established in literature, so this table is useful only in relative terms. In most cases forest farming occurs in forests and woodlands, the distinction being some form of canopy (adapted from Jacke & Toensmeier, 2005,[1] and Nelson et al, 1998[2]).

with a seasonal creek that is dominated by several 80-year-old eastern black walnuts as the overstory. These trees range from 50 to 80 feet in height and cast shade over the site while still allowing for a significant amount of secondary light to penetrate. This amount of light has proven to be acceptable for many species, including pawpaw and elderberry, while fruit production on the raspberries planted there has been inadequate from a cropping standpoint. This contrasts significantly with the forest at Wellspring Forest Farm, which is a 100 percent canopy of sugar maple. This type produces a very dense shade in which

Figure 4.2. Walnut Island is a portion of the MacDaniels Nut Grove where a seasonal creek defines its separation from the rest of the forest. To the surprise of visitors, many crops thrive here in the shade and "toxic" (juglone-affected) soils associated with the black walnut (see more on juglone on page 104).

none of the above crops would fare well. Of course, the conditions are perfect for mushroom cultivation.

GAPS, CLEARINGS, AND EDGES

As mentioned in chapter 3, another way to take advantage of a natural forest pattern (disturbance) is to utilize (or create) spaces in and around the forest that get some extra sunlight throughout the course of the day and season. In wild forests this happens all the time: A tree falls in the forest and creates an opening where more sun-loving species can thrive. In addition, utilization of forest edges (including old hedgerows) can be seen as a form of forest farming that also offers some flexibility in terms of light access. Table 4.2 offers some suggestions of plants in this chapter in relationship to light conditions.

Plant Selection Criteria

There are a score or more of potential plants to choose from in temperate agroforestry. Here are some criteria that guided selection for this book:

- Evidence exists (literature or grower or other expert advice) of successful cultivation in a forest niche.
- Natural habitat of the plant is in the forest.
- Potential exists for both domestic and commercial applications.
- Demonstrated cultivation techniques exist, even if underdeveloped.
- There are health and nutritional benefits for humans.

Table 4.2. Relationship of Food Species to Light Regime

Will Grow Well in Shade	Good Yields Possible in Partial Shade	Best on Sunny Forest Edge, in Gaps, or in Field Plantings	Canopy Species: Need Access to Sunlight (when mature)
Pawpaw	Schisandra	Staghorn Sumac	Hickory
Elderberry	Hawthorn	*Ribes* spp.	Sugar Maple
Ramps	*Ribes* spp.	Autumn Olive	Chestnut
Spicebush	Honeyberry	*Rubus* spp.	Walnut
	Hazelnut	Hardy Kiwi	
	Juneberry	Chokeberry	
	Groundnut	Sunchoke	

Table 4.3. Relationship of Food Crops to Commercial Potential

Food Plants with Clear Commercial Value	Plants with Possible Commercial Value	Plants Best on the Hobby Scale/ Experimental
Pawpaw	Schisandra	Sumac
Elderberry	Hawthorn	Groundnut
Maple and Birch Syrup	*Rubus* spp.	Honeyberry
Black Walnut	Ramps (Wild leek)	Sunchoke
Hickory (Pecans)	Spicebush	*Ribes* spp.
Chestnuts	Juneberry	
Hazelnut	Hardy Kiwi	
Chokeberry	Black Walnut Syrup	

Note: Some of the species, notably nuts, in the left column are limited by a lack of machinery for large-scale harvest. Species in the middle column may be limited by access to niche markets (hardy kiwi, for example) or competition with field cultivation (*Rubus* spp.).

Figure 4.3. A pawpaw fruit in the running for the "best tasting pawpaw" at the Ohio Pawpaw Festival (see insert on page 90). Fruits can range in color from bright orange to a dull yellow or white color.

With the exception of nut trees and pawpaw, elderberry, and tree syrups, none of the foods in this chapter are currently grown seriously as forest-farmed crops for commercial sales. *Ribes* (gooseberries and currants), *Rubus* (raspberries and blackberries), pecans, Chinese or hybrid chestnuts, and hazelnuts are available commercially but are usually grown in full-sun, orchard-style plantings. Landowner objectives should drive design to incorporate these species in the scale that fits personal goals.

Fruits for the Forest Farm

While light may be the limiting factor in getting good yields, there are nonetheless several species that have evolved to benefit from the shade and shelter that the forest offers. Pawpaw and elderberry seem the most prime for commercial production, while the other fruits should be approached on a smaller scale, at least at first. Most of these fruits are currently grown commercially in full-sun environments, but forest farmers can start to push the edges and bring these plants closer to where most of them originated—in the woods.

PAWPAW

Neal Peterson, a West Virginia farmer and owner of Peterson Pawpaws,[3] blames refrigeration at least partly for the demise of the pawpaw as a common eaten fruit in America. After World War II the widespread use of refrigerated trucks meant vulnerable fruits (and vegetables) could be stored longer and travel long distances from farm to table. An "artificial selection" began to take shape, in which those foods that could adapt to this new industrial food system survived, while others, such as the pawpaw, which has limited shelf life even under refrigeration, simply couldn't keep up.

A ripe pawpaw looks like it's borderline rotten. The fruit is really soft, the skin often dark and bruised. And it's best to get the fruit right off the tree this way, though you can pick somewhat unripe pawpaws and store them for three weeks at about 35°F (2°C); a refrigerator is closer to 41°F (5°C).[4] The poor shelf life of pawpaw has led to its demise as a mainstream commodity. Yet with the renewed interest in local, nutrient-dense foods, and consumer willingness to take time to process and preserve foods with a limited season, the pawpaw may just make a comeback.

The pawpaw boasts these unique qualities that none of the others, while wonderful foods, can match:

- It has the largest fruit of any fruit tree that can be grown in the cool temperate climate. Large fruits may exceed 1 pound in size.
- The fruit is a relative (same family, Annonaceae) of the custard apple (*Annona reticulata*) and both looks

Figure 4.4. The virtues of the pawpaw are many, from being the largest tree fruit native to the eastern United States to its ease of cultivation and aesthetic form; not to mention, the fruits are delicious.

and tastes very "tropical"—hints of vanilla, mango, banana, and avocado are common descriptors.

- It's both shade tolerant and, unlike many species, it can grow in association with black walnut; i.e., it's juglone tolerant.[5]
- The pawpaw is high in vitamin C, magnesium, iron, copper, and manganese. It's a good source of potassium and several essential amino acids and also contains significant amounts of riboflavin, niacin, calcium, phosphorus, and zinc.
- Studies from Purdue and other institutions have indicated the presence of cancer-fighting compounds in the fruit.[6]

Imagine all this good stuff from a tree that's native to twenty-six states in the United States, with a range extending from northern Florida to southern Ontario and west to the eastern portion of Nebraska. In addition to being a great food crop, the tree has ornamental value, with a beautiful form and leaf structure.

Ready for the kicker? It is deer resistant! It's even goat resistant. Grower Chris Chmiel of Integration Acres in southeastern Ohio grazes goats with his pawpaws, with no negative consequences, save the occasional small tree that gets trampled. Its deterrent effect on browsing herbivores, both wild and domestic, is attributed to toxins in the leaves.

From a forest farming perspective, the pawpaw is interesting, as it is found naturally as an understory tree, sometimes in very dense shade. The pawpaw thrives in floodplains where the soils are moist and rich in organic matter. While hardy to zone 4, in areas of the East north of Ohio, it is recommended that the

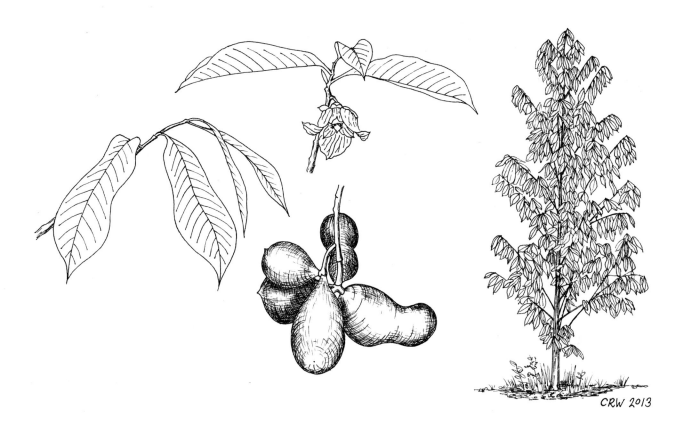

CRW 2013

Figure 4.5. Line drawing of the pawpaw tree. Note the wonderful form of the tree, which also lends itself to having ornamental value. Illustration by Carl Whittaker

trees be planted in warmer microclimates for good production. While many producers grow pawpaws in orchard-style plantings in full sun, the tree will fruit well in part shade and in forests with a full canopy, provided the overstory species are tall enough to allow for secondary light. Several pawpaws are thriving in the black walnut forest that is part of the MacDaniels Nut Grove at Cornell.

The basic planting requirements for the pawpaw are these:

- It likes loamy, well-drained soil.
- A pH between 5.5 and 7 is ideal.
- When young (years one through three), seedlings need to be protected from direct sunlight.
- At least two seedlings or different varieties or other genetically different trees (e.g., two different

cultivars) are necessary for cross-pollination (the trees are self-incompatible).
- If there are fewer than fifty trees in a planting, hand pollination is likely necessary.

Hand pollination may at first glance appear as a deterrent to growing pawpaws, yet all that is really required is to take a paintbrush out to the trees for one to two weeks in late spring (the last few weeks of May in New York) and transfer pollen from the mature flowers where pollen is loose. The key point is that this loose pollen must be transferred to the stamen of flowers in which the pollen mass is still solid. Of course, these trees must be genetically different (seedlings or different cultivars). Knowing both when pollen is ready to transfer and when the receiving flower is ready to accept is critical (see sidebar, Hand Pollination of Pawpaws, Step by Step).

HAND POLLINATION OF PAWPAWS, STEP BY STEP

For those planting fewer than a substantial number (thirty to fifty trees), hand pollination is recommended to ensure good fruit success. This equates to a minimum of about twenty minutes a day, two or three days a week, for two weeks. Of course, more than this can be done, as the flowering season can last up to a month.

The flower of the pawpaw is "perfect," meaning that it has both male and female flower parts. These parts do not mature at the same time, which is the biological mechanism called dichogamy, by which the tree avoids self-pollination. Thus, the task of hand pollination is bringing the pollen from the anthers of mature flowers (pollen will be loose) to the stigmas (female) that are receptive.

Follow the blossom development of the pawpaw through these stages:

Stage 1: The brown velvety bud expands and begins to open, revealing the outer petals, which are green in color.
Stage 2: The petals begin to turn color, showing a mixture of green and maroon as they further open and expand (figure 4.6 left). The stigmas are swollen and glossy or fuzzy.

Figure 4.6. As flowers mature they turn from green to dark red" (left). The color of the flower is blood red, and is know as a "carrion" flower, evolved to attract flies (pollinators) that are drawn to rotten meat (right).

Figure 4.7. The flower on the left has a bright green swollen stigma, and the pollen mass surrounding it remains hard. It is receptive to pollen from a flower with loose pollen, as in the one on the right. Once pollen has come fully loose, the stigma is no longer receptive to pollen.

Stage 3: The petals reach full size, with a dark maroon color. The flower is bell-like and usually positioned downward, with a small opening for admitting insects or a paintbrush. (figure 4.6 right)

Stage 4: The petals flare back substantially, exposing the ball of anthers, which become loose (figure 4.7 left).

Since the flowers develop at different rates over the course of many weeks, the hand pollinator can be on the lookout for male flowers, harvest the pollen from one tree, and deposit it on the ripe female flowers of a neighboring tree. For taller trees this will require a ladder, and in general it is best to manage pawpaws to a height of 6 to 8 feet if hand pollination will be a regular activity. This also keeps harvesting easy.

Figure 4.8. Coauthor Ken Mudge hand pollinating trees in the MacDaniels Nut Grove. The minimal work involved in hand pollination is well worth the rewards.

Pawpaw trees left unmanaged can grow to a height of 30 feet, so it is recommend that the trees be topped when they begin grow beyond harvest and pollination height, usually at eight to ten years.

In addition to the food yields of the pawpaw, the fruit tree deserves a place in every backyard butterfly garden, as it is the only host plant for the caterpillar of the zebra swallowtail. In fact, it is the toxicity of the pawpaw that gives the butterfly its protection from predators.

Propagation of the pawpaw is achievable from seed, which can be collected from fresh fruits. Seeds must be stratified (cold, moist pretreatment to overcome dormancy), either by letting them overwinter in the ground or by stratification in a refrigerator. See chapter 7 for additional information about seed stratification. However, like most trees that are cross-pollinated, pawpaw seedlings are not genetically uniform, and the outcome will be some degree of variability among different trees and between generations. For this reason many species are deliberately propagated with clones that have been selected for superior characteristics that will come true to type. In the case of the pawpaw some clones (cultivars) have been selected for early ripening, for use in more northerly locations. To maintain this early ripening characteristic from generation to genera-

Figure 4.9. Pawpaws are a taprooted plant, which makes them difficult to transplant. The root develops first, and a shoot from a planted seed may not emerge until the peak of summer. It's often recommended that seedlings be grown in containers for the first two seasons, then transplanted to field conditions.

tion, a cultivar must be grafted (or budded). See chapter 7 for more information on grafting and budding.

The bottom line is that almost anyone who tastes a pawpaw is hooked. It's a truly wonderful culinary experience—while pawpaw can be used in a number

CASE STUDY: OHIO PAWPAW FESTIVAL
ALBANY, OHIO

One way to build interest around a niche product is to throw a party. Chris Chmiel, along with his wife Michelle, owns Integration Acres, a small farm that sells a variety of forest and dairy (goat) products, most notably fresh and frozen pawpaw, and a range of jams and other value-added products. Chris started the festival in 1999, two years after the farm began offering pawpaw commercially. This was an enterprise that was not planned but discovered, when Chris and Michelle bought 18 acres to start a farm in Athens County, Ohio. After observing pound after pound of the native pawpaw rotting on the ground, Chris did some further research and found that the farm was situated in one of the best pawpaw growing regions in the world.

The festival, which marked its fifteenth year in 2013, has been growing in size and interest since the beginning. It features an educational pawpaw tent, competitions for the best-tasting and largest pawpaw, a pawpaw cook-off, and even a pawpaw-eating contest. Visitors have plenty to see, including craft and food vendors, who compete for best pawpaw food dishes and craft items, as well as local music and a beer tent, which features a surprising array of pawpaw brews for visitors to taste.

The site of the festival is Lake Snowden, an educational and recreational park owned by Hocking College. It is unique because, in addition to providing the grounds for a festival, including plenty of parking, camping, and open space, it hosts a grove of wild pawpaws, growing in their native habitat, an oak–hickory forest along a riparian creek zone. Extension educators brought visitors on tours of the grove and explained native flora and fauna of this ecosystem. Several organizations, including the newly renamed North American Pawpaw Growers Association, offered plenty of free tastings and even seedlings to those who joined as a member.

The event has connected growers and consumers with other farm- and history-related organizations, now sponsors, including Snowville Creamery and

Figure 4.10. One of the pawpaw beers on tap for tasting at the festival, in a pint glass displaying the 2013 festival logo.

Figure 4.11. A table full of pawpaws waiting to be tasted in the competition for best-tasting pawpaw. The fruit has an incredible range in color and associated flavor. Darker fruits are said to taste more like mango and lighter (white) ones more like avocado.

Ohio's Hill Country Heritage Area program. In 2009 Ohio senator Jim Stewart visited the festival and issued a proclamation to Chris and the festival that officially declared the pawpaw the state fruit of Ohio (a measure that had been proposed by a previous senator). The festival attendance has grown from hundreds to thousands of visitors, many of whom return year after year.

The success of Chmiel and others who have stepped in to organize the festival as it has grown is a good example for other forest farmers, highlighting that the good way to promote an obscure product or practice is to offer a fun and educational event to get the word out. Other locations may not start with a three-day festival (it took this one almost ten years to get there), but even a one-day affair highlighting the history, culture, and uses of a product offers a chance to connect with the community, as well as improve sales.

This same model has been used successfully by the maple industry, which supports open houses on a statewide basis each season. Chris has expanded on this notion, holding a one-day Spicebush Celebration (see page 136 for information on the plant) to highlight the virtue of this native forest shrub. The day includes tastings of spicebush beer and gin, educational talks on the spicebush swallowtail, music, and a cook-off.

For more info: http://www.ohiopawpawfest.com/.

of recipes, the best method of eating is simply to slice the fruit and eat it right out of the peel with a spoon. It's a fruit perfect for the backyard grower, as two to four trees would provide all the fruit a family could eat. The time is also coming for its reemergence as a commercial crop. Farmers looking for something different, especially if they are willing to process the pulp of fruit they can't sell fresh (easiest as frozen pulp) have a niche market just begging for attention.

ELDERBERRY

The elderberry (*Sambucus* spp.) has a long history of cultivation and excellent food and medicinal qualities. Especially compared to the pawpaw, the native shrub has a much better track record in terms of consumer awareness; in a 2011 study on consumer preference for the fruit by the University of Missouri, 1,043 households were surveyed throughout the United States.[7] Results showed one-third of respondents to be familiar with elderberry. The key to enjoying elderberry is in processing the fruit into syrups, jams, or other concoctions, including phenomenal wines made from both the elderflowers and the berries. The form it takes is shrubby, and it's a wonderful windbreak species, as well as good for wildlife. To top it off, propagation of elderberry from stake cuttings is almost as easy as for willow, and maintenance is minimal.

Figure 4.12. A developing flower set of the elderberry against a blue July sky in upstate New York.

Elderberries could be considered one of those plants that is a "food medicine," as it is both enjoyed in a wide array of dessert delicacies and beverages and equally valued for its benefits to human health. One study demonstrated that elderberry might have a measurable effect in treating flu symptoms.[8] It's also been cited to

Figure 4.13. Fruiting elderberry at a University of Missouri Planting. Photograph courtesy of Terry Durham

---

RECIPE: SIMPLE ELDERBERRY SYRUP

4 cups water
1 cup elderberries
1 cup raw honey

Pour water into medium saucepan and add elderberries. Bring to a boil and cover, reducing heat and simmering for 45 to 60 minutes, until half of the liquid remains. Remove from heat, and let sit until just warm to the touch.

Pour through a strainer into a quart mason jar. To the warm liquid, add honey and stir. Divide into pint jars and cover. Store in a cool and dark location. Standard dose: Daily: ½ tsp to 1 tsp for kids; ½ Tbsp to 1 Tbsp for adults

When showing signs of cold/cough/flu, take the normal dose every two to three hours instead until symptoms disappear.[11]

---

have positive effects on inflammation, allergies, and overall respiratory health.

Elderberry shows up in the foodstuffs and medicines of countries around the world, notably in Europe and in traditional Chinese medicine. One book from the 1930s called it "the medicine chest of country people."[9] Most often it is the flower, not the fruit, that is said to be most valuable medicinally. This isn't to diminish the excellent nutritional profile of the berries, which are very high in vitamin C and contain vitamins A and B6.[10]

The shrub grows rather fast and ranges from a height of 8 feet to 20 feet, though it responds very well to pruning and even coppicing, which makes it easy to maintain at harvestable height. It is tolerant of most soil types and even grows decently in heavy clay. Elderberries are said to be very shade tolerant and have been witnessed fruiting rather well in deep forests, though more notably along logging roads and other gaps in the canopy where they receive at least a few hours of sun in the peak of summer. They do well in light regimes with a high canopy. At MacDaniels Nut Grove recent plantings underneath black walnut appear to confirm resistance to the toxicity of the walnut.

Once they are planted, little maintenance is required. The root system sends up several new shoots (suckers) each season. These canes usually reach their full height in the first year, then develop lateral branches in the second year. Fruit set is best on the current-season growth of second-year canes but will often also fruit on first year growth. It's recommended that during the dormant season any dead or diseased canes be removed.

Elderberry doesn't bloom until late June, which means there is no danger of damage from frost. The berries develop and are ready to harvest in August or September. When ripe, most if not all the fruits will be soft and dark blue or purple to black in color. Berries should be harvested in the morning and put immediately in the fridge, as they soften easily. Remove the entire bunch by pruning, then strip berries for use. They can be frozen if desired, or processed in multiple ways, typically as juice (see sidebar, Recipe: Simple Elderberry Syrup).

Be aware that many publications refer to mild to medium toxicity of the raw berries. Most people can eat a few and are just fine. It's generally recommended to at least steam the berries, if not process them into

Figure 4.14. Elderberry is easy to propagate from cuttings, which can be taken almost anytime during the year. They should be rooted in a potting mix soon after cutting and watered consistently for best results, although a leafless hardwood cutting in the fall can be stuck directly into the ground.

Table 4.4. Suggested Commercial Prices for Elderberry Products

Product	Sale Price
Cuttings	$1–2.50 each
Plants	$5.00 each, 6 for $25.00
Fresh Berries	$.50/lb, with stems, to winery
	$1.25 pick your own
	$3.00/lb destemmed and sold retail
Wine	$10–14/bottle
Juice	$12–17 per 11 oz bottle retail
Concentrate	$25 per 375 ml bottle retail

Source: University of Missouri Factsheet[12]

other products—which is what most people end up doing. For most people the berries are too sour to be eaten raw.

Elderberry is a pioneer species that is fast growing, easy to propagate, tolerant of a wide range of conditions. It can effectively be employed in almost every light regime of the forest farm and will offer great benefits to people and the ecosystem. Windbreaks, hedges, and riparian plantings are good places to consider including elderberry.

Propagation of elderberry ranks up there with the easiest of plants. Cuttings can be taken anytime. Cuttings from the previous season's growth, taken in winter to early spring before budbreak, can be treated with a rooting hormone to accelerate rooting. Cuttings of half-ripe wood can be satisfactorily rooted even in summer. For more information about rooting of cuttings, see chapter 7.

As a final note, the economic potential of the elderberry is quite high. Of the forest-farmed fruits it is perhaps second only to the pawpaw in its economic potential. Some of the more lucrative uses are juicing the berries and either marketing medicinal syrup or fermenting the berries into a wine or cordial for sale to niche markets. Commercial growers may want to consider both the native American elderberry (*Sambucus canadensis*) and its European relative (*Sambucus nigra*).[13]

RIBES SPP.: CURRANTS, GOOSEBERRIES, AND JOSTABERRIES

Ribes species are abundant in the understory of many temperate forests. They are recognizable by their serrated, maple-like leaves and low growing form. Native wild populations are highly variable in their production of fruit. If a forest farmer is lucky enough to stumble across a specimen fruiting well in the shade, he should take note and make plans to propagate from the mother plant.

Though native to the continent, cultivars of gooseberries and currants were brought to America early, introduced in the Massachusetts Bay Colony in 1629. By the end of the nineteenth century commercial acreage was common and totaled almost 7,000 acres,[14] until 1909, when the white pine blister rust fungus

(*Cronartium ribicola*) first appeared and spread from an import to New York State. All *Ribes*, but notably black currants and gooseberries, are considered an intermediary host for the fungus, which needs to spend time on the plant to complete its life cycle. A federal ban on planting was enacted, but still much damage was done, with half the pines in New Hampshire infected as one example of the destruction.

The federal ban was lifted in 1966, though several states, including Delaware, New Jersey, and North Carolina, prohibit the importation and culture of all currants and gooseberries. Maine, Massachusetts, Michigan, New Hampshire, New York, Rhode Island, and Vermont allow planting only in certain areas. Culture of black currants in Massachusetts is prohibited. A forest farmer planting this species should be aware of any neighboring white pine forests, nurseries, or plantings that could be affected. In 2011 a mutant form of white pine blister rust was discovered by Cornell University researcher Kerik Cox in Connecticut. After two years of study, some scientists now believe a large number of cultivars previously thought to be immune to the fungus may be susceptible.[15]

Propagation of *Ribes* is rather easy; layering of the canes in the summer or fall after fruiting is best, although they also root easily from cuttings. Layering involves bending side branches down to ground level and covering about 4 inches worth of the plant with a healthy scoop of a mixture of wood chips, compost, and soil. Place a small stone over the pile to weight it down if necessary. The following spring uncover and prune at the lowest point. All parts of the plant in the soil mixture should have developed a healthy root system. See chapter 7 for additional information on propagation by layering.

The berries come in a variety of sizes, shapes, and colors. Cultivars have been selected for resistance and should be the first choice of forest farmers and include:

- Red currants (*Ribes rubrum*, *R. sativum*, and *R. petraeum*); characterized as dependable, vigorous, and very productive, the fruits can range from dark red to pink, yellow, and white. Good

Figure 4.15. Jostaberry fruit, a cross between currant and gooseberry. Photograph by Zualio, Wikimedia Commons

cultivars include 'Cascade', 'Wilder', 'Rovada', and 'Champaigne'.

- Black currants (*Ribes nigrum*): These are less susceptible to white pine blister rust and often boast larger and more flavorful fruit. Recommended cultivars include 'Consort', 'Crusader', and 'Titania'.

- Gooseberries: There are two types: American (*Ribes hirtellum*) and European (*Ribes uva-crispa*). American fruits are smaller but more resistant to mildew, and the plants tend to be more productive overall. European cultivars have larger and better-flavored fruits. Recommended cultivars are 'Oregon Champion', 'Poorman Captivator', 'Pixwell' (American), and 'Careless Clark Tixia' (European).

- Jostaberries: These berries are a complex hybrid of black currant (*R. nigrum*), American black gooseberry (*R. divaricatum*), and European gooseberry (*R. uva-crispa*). The plants are thornless and resistant to many diseases, including white pine blister rust. The berries are large and delicious.

The main cultivar, 'Josta', is widely available in the United States.

Unlike many berry crops, currants and gooseberries tolerate partial shade and grow best in cool, moist environments. Northern slopes with some protection from direct sun are ideal. Planting sites should have good air circulation, as all *Ribes* species are susceptible to powdery mildew. This group of plants are also heavy nitrogen feeders, so interplanting with nitrogen fixers may be a good strategy. Plants should be spaced 3 to 4 feet apart.

Remove any flowers to prevent fruit during their first season of growth so that root development is favored instead. Expect a light crop the second year and a full crop by the third. Currants and gooseberries ripen over a two-week period in June. Berries do not drop immediately upon ripening, so they usually can be harvested in one or two pickings. Currants can be picked in clusters, and gooseberries are picked as individual fruits. Expect mature plants to yield 90 to 150 pounds per 100 feet of row. Wait for fruit to turn color before picking. Gooseberries come off easily when they are ripe; currants require some trial and error to determine the right time.

RUBUS SPP.: RASPBERRIES AND BLACKBERRIES

For many, the first encounter with forest foods is picking blackcaps or raspberries along a hiking trail or creek bed in the woods. The wild fruit, while small, can be abundant at the right time of year and location. This might lead one to conclude that growing *Rubus* fruits in the forest farm would be very easy. The genus *Rubus* is a large collection of flowering plants in the rose family, most of which have semiwoody stems with thorns. Some of these are commonly referred to as brambles.

Attempts to grow bramble cultivars in the MacDaniels Nut Grove have failed to produce a fruit crop for several years. Notable, however, is the exceptional vegetative growth when grown in shade. Beds that were formerly intended for fruit production have been renovated for propagation, the idea being that

Figure 4.16. The flowers of the salmon berry (*Rubus spectabilis*) are gorgeous, and the plant grows well in shade, with tasty pink berries, albeit small. Photograph courtesy of David McMaster

canes can be layered in a shady nursery, then planted on forest edges or fields or sold as a nursery crop. See the section on propagation by "layering" in chapter 7. At this point it cannot be recommended that forest farmers plant *Rubus* as a reliable income-generating crop, even under partial shade conditions. The fruit sizes, and yields, are simply not as good as when grown in full sun.

Yet forest farmers should consider the plant as part of a possible nursery operation and also if it's desired to take advantage of some of the other yields the plant offers. Black raspberry leaves (*Rubus occidentalis, R. leucodermis, R. coreanus*) when dried make a wonderful tea. Given the prolific nature of these plants, forest farmers who wish to experiment should acquire many cultivars and plant in varying locations around their site. Later, undesired plants can later be sheet mulched.

STAGHORN SUMAC, *RHUS TYPHINA*

This native pioneer tree species is one of the more misunderstood and underappreciated of forest and hedgerow plants in North America, considered by many to be a persistent and aggressive "weed" species. Yet beauty is in the eye of the beholder, and this species, not to be confused with its somewhat similar-looking

Figure 4.17. The staghorn sumac is distinguishable by its fuzzy green and red stems and shape of the flower head.

Figure 4.18. Ripe sumac berries. Ripeness is determined by sampling the berries until they are soft and fragrant and don't have any bitterness to them.

relative, poison sumac (*Rhus vernix*), has some very appealing qualities. It may not be planted by the forest farmer, but perhaps he or she will pause a moment before automatically removing the species.

Sumac is a sun-loving species, yet it thrives along forest edges and hedgerows, which are important habitats to maintain for a diverse array of wildlife. It is a species that also thrives in disturbed sites and on dry soils. The fruit, long and pointy clusters of fuzzy red berries, is certainly beautiful as it develops fully in the later parts of the summer (August/September in New York). The fruits, unless harvested, usually persist on the tree throughout the winter.

These characteristics will ensure identification of the "right" sumac (*Rhus typhina)*:

- A small tree/large shrub with stout twigs that are fuzzy (like deer antlers)

- Large, slender, feathery compound leaves
- Upright spikes of multiple small yellow or green-yellow flowers with five petals
- Thick, upright clusters of tiny, round, fuzzy red or red-orange berries

Contrast with the stark differences in the poison sumac (*Toxicodendron vernix* or *R. vernix*):

- A medium shrub or small tree with smoother branches
- Oval-shaped leaves with smooth edges
- Yellowish green flowers in small, branched clusters
- Fruit is grayish white berries

Determining ripeness of the berries is important, because harvesting too early leads to astringent qualities and harvesting too late means the berries are usually

Figure 4.19. Poison sumac. Since the practical interest in sumac mostly has to do with the fruit, proper identification is very easy. If unsure about the sumac you are observing, then simply look, don't touch. Photograph by Freekee, Wikimedia Commons

Figure 4.20. Spiels, or spouts for tapping sugar maple and other trees, can be made from one-year-old sumac wood, because of its soft pith that is easy to remove with a wire rod.

infested with insects. The best way to test is to allow the berries to change from light to a very dark maroon, then begin sampling. When sweetness overwhelms any astringency, it's a good time to harvest. Ripe berries are harvested by clipping the entire flower head (inflorescence), which won't affect future fruiting.

Traditionally, flower heads were soaked in cold water for from thirty minutes to several days to make a "sumac-ade," which is high in vitamin C and tastes very much like a pink lemonade. Soaking in a glass container in the sun is said to further enhance the flavor. It is not recommended that the mixture be boiled because it can result in a bitter beverage. The berries are also easily dried and can be used as a cooking spice, popular in Middle Eastern cooking.

The leaves, twigs, and fruit also make attractive dyes, and the berries can be easily dried for fall and winter decorations. The wood, with its very pithy core, can be used to make a spiel for tapping sugar maple trees (see figure 4.20) by simply fashioning a coat hanger or rigid length of wood around a short section of wood and plunging it in and out of the pith, which will remove the inside and leave a hollow core.

Above all, the berries provide a late winter food source for birds and are perhaps most valuable in this regard if at least a portion of berries is left. While forest farmers may not want as extensive a population as is often found in neglected forests and hedgerows, leaving some portion of the plants will provide an easy annual harvest.

HAWTHORN (*CRATAEGUS* SPP.)

Often considered a weedy species by many, hawthorn is a small tree that the judicious forest farmer may decide to keep, and for good reason. Common in thickets and hedgerows and disliked mostly because of its persistent

Figure 4.21. Flowers of hawthorn show its close relationship to apples and pears (same family). Hawthorn is a great food for pollinators. Photograph by Wikimedia Commons

weedy growth habit and large thorns, the species turns out to be one of the more useful medicinal foods. It grows abundantly in the woods and doesn't need any care whatsoever.

Many herbal and traditional medicines are often overlooked by the mainstream medical establishment, but not hawthorn. Research has noted that antioxidants found in the berries may be very successful in the treatment of several cardiac, respiratory, and circulatory conditions.[16] In 2008 a German study provided evidence that a drug containing hawthorn berry juice was safer and just as effective when tested against other heart and blood pressure medications.[17] But more research is needed.[18]

In addition to their use as medicine, harvested berries can be used for jams and preserves and can also be dried. Ground berries then can be mixed with flour and will add an interesting dimension to breads and

Figure 4.22. A native hawthorn rootstock grafted with pear, a compatible species within the Rosaceae family. Initial results of this by Sean Dembrosky of Edible Acres show promising potential. The hawthorn was cut at chest height to limit browse pressure from wildlife.

baked goods. Dried leaves can be used as a delicious and medicinal tea. Flowers have traditionally been used in syrups and puddings. The berries are also an important food for birds, as well as food for the caterpillars of many lepidopteran (butterfly) species.

MAKING TINCTURE FROM HAWTHORN BERRIES[19]

1. Harvest soft, plump berries.
2. Take a handful of berries and roll gently to remove the many small stems.
3. Wash the berries and place in a food processor or blender, grinding the berries only until they are crushed.
4. Place crushed berries into quart mason jars, leaving one inch of headspace. Top off with 80 to 100 proof vodka, gin, or brandy. Be sure to fill and press on the berries with a spoon to make sure they are all fully submersed.
5. Put the jars in a warm place, and leave for at least four to six months. Shake daily (or as often as you can remember!).
6. Strain berries through a stainless steel sieve, and send the berries to the compost. Store in mason or tincture jars in a cool, dark place. The tincture will keep indefinitely.

The tincture should be carefully used under advice from a health care professional or herbalist. This powerful medicine has some interactions with other medications. Dosage depends on the individual.

Figure 4.23. *Schisandra* berries ready for picking in New York in late August. Vines should be trellised to make harvesting efficient.

As a plant hawthorn is extremely tough, tolerant of cutting and neglect. Because it tends to be ignored by deer and other browse animals, some forest farmers are experimenting with using the rootstock to graft more valuable species such as pear (see figure 4.22). Its thorny wood makes a good living fence for livestock, and it was traditionally used in layered hedges in European grazing systems. The wood is very hard and tough to work, often historically used for tool handles. Its fuelwood value is excellent, and it coppices readily. A forest farmer might consider keeping the trees low to make for easy harvesting.

SCHISANDRA BERRY
(SCHISANDRA CHINENSIS)

Sometimes called magnolia vine, *Schisandra* is native to the forests of northern China and Russia. In Chinese *schisandra* literally means "five-flavor berry," as it contains all the tastes of salty, sweet, sour, pungent (spicy), and bitter. *Schisandra coccinea*, also known as southern magnolia vine, is a rare species found growing in undisturbed streambeds in North Carolina, Tennessee, Georgia, Florida, Arkansas, and Louisiana. It is reported to grow in a range of light conditions and soil types and to grow quite rapidly. The plant is used as a common remedy for many ailments, including infections, insomnia, and coughing and is considered one of the fifty essential plants in Chinese herbal medicine.[20]

Modern medicine has acknowledged its potential; it was named one of four "well-established adaptogens" and confirmed through research of its ability to increase immunity. Part of this may be the exceptional vitamin C content, which is more than six times that of an orange and twenty times the amount in an apple. It also readily promotes iron intake and provides eight essential amino acids.[21]

Figure 4.24. Canned juice from *Schisandra* is easy to make with a steam juicer at home. It's astringent but good on its own watered down a bit or as a tasty addition to smoothies and mixed alcoholic drinks.

Schisandra is another plant that finds itself in the margins of forest/field ecotones, often in sandy, well-drained soils and by streams and brooks. It is a moderate grower that can reach heights of 30 feet, though it is easily pruned to keep it under control. Some sort of trellising system is recommended for production. Younger plants especially enjoy partial shade.

Flowers emerge in April or May, with the fruits ripening in late summer. Berries form in clusters somewhat similar to grapes. Plants produce both male and female flowers, which may be susceptible to frost and may need to be protected by covering with agricultural cloth.

Propagation of *Schisandra* is best through seeds or hardwood cuttings. Prune the vines of dead wood after the harvest, given that the following season's flowers emerge on the previous year's wood. After ten years productivity may decrease. So far, it appears that neither deer nor birds are interested in the vegetation or fruit.

SOME OTHER FRUIT POSSIBILITIES

A few other species are listed as having some shade tolerance and should be considered as possible additions to the list of the forest farmer. Further research is recommended on these, and the authors cannot

fully vouch for their full potential for the forest farm, though these seem to be likely candidates.

Juneberry

Juneberry (*Amelanchier* spp.) actually refers to a number of species of shrubs, many native to various parts of North America. Notable are the more treelike species *A. canadensis*, which is also called eastern serviceberry and often found in forest edges as well as in deeper woods, and the species *A. alnifolia*, which has been more developed for commercial production and offers large berries, about blueberry size. The berries come into production in early June, as the name would suggest. They are phenomenal fruits with a sweet-tart taste and a slightly nutty undertone.[22] It should be noted that the species *alnifolia*, pictured in figure 4.25, is a

Figure 4.25. Cultivated juneberries such as this *A. alnifolia* produce large clusters of blueberry-size fruits that have an almond-berry flavor. Productivity of this species may not be as dramatic in forest farm settings with limited light, where *A. canadensis* should be considered. Photograph courtesy of Jim Ockterski

species from Canada that is adapted to growing and fruiting in full-sun conditions. The *Amelanchier* native to the eastern United States, *A. canadensis*, is likely a better candidate for forest farms because it is adapted to forest edges and understory habitats. The challenge is that this species has not been bred for large fruit size or production, a task that a passionate forest farmer could take on.

Hardy Kiwi

The hardy kiwi (*Actinidia arguta*) is a woody vine that can bear prolifically once established. It is critical to select the species *A. arguta* and not its Asian relative, *A. chinensis*, as the latter will not survive harsh winters. Kiwis need both male and female plants, at a ratio of one male to every nine or ten females for good pollination and fruit set. Its potential in forest farming stems from the fact that while the vine does well in full sun, too much exposure can result in early breaking of dormancy and young growth, and flower buds can be easily damaged by a late spring frost. Early fall frosts can also be a threat, as well as heavy winds and hot, dry conditions.

Planting in areas without any potential frost pockets and moderate microclimates ensures better success.

Honeyberry

Honeyberry (*Lonicera caerulea*), sometimes called haskap, is a shrub originally from Asia that grows to 4 or 5 feet tall. It produces sweet, tangy blueberry-like fruits very early in the season, before or at a similar time to strawberries. This quality offers very promising combinations of honeyberry in forest farms with species that leaf out late—permitting sunlight to penetrate the canopy and into the forest floor. The berries are being developed for markets in Canada and are a good possibility for cool temperate climate producers in many regions.

Autumn Olive

Autumn olive (*Elaeagnus umbellata*) is a nonnative Asian species that is considered invasive in many states. Don't plant in your area if it's illegal or not already established. It was originally brought to North America in 1830 by the US Soil Conservation Service for landscape conservation purposes. It was widely planted

Figure 4.26. Hardy kiwi fruit are not like the tropical kiwi but are smaller without a fuzzy skin, making them ready to eat right off the vine. Photograph courtesy of Björn Appel

Figure 4.37. Fruits of the autumn olive shrub. Photograph by VoDeTan2Dericks-Tan, Wikimedia Commons

for windbreaks and to attract wildlife. Birds enjoy the speckled red fruit and have spread it throughout the scattered landscape of the United States, most notably in abandoned fields and forest edges.[23] It is a vigorous nitrogen-fixing shrub, and recent research discovered extremely high lycopene content in the fruit, at a rate more than ten times that of tomatoes, which are often highlighted for this trait. Lycopene is widely considered an important phytonutrient, thought to prevent or fight cancer of the prostate, mouth, throat, and skin and to reduce the risk of cardiovascular disease.[24] Yields can average between 5 and 35 pounds per bush. The fruit has been harvested with conventional blueberry harvesters. A delicious use of the berries is in dried fruit leather.

Black Chokeberry

Black chokeberry (*Aronia melanocarpa*) is a small (5 feet or less) tree well suited to fruit production. A native member of the rose family, it is a prolific suck-erer and can fill in large areas. The shrub is well adapted to grow on a wide range of soils, from very wet to very dry, and can grow well in partial shade. *Aronia* is one of the few forest farming crops that already has a large commercial market. Especially in Europe, the juice of the half-inch-diameter dark black fruits can be found in many mixed berry juices and is used for jellies, pies, and yogurt. The berries are very astringent when eaten raw (hence the name).

Nuts for the Forest Farm

As discussed in chapter 2, when J. Russell Smith published his seminal work *Tree Crops: A Permanent Agriculture*, he envisioned a tree-based agriculture in which nut trees would play an important part. Smith's main focus was on the idea of using trees for marginal lands and instead of dealing with the challenges of harvesting and processing for human use, he thought the easiest application was in feeding animals and thus reducing the need for grain inputs to feed:

> When tree agriculture is established, chest-nut and acorn orchards may produce great forage crops and other orchards may be yielding

persimmons or mulberries, crops which pigs, chickens, and turkeys will harvest by picking up their own food from the ground. Still other trees will be dropping their tons of beans to be made into bran substitute. Walnut, filbert, peach, and hickory trees will be giving us nuts for protein and fat food.

In this section, four main types of temperate nut trees native to the eastern parts of North America are discussed. These include walnut, hickory, chestnut, and hazelnut. Commercial production of each of these four, with the exception of pecan (a type of hickory), is mostly in the Midwest and central United States, especially black walnut and Chinese chestnut. Most of these are planted as orchard-style plantings and not integrated with other forest farming elements, though the potential is certainly there.

While nuts are a crop with wide appeal, there is less production going on in the cooler areas of the temperate climate. Interest has mainly been left to hobby growers, with the Northern Nut Growers Association leading the charge in the development of improved varieties and better cultivation strategies. In addition, many states and provinces have their own nut growers asso-ciations (unaffiliated with NNGA), including Indiana, Iowa, Kentucky, Michigan, Missouri, Nebraska, New York, Ohio, Pennsylvania, and Ontario.

With respect to cultivating temperate nut trees for nuts, as nontimber forest products for forest farming in the East, keep in mind that each of these species are canopy trees, meaning they might grow initially in the understory of the forest but ultimately need access to full sun to bear nuts. Often nut trees start out plantation style but eventually grow into a mature forest before the forest farmers "move in" with other shade-tolerant nontimber forest products. Gaps in the canopy of a mature forest are also suitable locations for planting nut trees.

MASTING OF NUTS

Important background for anyone interested in nut cultivation is understanding the dynamics of masting, which is the pattern by which nut trees produce prolific

Figure 4.28. Masting of nut trees at MacDaniels Nut Grove in 2013 made harvesting easy; this bin of hickory nuts was collected in under ten minutes! It is unlikely a harvest of this size will happen again for several years.

amounts of nuts one year and very few the next, in synchronicity with other neighboring trees. This habit is the result of an evolved relationship with nut-loving birds and forest rodents, a relationship that appears to have originated as early as the Paleocene era, around sixty million years ago.[25]

Most of the nut trees produce crops at intervals of two to five years, with a large crop rarely occurring in two consecutive years. When a large mast does occur, the yield can be a hundred to a thousand times larger than that of an off year within the masting stand. The fact that trees within a local population and sometimes even a region mast in synchronicity suggests that the environmental conditions (in addition to genetic traits) play a role in setting the tone. Warm spring

temperatures during pollination periods, dry seasons, and late frosts are all thought to have a direct effect on masting in a given season.

Three main theories exist about the adaptive nature of masting in nut trees. The first is that mass production of flowers in a species increases the probability of cross-pollination. The masting also serves to reduce the loss of nuts to insect predators, as populations cannot often rise to increase populations proportionately. The third aspect of this dynamic is that masting increases the activity of "hoarding" animals that tend to harvest and store as many nuts as possible. Thus, a larger population of seeds is left in the ground to germinate, thereby exploiting the tendency of squirrels, jays, and other animals to overstore nuts by burying (planting) them when available. These three hypotheses are the most tested and commonly held, though others do exist. None of the three is mutually exclusive, and all likely play a role in nut tree evolution over a long time frame.

Breeding of cultivars has to a degree reduced the disparity in mast years versus nonmast years, but nut growers should be aware that this would be a normal occurrence as trees produce over their lifetimes. In the long term, even if possible, it would not be wise to breed the masting trait out entirely, as this would begin to lead to other problems, such as overpopulation of rodent species in the woodlot.[26]

WALNUT

The stately walnut tree (*Juglans* spp.) has long been regarded as valuable for a variety of different reasons. Lore, storytelling, and even financial investors joke half-seriously about the idea of planting a walnut orchard as a biological "retirement fund" when a child is born, in order to pay for college.[27] This is based on the notion that walnut lumber is extremely valuable and has only increased in value over time. While this is an appealing anecdote, the reality is that harvesting a well-tended grove of walnut at maturity would be an emotionally difficult task. If the trees are grown instead for nuts, the form and function of the tree can be enjoyed for generations by the planter and his or her descendants. In practice the cultivars and management for these two different yields (lumber vs. nuts) are different in form. In other words, trees are usually grown for timber or nuts, but not both.

The wood of black walnut is some of the most sought after and valuable in North America and is notably heavy, strong, and shock resistant, ranking among the more durable hardwoods including the cedars (*Thuja* spp.), chestnuts (*Castanea* spp.), and black locust (*Robinia pseudoacacia*). It is straight grained and can be worked easily with hand tools. The heartwood is a stunning rosy brown and is often used for fine furniture and for interior finish wood.

The common name "walnut" refers to several species in the genus *Juglans*. The best known is the exotic (imported) English walnut (*J. regia*). English walnut production is a huge industry (for a nut) and is located almost exclusively in California, in Sacramento and the San Joaquin Valley. Of next importance is the native eastern black walnut (see figure 4.29). Trees can grow as tall as 125 feet but usually grow to 80 feet at maturity, with a typical canopy width of 40 or more feet. It is a taprooted species that prefers loamy, well-drained soils and thrives in wet bottomlands. The walnut strives to be a canopy tree, though it can often be found sharing and thriving underneath the faster growing black locust.

Walnut Toxicity

One caveat of these species that troubles gardeners and farmers alike is that black walnuts, and to a lesser extent other *Juglans* species, are allelopathic, which means the roots exude a toxin that inhibits the growth of other plants. This is sometimes referred to as walnut toxicity. In the case of walnut the allelochemical is named juglone, which inhibits the growth of many other plants. It occurs in many parts of the tree, including the leaves, but these contain lower concentrations than in the roots. The chemical produced by the roots, called hydrojuglone, is actually nontoxic and colorless. When this is exposed to air, it oxidizes into the highly toxic juglone. Several of the *Juglans* species, including English walnut, hickories, and pecan also produce juglone, but in much smaller amounts when compared to the eastern black walnut. Other common trees that also have allelopathic properties (with a wide range of intensity) include sugar maple, hackberry, sycamore, black cherry, red oak, black locust, and sassafras.

Many feel that this walnut toxicity means the elimination of possibilities when designing walnut plantings or working within existing conditions. But of course, as with any ecological dynamic, there is more to the story. The toxicity of juglone fluctuates both with the age of the trees and within the seasonal flow. Much appreciated is a review of walnut literature conducted in 2006, which looked at the overall research on the walnut juglone issue and provided a relevant and prudent analysis.[28] (Researchers, including farmer researchers, take note, as this type of synthesis work is an important step for furthering agroforestry development.) In the paper the authors provide a relevant

Figure 4.29. A healthy, middle-aged eastern black walnut tree at Wellspring Forest Farm. This is one of around half a dozen on-site that produce good nuts and were tapped for syrup for the first time in 2012.

framework for considering the design of systems that work with juglone toxicity.

Age and Light

The first part of the puzzle is that allelopathy is highly variable with age, being almost completely absent from years zero to ten, then increasing with intensity over time. Plant considerations can thus be lumped into short (zero to fifteen years), middle (fifteen to thirty years) and long term (thirty-plus years). The second major variable over this time scale would be light; obviously, young seedlings would cast very little shade, whereas mature trees would become canopy trees and therefore restrict the growth of plants in the understory.

If the forest farmer is starting with established groves, the choices will of course be limited by the combination of juglone and low light. But if planting in a field, a spacing of 6 feet between trees and 30 feet between rows would allow for juglone to accumulate less quickly. Based on this, the authors of the previously mentioned paper classify the following terms, with notable change in the possibilities for intercropping with black walnut. While in many forest farms the latter stages are most likely to be of interest, the full range is presented here because it illustrates the dynamics of juglone very well.

Short Term: Ample Light and Little to No Juglone (Years 0–15)

For the first decade after planting, the abundance of light and lack of any toxicity means that the possibilities are basically wide open for planting. It also suggests a transition from pasture crops to walnut forest that might present a compelling case to cropping farmers. Soybeans, corn, and wheat can be grown in association with black walnut for up to seven years, until shade begins have a negative effect.[29] Most annual vegetables could be planted for five to ten years, but probably not longer. Playing it safe with field crops would mean planting those that exhibit resistance to juglone, including alliums (onions and garlic), parsnips, beets, and sunchokes (*Helianthus tuberosus*), and could stretch this phase beyond the first ten years.

Grazing could also be a viable option in early stages when sunlight is ample, and it might even persist indefinitely as the trees age, if the spacing between rows was increased to allow between 30 and 50 percent light to penetrate and allow grass regeneration. Poultry and ruminants such as sheep, cows, and goats have excellent potential (for more on animals in tree crop systems, refer to chapter 9). Forages that thrive in sun, such as alfalfa, ryes, and fescue, would be appropriate to this early succession phase.

Specifically, researchers have explored black alder (*Alnus glutinosa*) and Russian olive (*Elaeagnus angustifolia*) interplanted with black walnut, and results suggest that the walnut growth and yield can increase because of the ability to increase available nitrogen in the soil.[30] Other species to consider include the red alder (*Alnus rubra*) and black locust. These trees could be interplanted within rows or between rows of existing stands and coppiced periodically.

Midterm: Medium Light and Medium to High Toxicity (Years 15–30)

The midterm phase is potentially where a forest farmer might not be dealing with a planted forest but is starting to get back into the forest and work with existing stands of young black walnut. During this middle term of walnut stand growth, toxicity begins to play a factor, as does light. Many marketable fruits mentioned in this book are juglone tolerant, including currant, elderberry, mulberry, pawpaw, and persimmon.

Grazing could continue, though pasture composition would likely shift to cool-season and more shade-tolerant grasses, such as fescue, Kentucky bluegrass, clovers, and timothy. If dealing with an existing forest, some thinning may need to occur, as stands at this age are often too dense. This effort is often a boon to forest growth, as efforts can be focused around removing diseased or dying trees and otherwise undesirable species from the woods.

Of course, if the trees are being grown for nuts, this phase is when nut trees come into maturity and full bearing. While the full yield potential is not known, some estimate that under ideal conditions an acre of good cultivars may be able to produce upward of 1,000 pounds of nut meat.[31] This will vary widely, of course, depending on whether the trees are wild seedlings or selected cultivars. In addition, many cultivars are selected to favor either high yield *or* ease of cracking, but generally not both. When thinning a woods (or a plantation) keep in mind that any young saplings can potentially be cut to above browse height (6+ feet) and selected cultivars could be grafted onto the native rootstock.

Long Term: Low Light and High Toxicity (30+ years and existing stands)

As a planted grove matures, light and toxicity are both important major factors in site use. If the forest farmer has an existing grove of walnuts that have reached this phase, then trees should be evaluated for health and vigor. Walnut is highly susceptible to nectria canker, a fungus that infects the bark and diminishes the tree, often at weak points such as crotches, where growth splits into multiple directions, or areas where branches have broken and exposed inner bark. Nectria cankers may reduce or eliminate the timber value of a tree. A thinning to remove any of the highly diseased or otherwise unfit trees would be a good thing to consider. This might also open up some light for additional cultivation.

Of course, at this phase nuts will be prolific. The challenge will be collecting them, as large trees drop nuts all over the place. Keeping the understory open to access will greatly aid in the harvest. Additional crops should accommodate this, and with a closed canopy the potential for shade-loving crops such as ginseng and mushrooms is great. The walnut trees can also be tapped for syrup in the spring, much like the sugar maple (see sugaring section later on in this chapter).

Where Does the Juglone Go?

Walnuts produce a toxin and exude it into the soil, so it must go somewhere. A common question is if the toxin becomes more intense over time, or if it is dissipated. As one might expect, multiple factors such as soil type, drainage, temperature, and microbial action combine to make this question a difficult one to answer. The most important factor appears to be maintaining a healthy, aerated soil, which in turn supports a healthy population of aerobic microorganisms that can degrade the chemical and render it nontoxic.[32]

A Final Thought on Juglone

A considerable amount of writing has been devoted to this topic, mostly because it offers forest farmers some important items to consider in the context of the forest farm. One is that when something at first appears to be a severe limitation it might also be an opportunity. As the authors wrote in their research synthesis,

> Black walnut could be re-cast as an ecological resource if its beneficial interactions were emphasized as often as its detrimental interactions. Black walnut's allelopathic chemical juglone is lethal to some popular flora, but the remainder of the plant kingdom may effectively receive a selective advantage due [to] reduced competition when grown with black walnut.
>
> Additionally, black walnut is not likely to attract insect pests, as the leaves have been observed to be amongst the least popular for forest insect pests (Shields et al. 2003), which holds consistent with their reputation as folk insect repellant (Walker 1990).
>
> . . . This assessment suggests the need to destigmatize juglone and recast it as a resource in need of management. Just like shade, low pH soil, or a dry moisture regime, juglone will favor certain species and disadvantage others. The logic of short-term gains is a major argument against tree crops in the Midwest, and the potential for multiple yields in an ecosystem mimicking polyculture could become its counterargument.

Fortunately, forest farmers are often willing to think "outside the box," and this type of approach to growing will be critical for the success of forest farms and agroforestry plantings as a whole. Ultimately, the combination of species, the timing, and the management needs of the system will have to be considered for each unique site and context.

Harvesting Black Walnuts

Arguably, of the nut trees examined in this chapter, walnut is the hardest to deal with. Its husk is hard to peel and can stain skin and clothes for days after harvest, and the nuts, especially the wild ones, don't often crack out easily. Many of the nuts are so small that it's hard to justify the work. That said, for those willing to work within these limits, the reward of the tasty, meaty nut is worthwhile. Black walnuts are low in saturated fats and high in the "good fats" that are said to lower detrimental cholesterol. The nuts are also a good source of iron, fiber, and minerals.

Harvesting at the right time is key to maintaining optimum color and taste in the kernel. Nuts are ripe and ready when a thumb pressed into the husk leaves an indentation. For commercial harvesting a pecan harvester can be modified to get nuts off the ground. At the small grower or hobby scale, a Nut Wizard is a simple tool that will save time and effort. Ground-harvested nuts should be harvested regularly to avoid the darkening of the kernels that occurs over time and to avoid problems with husk fly.[33]

Removal of the husk (hulling) should take place as soon as possible if nuts will be sold to markets. Homemade tire and cage hullers or commercially available dehullers are a worthwhile investment for commercial growers, while hobbyists can either hull by hand or simply lay nuts on the driveway and drive back and forth over them with the car or tractor. When markets are less of a concern, the harvested nuts can be laid on the drive, and the action of coming and going for several weeks makes easy work of dehulling. After hulling, nuts should be washed in the shell (some recommend in a 1,000 ppm chlorine bath) and agitated to remove and disinfect the nuts. Any nuts that float to the top are not good and should be discarded (this is true for most tree species). Place washed nuts on screens, and allow them to air-dry. A fan can speed up the process.

There are ample markets for the nuts, if harvesting and processing can be streamlined. Processed nut meat can sell for around $15 a pound, while nuts still in the shell can be sold for $15 per 5 pounds. Considering even a low yield per acre of 1,000 pounds, a $15,000 per acre

Figure 4.30. One of the best tools for collecting nuts that have fallen to the ground is the Nut Wizard.

payoff isn't something to balk at. Since a national market is not well developed in cold temperate regions, the best bet probably is to market nuts locally. Additional value can be added to the harvesting process as well, if only for recreation. The green husks can be boiled to make a pleasant yellow-green dye, while older husks can be soaked in water for up to two weeks, then used as a dark brown dye.

CHESTNUT

The American chestnut (*Castanea dentata*) is an impressively large tree—some call it the "redwood" of the East. The natural range of chestnut stretches through much of the eastern United States and Canada, with the heart of the range being the Appalachian mountains, where, at its peak, one in every four trees was an American chestnut. In addition to the delicious nuts, the tree provided the primary source of tannin for treating leather and was a highly valued timber species, as the rot-resistant wood is dense yet easy to work.

Figure 4.31. A hybrid American chestnut in a spiky hull a few weeks before harvest. At New Forest Farm, Viola, Wisconsin.

The chestnut is a shade-tolerant species in its younger years, slowly emerging from the understory and eventually becoming a canopy tree when mature. It is a prolific producer, with mature trees able to produce as many as six thousand nuts in a season. Yield rates are not as extreme as with the masting pattern of many other trees; usually the trees offer reliable nuts each season, with some years better than others. Trees flower late in the spring, past the danger of frost, which could damage the blossoms.

Compared to other temperate nuts, the biggest virtue of chestnut is that the shell is quite thin and can easily be "scored," then microwaved, roasted, or boiled to help remove the leathery shell and papery seed coat, revealing the nut meat. The husks, of course, are another story, as their spiky nature makes leather gloves

a necessity for harvest. The food value and nutrition of chestnut is phenomenal; it is one of the only nuts to contain vitamin C. It is low in fat and has nutritional qualities comparable to brown rice or wheat. Because the nut has a high moisture content (49 percent), measures need to be taken to store nuts carefully; because of the high moisture content, they are more vulnerable to molds and rot.

Historical Use

Native Americans valued the chestnut both as a direct food source and as a feed for game, which they relied on heavily as a primary food source. Wildlife such as bear, deer, squirrels, and wild turkeys all enjoyed the abundance of nuts the trees produced. The now-extinct passenger pigeon, whose extinction invokes a sense of the real loss of biodiversity that came with the industrialization of the natural landscape, was also a heavy feeder of the nuts. To stimulate chestnut production, tribes would often develop crude orchards by burning out competitive understory species.[34]

For settlers, the chestnut proved to be yet another source of wealth that helped with the rapid colonization of the American landscape. The cutting of just a few giant trees could support construction of a home, and the nuts made it virtually unnecessary to graze livestock on pasture, which often meant time and labor-intensive clearing of the forest for pasture. The large populations of game that thrived on the nuts were of course also important food sources for settlers and Native Americans alike.

By the mid-nineteenth century, expanding transport networks of steam-powered boats and trains carried nuts and lumber north to rapidly growing cities. For example, Patrick County in Virginia boasted an incredible 160,000 pounds of harvested chestnuts, according to the 1910 census. Overall, the state was producing for sale around 700,000 pounds of nuts a season. Reports abound of storage facilities being knee-deep in nuts and trainload after trainload taking nuts from the rural counties to the cities.

Though lore focuses on the nut of the chestnut, in economic terms the wood did much more to shape economic and social orders of the time. While it doesn't possess the same charming heartwood as black walnut or black cherry and isn't as strong as oak or as rot resistant as black locust, the chestnut could adequately meet all of these needs. Chestnut is a versatile tree, a "jack of all trades," which is perhaps its greatest virtue. During the rapid expansion of settlers in the twentieth century, chestnut was used for telephone poles, railroad ties, house and barn construction, furniture, and even pianos and packing crates. By 1909 roughly 600 million board feet of chestnut was cut each year in the United States, about one-fourth of all the lumber cut in southern Appalachia.

Unlike other resources that appeared abundantly in the North American landscape, such as beaver and white pine, the ultimate downfall of chestnut was not from overexploitation per se but from a fungal blight

Figure 4.32. A blighted chestnut trunk at Badgersett Research Farm, Canton, Minnesota. Researchers at this site are allowing the blight to spread so they can test resistance of their hybrid varieties.

that showed up in New York City in 1904. Various efforts were undertaken to stem and constrict the spread of the blight, mostly notably in Pennsylvania, where a small army of young men was employed to scour the forests of the state and cut down and remove infected trees. But efforts large and small proved futile, and by 1920 the official policy of the Forest Service came to be that remaining trees should be harvested as quickly as possible, essentially giving up hope of any recovery of the population.

The fact that so many American chestnut trees had been harvested at this time certainly played a part; as with much of modern forestry, loggers tend to take the largest, healthiest trees in a given stand, since these will fetch the best price for their efforts. While logical in one sense, the problem with this approach is that the best trees are also the ones that have the best genetics; that is, are most well suited to the environmental factors and growing conditions of a site. This vigor and health may have offered some resistance to the blight. Further, since the best trees were long gone, seedlings were growing from the genetically inferior trees, which contributed to a downward spiral, with those trees more vulnerable to the blight.

The most unfortunate part, however, was the attitude of the Forest Service and governments, which more or less gave up on the tree. An assumption was that all hope was lost. And in this assumption the search for possible resistance was set back in many ways. There are still said to be roughly one million surviving chestnuts around North America, many of which have stump sprouted and lived to at least bearing age. This is one pathway for efforts to restore the chestnut, through the identification of surviving trees and propagation from the nuts, which inherently have at least some degree of blight resistance.

Bringing Back the American Chestnut

If nothing else, the impressive and wide-ranging efforts of scientists, farmers, and landowners to "bring back" the American chestnut demonstrate the mark its legacy has left on a culture that both enjoyed its ample abundance and witnessed its terrifying demise. Today the social dynamics around restoration offer many approaches to—and conflicts about—the "best" way to restore the American chestnut to full glory. Others question if this is really the best goal anyway, that to some degree it should be accepted that what is gone is gone.

Most notable of organizations devoted to the cause is the American Chestnut Foundation, which operates as a national organization with several state chapters. The expressed goal of the organization is to "restore the American chestnut tree to our eastern woodlands to benefit our environment, our wildlife, and our society." This has primarily been done using the backcross breeding method, in which Chinese chestnut trees, which are naturally resistant to the blight, are crossed with their American relatives, resulting in trees that are 50 percent American and

50 percent Chinese. The trees are then backcrossed with American species, which results in trees that are 75 percent American and that should hold some resistance to the blight. This procedure is replicated multiple times, which has successfully resulted in trees that are $^{15}/_{16}$ American chestnut. Multiple trial orchards maintained by the foundation currently boast over thirty thousand trees in various stages of the breeding program.

A second approach by the American Chestnut Research and Restoration Project (SUNY–ESF in Syracuse) focuses on genetic transformation (genetically modified organisms [GMO]) to provide a solution that is highly sophisticated and has made some significant progress. The method involves a multistage process in which American chestnut is genetically transformed with genes that confer some resistance to chestnut blight.

A third approach has been to identify the "remnants" of blighted trees that naturally showed resistance; in other words, trees that stump sprouted. Some estimate that around a million of such trees remain, but many succumb to the blight at varying ages: some before setting any nuts, some just into bearing age, and some last considerably longer. A minor faction of enthusiasts collects and grows trees from their harvest, yet not on the scale of either of the previous options, as the practice of seeking out isolated remnant trees is naturally limiting in its potential to be done on a large scale.

These conversations around chestnut restoration have largely been aimed at the broad goal of "restoring" the American chestnut to some semblance of its former place in the eastern temperate forest. There are two main challenges in this approach. One is the assumption that it would be possible to have a modern chestnut forest that looked anything like those forests of the early twentieth century. This is simply impossible, because there are so many dynamics in species composition, management strategies, and other factors that mean the forest will always be a new iteration. This is an important consideration, especially with the rapid onset of climate change, which will further change the "normal" conditions of forests around the globe. At a 2004 conference held in

CHESTNUT IN THE MACDANIELS NUT GROVE

Today one lone Chinese chestnut tree survives at the nut grove. This is but one of sixty blight-resistant Chinese chestnuts that MacDaniels planted in the orchard in 1925, at the tail end of the American chestnut blight epidemic. MacDaniels wrote at the time that this artifact as a caption to a crude map of the area now referred to as the nut grove. Thus, the remnant tree boasts a considerable cold tolerance, a trait that has potential in offering some genetics useful for moving the range of chestnut northward.

North Carolina titled "Restoration of the American Chestnut to Forest Lands," Steve Oak of the USDA Forest Service noted that "these forests are nothing if not ever-changing as a result of the way that people interact with the forest . . . there's never been anything like what you see here today. And the forest that will result in decades hence from what is here today will be like nothing else that has ever existed in the past."[35] This sentiment hits at the core of what forest farmers should consider as a core approach to thinking about the "purity" of species and cultivars; there is simply no going back to the glory of what once was. The question becomes how to move forward.

The second assumption in the restoration of the chestnut would be that our modern culture has an abundance of time and resources to be able to devote time to a pursuit of breeding pure American chestnuts, for the sake of "preserving" the purity of the species composition. This is a perspective highly unique to the human condition. In natural systems species are constantly evolving and changing. Further, while the American chestnut offers some advantages as a forest tree species, it does not contain all the qualities one would look for if food production was a main goal. Compared to its relatives, the American nuts are not as large and the form is not as amenable to orchard-style production. For these reasons, and most importantly because of blight concerns, commercial production in the US has mostly centered on cultivars

of the Chinese chestnut, which boasts a shorter form and larger, more prolific nut production. Worldwide, other species that are used for food production are the European chestnut (*C. sativa*) and the Japanese chestnut (*C. crenata*). These have not been seriously considered for North American production, because both are blight susceptible and lack the cold tolerance of the Chinese variety.

Forest farmers should consider planting hybrid chestnuts as part of their operation. These can be the backcrossed American chestnuts or a mixed hybrid, such as those offered by Badgersett Research farm (see the case study at the end of this chapter). Chestnut trees are rather tolerant of a wide range of soils and can even grow decently in heavy clay.

The University of Missouri offers some promising data on the production potential of (Chinese) chestnut. As of a 2006 report, production was minimal, at around 1.5 million pounds. Yet publications note that, as with many agroforestry crops, demand exceeds supply. Grower retail prices range from 75¢ to $6 per pound at farmers' markets; $1.50 to $6 a pound at on-farm sales; and $2 to $7 per pound at restaurants. According to market research, demand for fresh chestnuts is expected to continue to increase by 10 to 25 percent over the next five years.[36]

HICKORY

Many believe that hickory is the most delicious of the nuts capable of growing in the cool temperate forests of North America. It is also one of the smaller nuts, often the reason it is overlooked, along with the fact that masting cycles of wild hickories are rather extreme; that is, when it rains it pours, so to speak. The year 2013 was the best nut-producing year at the MacDaniels Nut Grove since it was "rediscovered" in 2002. During a recent open house, hundreds of visitors were treated to the terror and the thrill of walking around and dodging the multitude of nuts falling with each gust of wind on the blustery fall day when staff and students offered tours, tastings, and information on forest farming.

Considerable time was spent gathering nuts during the harvest season, and it was a dramatic comparison

Figure 4.33. This large hickory nut from the MacDaniels Nut Grove shows the power and potential of breeding. The nut is in a large-size hand and is more than 3 inches in diameter.

to see the difference in the nut size between some of the grafted cultivars planted over seventy years ago and their more recent seedling offspring. Edible hickories include the pecan (*Carya illinoiensis*), shagbark hickory (*Carya ovata*), and shellbark hickory (*Carya laciniosa*). Pecan is the only hickory at present that has major commercial importance in North America. Its range is mostly southern, extending only as far north as southwestern Ohio, although there are some northern (cold-hardy) cultivars. The state with the greatest commercial production is Georgia. In an agroforestry context pecan lends itself more to alley cropping than forest farming.

The "true" hickories worthy of consideration in the northern quarter of the United States are shagbark hickory (*Carya ovata*) and shellbark hickory (*Carya laciniosa*). Shellbark hickory is likely to be found in the same general habitat as black walnut— low flood plains (shellbark hickory grows best on deep, fertile, moist soils), streamside, and its range is greatly restricted. In New York State there is only one population (inlet to Owasco Lake) compared to shagbark hickory, which is found on dryer hillsides, often in association with oak. As with walnut, both of these hickories are typically tall canopy trees. As such, they will survive but not thrive when planted beneath an established forest canopy. The range of shagbark hickory is considerably more extensive than shellbark hickory.

Hickories require at least 2,250 growing degree days (GDD), but experience in Ithaca, New York (zone 6), is that most hickory selections that are from the Midwest

NUTS: FOR THE NEXT GENERATION

As I harvested the benefits of the MacDaniels nut orchard planting this past season, I couldn't help but think about the fact that nuts offer in some ways the ultimate vision of forest farming. Here we were, in a mature forest canopy that was once completely deforested. Human intention had planted this grove, and it wasn't until the planter (Dr. MacDaniels) was gone that other people reaped the benefit of the harvest.

Ultimately, forest farming is about growing and preserving forests, and getting into nuts is in many cases a practice that is mostly a gift for the future. This harks back to past generations, when families passed orchards and farmland from one generation to the next, with the idea that the kids and grandkids would be able to work less and reap more. Unfortunately, this cycle has largely been broken in the United States, and we are forced in many cases to start over, to plant the trees not only for our own yields but also for the benefit of future generations.

This is one aspect of forest farming that is really critical: that short-term yields are balanced with long-term benefits, often to others down the line. If the farmer who had our land before us hadn't decided to leave the 1-acre sugar grove, we wouldn't be able to make syrup today. It is both a humbling and an eye-opening experience to consider that much of the trouble in the world today is due to disconnection from these larger cycles and a focus on short-term gain. Forest farming invites us to change these cycles and to offer a gift for generations to come.

— Steve

and the south of New York State fail to ripen except in years when the period of summer warmth and the length of the growing season (i.e., Growing Degree Days) is slightly longer than average. Climate change is bringing the conditions to the Northeast that will favor better hickory production.

Hickory nuts are excellent nutritionally, but wild hickory nuts are almost too small to crack out to retrieve the nut meat. The Creek Indians, however, solved the problem without resorting to selection or breeding. An eighteenth-century explorer, William Bartram, reported that

> they pound them [hickory nuts in the shell] to pieces and they cast them into boiling water, which, after passing through fine strainers, preserves the most oily part of the liquid; this called by the name which signifies hickory milk [powhicora]; it is as sweet and rich as fresh cream, and is an ingredient in most of their cookery, especially homony and corn cakes.[37]

In general hickories are slow-growing trees with deep taproots, making them difficult to transplant.

Planting of clonal (grafted) trees of selected cultivars is recommended if your goal is to produce salable-quality nuts. Seed or seedlings should only be propagated as rootstocks for grafting. If you prefer to do your own grafting onto wild hickory seedlings growing in your woodlot, grafting stock and scion of the same species is recommended because delayed incompatibilities have been observed between shagbark and pignut hickory, and possibly other combinations, although grafting shagbark onto shellbark appears to be a good combination. In more modern times, the problem of small nut size was greatly improved genetically through selection and breeding of trees that produce larger-size nuts. The details are discussed in chapter 7, but suffice it to say that dramatic increase in nut size was achieved by selection and/or breeding. Figure 4.34 illustrates increases in nut size compared to the wild type by selection (bottom) or breeding (top).

Another option worth experimenting with is the naturally occurring hybrid cross of pecan and one of the various types of hickories, which results in what is called a "hican." These crosses have a high variability but can offer the cold tolerance and rich flavor associated with hickory while gaining the productivity and crackability of the pecan. This tree is reported to survive in zone 5 or 6 but isn't ready for primetime.

As for the future of hickory nuts as a food crop suitable for forest farming, there is much work to be done. Aside from pecan, for which many cultivars have been

Figure 4.34. Comparison of nut sizes from selected wild nuts (bottom) and breeding combined with grafting (top). All are from the MacDaniels Nut Grove.

selected, relatively few cultivars have been selected of shagbark and shellbark hickory and several interspecies hybrids. Some of these thought to be better suited for northern growing conditions include:

- Shagbark: CES 26, Fox, Davis, Wilcox, Porter, Neilson
- Shellbark: Fayette, Henry, CES 24
- Hybrids: Weschcke (*C. laneyi* = *C. cordiformis* × *C. ovata*), Weiker (*C. dunbarii* = *C. ovata* × *C. laciniosa*)

Genetic improvements and selection have come a long way toward achieving the goals of J. R. Smith and L. H. MacDaniels to improve human nutrition. It is ironic then that there are few who care to take advantage of these advances. Hickory nuts are virtually unknown to the public and not grown commercially anywhere. They are the orphan of nontimber forest products.

HAZELNUT

The hazelnut grows wild in many locales north of the equator, with ten species in the genus *Corylus*, which is the birch family. The species found in North America take a shrub form and are the American hazel (*Corylus americana*) and the beaked hazel (*C. cornuta*), while the hazel of commerce is the European hazelnut, *Corylus avellana*. It is grown in Oregon and Washington and

Figure 4.35. Hybrid hazelnuts from Cornell plantings. Note the variability of sizes, which is a consistent trait in the relatively young breeding process.

produces a rather large nut that is known far and wide for hazel butter (Nutella), sugary confections (chocolate truffles), and coffee flavoring.

The hazelnuts of the eastern forests are just too small to be very useful, except when it comes to hybridizing with the European hazelnut, which, despite its size, is highly susceptible to a fungal disease known as eastern filbert blight. This disease is caused by the fungus *Anisogramma anomala,* a species indigenous to the northeast United States. On the American hazelnut (*Corylus americana*) it appears as an insignificant canker, while on the European hazelnut (*Corylus avellana*), it is lethal. Resistance to the blight is inherited in European × American hazel hybrids. This has been the basis for breeding attempts to develop hazelnuts more suitable for North America. The earliest breeding programs began in the early twentieth century in the East. Even today "amateur" breeders, many of whom are active in the Northern Nut Growers Association, continue to seek the perfect hazel, and a prominent breeding program is under the leadership of Tom Molnar at Rutgers University in New Jersey. Similarly, hazelnut breeding attempts have been conducted in the Midwest by Weschcke and by Ferris in the 1930s, and private breeders, particularly Phil Rutter[38] (Badgersett Research Corporation, Minnesota) and Mark Shepard (New Forest Farm, Wisconsin) continue to carry the torch. Their efforts have been directed toward hybrid varieties suitable not only for the nuts themselves but also for a high-quality oil that can be pressed from them.

Propagation runs the same continuum as with chestnut, from controlled hybridization to tissue culture to Rutter's "hybrid swarm" approach (see the case study at the end of the chapter).

In 2004, Melissa Madden and Ken Mudge obtained about two hundred bareroot hybrid hazelnut seedlings from Mark Shepard's breeding program at New Forest Farm, in Wisconsin. These were transplanted at the Dilmun Hill Student Farm at Cornell University, and over the next four years growth and survival were monitored annually. The first nut production from this planting began in 2008. Over the next four growing seasons nut production overall steadily increased. Not surprisingly, there has been considerable variation in nut yield among the 180 surviving plants in the trial. This is typical of the genetic variability expected of a population of seedlings.

The goal of this research project is to select and clone those individual plants grown from these seedlings that have consistently high nut yield from year to year. These selections could then be cloned and distributed as named cultivars to forest farmers for production of hazelnuts as a perennial crop. By selecting the best producers and eventually clonally propagating the best performers, we hope to achieve some measure of genetic improvement over unselected seedlings. Since 2009, students in the class Practicum in Forest Farming harvested nuts from each plant and recorded the number of nuts per plant, the weight of all the nuts per plant, and the percent kernel, which is the ratio of the weight of the nut meat to the weight of the entire nut including the shell.

The nuts are high in protein (19 percent) and oil (79 percent), the oil being monounsaturated. While hazelnuts are smaller than many of the other nut types, their shells are thinner, which makes them significantly easier to crack than walnuts or hickory nuts.

Growing the hazelnut as a forest farming species is pushing the edge of the concept, as they are a species adapted mostly to midsuccession "shrubland"-type woods, and they thrive in edge environments. While

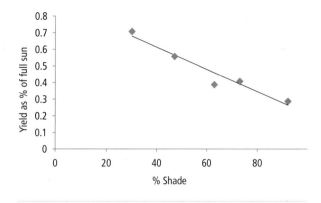

Figure 4.36. This graph shows the shade tolerance of hazelnut in relation to yield potential. Depending on the goals of the forest farmer, hazelnut is a species to try in various light conditions (adapted from Hampson et al., 1996).

cultivated hazelnuts are almost universally grown in the open (full sun), they deserve a trial as a nontimber forest product for the temperate forest farm. After all, American hazelnuts have evolved in lightly shaded forest edges and gaps. One can never expect any crops that are grown under full sun to perform as well in the shade, but studies show that hazelnut can perform tolerably under such conditions.

In one study, hazelnut grown under 30 percent shade yielded 70 percent as much as it did under full sun. This may not be acceptable to someone whose prime directive is commercial producing of high-quality hazelnuts, but in a multifunctional forest farm, hazelnuts may make a worthwhile contribution to the overall productivity of the system when grown on a sunnier perimeter of the woods or under moderate shade of an overstory such as black walnut or locust.[39]

Hazelnuts are easily grown from seed (they require cold stratification) or by layering an established plant. Hybrid seedlings are often predictable in expressing many of the desired traits of the parents. The massive, fibrous root systems of the shrubs can be quite tolerant of a wide range of soils. One of the best potential uses is as a windbreak or snow break, in open fields or along the edge of the forest farm to minimize impacts. Plant spacing for windbreaks is 3 to 5 feet, while for nut production 4 to 10 feet is best. They will begin to produce nuts after three or four years, peaking at around eight years.

Figure 4.37. One of the clear benefits of breeding through selection: a clean break of the nut from the husk. This is one quality important to the potential of machine harvesting of hybrid hazels, one aspect being researched at Badgersett Research Farm. Photograph courtesy of Philip Rutter

Each seed (nut) is enclosed in a husk (two bracts), and there are usually two or more husks in a cluster. Unfortunately, these do not always ripen at the same time, so it is necessary to harvest the plant over several weeks. Inevitably some husks or seeds will fall to the ground before harvest, and if the plant has been grown as a multistemmed shrub, recovering these is very difficult. On the other hand, if the plant is trained to a single stem, collecting the wayward burrs from the ground is much easier.

Nuts mature in August or September and are ready when the husks are mostly browned and the nut releases easily from the husk. Often nuts will hang on the bush for up to two weeks when ripe, and it's important to try to get them before they hit the ground. Depending on the site, squirrels, chipmunks, mice, crows, and jays can be a challenge to control. Best management practices include keeping vegetation mowed near the plants, installing roosts for hawks and owls, and maintaining active human (and perhaps dog) presence. Badgersett has in recent years experienced crop loss to crows, which are particularly intelligent (for a bird). Hawk-shaped kites set up on lines and moved periodically seem to work in offering some deterrence in the field.

Once harvested, the nuts should be allowed to mature further in shade for days and weeks, until the husks come off easily. This can easily be achieved by spreading nuts out on screens to air-dry in a sheltered and ventilated space. Keep in mind it is best to dry the nuts as quickly as possible in a passive manner. For commercial operations a grain dryer would work well for this task.

THE FUTURE OF NUT PRODUCTION

Looking back at the nut crops as a whole, a simplified summary may be that the ones ready for primetime across climate variation would be chestnuts and hazelnuts. Trees that need further development but have some good beginnings are walnuts and hickories. Of course, in southern states nuts, particularly pecans, are already somewhat common as a commercial venture. The take-home message may be best summarized by the Chinese proverb, "The best time to plant a tree is yesterday." Northern temperate states especially will become prime nut-producing regions in the coming decades, which is convenient, considering the high nutritional value and stress tolerance of many species. Finally, planting nut trees connects forest farmers to a longer time scale and is one of the ways an interest in supporting the vitality of future generations can be demonstrated on the forest farm.

Tree Syrups: Maple, Birch, and Walnut

One of the oldest forms of forest farming comes in the tapping of tree sap for delicious and nutritious products that arrive as the seasons change (thaw) from winter to spring. By far the most common practice is with sugar maple, though there are several other trees that warrant attention, depending on the location of a forest farming operation.

MAPLE

Maple sugaring is the first act of spring for farming in the Northeast. It signals the awakening of the plant kingdom, with copious amounts of sap flowing up from the roots of the sugar maple (*Acer saccharum*), awakening dormant buds and pushing forth flowers and eventually leaves that will be the solar array for the trees and the forest. Even though humans have harvested sap since pre-Colombian time, mainly in North America, the entire physiology of the sap run is not fully understood.[40]

The process of collecting and boiling sap has barely changed over time. Innovations have mainly come in how sap is moved from the tree to the fire and how quickly the boil is conducted. But the main process is both simple and timeless. Native Americans used sharpened stones and later hatchets, hacking a V into the trunk of the tree and collecting sap in a wooden trough. Sap was boiled by cooking rocks in a hot fire, then placing them into the sap and constantly replacing rocks throughout the night. The natives also relied more heavily on letting sap freeze, which naturally separates water from sugar. The remaining liquid was then boiled off, but it took a lot less time than boiling alone. Some sugarmakers still take advantage of this freeze/boil strategy today.

When European settlers arrived, so did metal. Buckets were easier to make, maintain, and store. Since metal can come into contact with fire, the boiling process was revolutionized. Large cast iron kettles over fires worked but also wasted a lot of heat. Eventually metal tanks were fabricated to fit perfectly over fireboxes, which channeled fire toward the pan, thus making for a more efficient boil.

Modern life brought plastic, making sugaring cheaper to set up and maintain. Tubing lines can be cleaned out and reused for several seasons. They are particularly helpful on steep sites and those challenging to access. Tubing systems are now the maple industry's standard, mainly because agricultural practices have all tended to evolve to replace human labor with technology, which often equals efficiency. The addition of vacuum systems has also led the yield per tree to increase. Fuel to boil sap is equally diverse, with many sugarmakers abandoning wood firing, choosing to boil sap with gas or oil as the fuel. Today the choice of equipment is a combination of the sugarmaker's desired scale and personal goals for the farm.

Figure 4.38. Liz Falk of Wellspring Forest Farm adds fresh sap to the boiling pan. Sap is 98 percent water, with a slight taste of sweetness. One will never have cleaner water than the water filtered through a tree!

Sugaring Basics

Following are some of the key points to keep in mind with regard to sugaring. Some vocabulary is useful, too; a "sugarmaker" is one who collects and boils sap into maple products. A "sugar bush" is the collection of trees that are tapped for sap, while a "sugar shack" is the place sap is brought to boil. And finally, a "run" is the period of time when trees are producing sap for collection.

Sap Is 2 Percent Sugar

What comes out of the tree is overwhelmingly water, and likely the cleanest water one will ever drink. Sap is an amazing tonic and will keep in the fridge for up to a week (a similar shelf life to milk). The 2 percent sugar content is an average, with some trees occasionally giving more. Sugarmakers often assume that 40 gallons of sap make a gallon of syrup. If the weather is cold enough to freeze the collected sap, it is worth removing any frozen chunks, which are almost entirely water. It will significantly reduce boil time.

Figure 4.39. The spot from last year's tap will fully heal over by the following season, as shown in the photo. New taps should be drilled far away from the previous year's wound. Photograph courtesy of Jen Gabriel

JUST DRINK THE SAP!

In many cases, the amount of time required to boil sap into syrup makes this process impractical for the homeowner on a small scale. Thus many people do not tap trees, choosing instead to support a local sugarmaker for syrup. One thing that anyone with a few healthy sugar maples should consider is tapping for the sap alone; it offers a chance to connect to the seasonal change of nature, as well as enjoying some potential health benefits.

Maple sap, along with other tree saps, has long been viewed as a spring tonic by many cultures around the globe. It is usually about 98 percent water and 2 percent sugar, but little known is that it is also loaded with minerals, nutrients, enzymes, antioxidants, phenolic compounds, and more. In Korea specifically there is a long history of sap consumption, and most comes from the *Acer mono* species, a maple that is called *gorosoe*, meaning "the tree that is good for the bones" in Korean. This is likely because of the high mineral content in sap, most notably calcium, magnesium, and potassium.

There are even places in Korea where people can take weekend retreats, visiting the mountains and consuming as much as 5 gallons of sap per day while sitting on heated floors with conditions similar to a sauna. The idea is to detox the body of the bad stuff and unclog the body from a long winter. In Korean markets maple sap usually sells for $5 to $10 a gallon.

Most analysis of the health benefits of sap has been done on the basic content, which has over fifty vitamins and minerals and a number of probiotics similar to those found in yogurt and other dairy products. More research would be useful, but it's hard to argue against the idea of drinking sap as a healthy option in the springtime; after all, it is water filtered in a tree and loaded with a bunch of nutritional compounds.

If you are interested in collecting and enjoying sap, its important to note that while sap is essentially sterile when inside the tree, it can quickly become contaminated. The choice of container for collection is thus very important. Maple buckets and jugs (a milk jug can make a great collection vessel) should be thoroughly cleaned before use. The best sap runs during the beginning and middle of the season, but as the temperature warms toward the end of March and into April, it's best to stop drinking it straight. Sap can be stored in the fridge (or outside if it's below freezing) for several days and should generally be treated like milk; it's best consumed within one week of its coming from the tree. And while some of the good bacteria may be killed, to be extra safe some choose to boil the sap to effectively pasteurize it and render it completely safe.

One Tap per 10-Inch-Plus Diameter Tree

While opinions vary, it is best to tap trees that are 10 inches in diameter or greater. Because of the increased environmental stress from climate change and other factors, most Extension agents recommend only one tap per tree, regardless of how big it is, though some choose to add a second tap to trees 18 inches or greater. This is a choice based on short-term yields versus long-term health. When in doubt, tap less.

A Quart per Tree

In any average season one can expect each healthy mature tree to produce enough sap (about 10 gallons total) to boil down to a quart of syrup. This is a rough figure, although seasonal variables make this a very flexible number. So ten trees would yield 2.5 gallons, fifty trees about 12.5 gallons, and one hundred trees about 25 gallons.

Runs Highly Variable

A "run" is a period of time when the temperatures are warm enough for the sap to flow, then cool down, stopping the flow. No two seasons and no two runs are alike. The basics are that sap runs when temperatures rise above freezing (32°F) and stop when they drop below freezing. Yet things quickly get more complicated. Sap flow is much like a faucet. It can run slowly or rather quickly, depending on conditions. A day with temperatures that barely get over 32°F is a slow run, but it seems that over 45°F the run also slows.

Figure 4.40. Sap boiling on its way to maple syrup at Sapsquatch Pure Maple Syrup in Enfield, New York. Each division in the pan helps move the sap along, and finished syrup is drawn off about once per hour if everything goes right. Photograph courtesy of Jen Gabriel

Table 4.5. Comparison of Various Scales of Sugaring

Scale	Personal Use	Surplus to Share	Commercial
No. of taps	5–50	100–300	500–1,000+
Approx. gals of syrup	1–10	20–60	100–400+
Collection method	Buckets	Buckets or tubing	Tubing, vacuum system
Boiler	Woodstove or turkey fryer	Outdoor homemade or purchased evaporator 4' x 8'	Commercially purchased or fabricated system
Pros	Low-cost system with minimal inputs and low stakes. Use of existing buckets, pans, etc., is possible.	For the time and energy, this can produce a decent amount of syrup. Bad years aren't of much consequence.	Over the long term (10+ years), sugarmakers can yield a profitable harvest and complement other farming and forestry work well.
Cons	Question of scale; at this level much time and energy is spent for little yield. Since there is a base amount of investment of resources and energy, it may be more efficient to step up one scale.	This scale could otherwise be called an expensive hobby. There are ways to keep it cheap, but it's easy to spend money to make the work easier, yet recoup little from any sales/trades.	Sugarmakers are almost entirely devoted to the process from Dec/Jan through March. Long days and nights are a test of endurance. At this level it's necessary to get good yields, and with seasonal variability, there is no guarantee of a profit.

The ideal run starts with overnight temperatures dropping down into the mid-20s. Then a quick warmup in the morning follows, with temperatures reaching around 40 degrees before a long descent back into freezing temperatures. Of course, the backyard sugar farmer cannot control the weather, so sugaring requires patience and flexibility. It also forces sugarmakers to pay attention to the subtleties of the natural world.

Syrup Is 67 Percent Sugar

Sap is most often boiled down to a syrup that is 67.7 percent sugar. This ratio is shelf stable, requiring no refrigeration if bottled properly. Any less sugar and the product will mold at some point, whether in weeks or months. At a higher sugar content, the syrup begins to crystallize.

Scale: Where to Begin?

It's easy to tap trees: The key is to consider ahead of time how many are appropriate. One approach is to consider a goal of how many gallons of syrup to produce. Since a tree provides on average a quart of syrup per season, a starting point would be that tapping four trees would provide about a gallon of finished syrup. Those who heat their home with wood can easily keep up with this—adding a bit of sap with each run to a pot on the woodstove, then finishing a gallon at the end of the season over the main stove, keeping in mind that when syrup gets close to finished it can easily burn.

Once you get over five trees it begins to get more complicated. Boiling will need to happen outside, as the amount of steam coming off a boil could do damage to a home. Besides prefabricated rigs that you can order from supply companies, the easiest (and cheapest) backyard rig starts with cinder blocks; it's easy to create any dimension you want. Backyard sugarmakers usually tap somewhere between ten and two hundred trees.

As for boiling pans, a restaurant supply company can provide a 6- or 8-inch-deep "hotel pan" in a variety of sizes, which can work quite well on cement blocks. Multiple pans allow for you to begin tapping ten to forty trees with ease. Beyond that a local welder can be contracted to fabricate a pan, or a used pan can be purchased from a sugaring supplier.

Commercial sugarbushes usually start at a thousand taps or even as big as sixty thousand, though a few folks that sell syrup still maintain smaller operations. The experience of sugaring at this scale is both intense and highly rewarding, with long days and late nights around the boiler becoming the norm for several

Figure 4.41. A medium-scale boiler setup at Wellspring Forest Farm, designed to handle the sap from about a hundred trees. The pan was bought used from a maple producer, the firebox is made of cinder blocks, and the chimney stack was reclaimed from neighbors when they replaced their chimney.

months. Successful operations usually combine syrup with other strategies. Some host pancake breakfasts on weekends during sugaring season, combined with tours of facilities and the woods. Others engage in "community sugaring," working with school groups

and families interested in the process. Festivals, statewide open houses, and mail order to places around the globe that don't have access to maple syrup are also good strategies.

Buckets vs. Tubing

Once the number of trees for tapping is determined, the next step is to decide on the method you will utilize to collect sap. This decision includes thinking of what materials are preferred (metal vs. plastic), the aesthetic desired in the sugarbush, and how you want to use your limited labor.

Buckets

This choice involves setting up buckets to collect the sap, which will need to be emptied and the sap moved to the sugarhouse. This can be labor intensive, though on a gently sloped piece of forest with good access, a single person can easily harvest hundreds of buckets in under an hour.

Traditional buckets are made from either galvanized tin or spun aluminum. The bucket contains a hole that hangs on the tap, which is pounded into the tree. Taps, also know as spouts or spiels, come in two sizes; ⁵⁄₁₆ inch and ⁷⁄₁₆ inch. It is widely recommended that the smaller ⁵⁄₁₆ be used; these are sometimes called "health spouts," since you are wounding the tree in tapping it, and though the smaller size won't affect the amount of sap collected, it will allow the tree to heal faster.[41] There are many types of spouts made of metal and plastic. Old and rusty spouts should not be used.

A lid is also part of the setup, to keep debris and rain from falling into the bucket. Because the maple industry has largely adopted tubing systems, there is currently very limited production of new buckets. A set of bucket, spout, and lid could easily cost $20 to $30. Luckily, since so many commercial sugarmakers have abandoned buckets, used ones are available, selling for $8 to $15 a set, though it does take a bit of hunting around to find them for sale. Over the last few years, these have become harder and harder to come by.

A bucket system can be easily constructed from locally available materials, utilizing food-grade 5-gallon

Figure 4.42. A spun aluminum bucket and lid, the traditional Canadian-style bucket. Used metal buckets can be purchased for a reasonable price but are increasingly hard to find.

Figure 4.43. Maple-specific tubing strung between trees at Sapsquatch Pure Maple Syrup. Smaller 5/16-inch-diameter tubing comes from the trees as "drop lines," which feed the larger ½-inch line running from left to right toward the sugar shack. Photograph courtesy of Jen Gabriel

buckets. These buckets are either hung on the tree or placed on the ground beneath the tree, and a plastic tap and short piece of line connect the tree to the bucket through a hole drilled in the bucket lid. A 5-gallon capacity means sap won't need to be collected every day—a nice side benefit. Sometimes if two or three trees are really close together they can even share the same bucket. There are many variations on bucket systems that deviate from these basic examples; the key is to keep the sap clean and isolated from outside elements.

Tubing

Where access, steep slopes, or limited time prevails, tubing systems are the natural choice. Tubing is available from sugaring supply companies at a low cost, around $.10 per foot. Tubing can be used to drop into buckets or tanks as mentioned, but it is best utilized by collecting trees from the top to the bottom of the slope along one length.

"Drop lines" consist of a spout and short length of tubing that connect to the main line, which runs downhill. Most tubing lines are 5/16 inch, though when multiple lines come together a "mainline" can be added, usually ½ inch or ¾ inch, to accommodate the increased volume. A tank at the bottom of the line allows for a single collection point and ideally is right at the point where sap will be boiled. If a sugarmaker decides to go with tubing, it is highly recommended she read up on the topic (see inset) and work with a local experienced person to learn the technique. Most modern tubing systems employ a vacuum system, which pulls sap from the trees and can boost yields while likely not damaging the tree.[42]

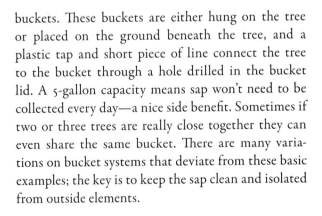

Figure 4.44. Tapping trees at Wellspring Forest Farm. Modern cordless drills hold a decent charge and can tap dozens of trees before needing a recharge. Be sure to drill straight into the tree to avoid inflicting more damage than is necessary.

Tapping the Trees

This part of sugaring is the most important to take extra care and detail with, as tapping is essentially wounding the tree, and the goal is to create a clean wound that will heal quickly after the tap is removed. It is worth purchasing a special tapping drill bit from a maple supplier, as it will last forever and it is specially engineered to leave a cleaner hole, critical to helping the wound heal itself. Tapping can be done anytime from early to late February (at least in Upstate New York) through the end of the season in late March to even mid-April some years. The key is to avoid leaving taps in longer than six to eight weeks; otherwise the tree will naturally heal itself around the tap hole and the tap will be almost impossible to remove. Taps should be removed as soon as trees break bud, if not sooner.

To select a spot for tapping, first examine the entire tree. Is the crown complete, and does it appear healthy? Avoid tapping areas that appear diseased or damaged. If the tree has been tapped in previous years, tap the opposite side from the most recent hole. No trees smaller than 10 inches should be tapped.

Drill with a high-speed cordless drill about 1.5 inches into the tree. Take extra care to keep the drill straight in and out of the tree, to avoid an "oval" hole. The goal is to get through the sapwood and slightly into the hardwood center of the tree. Insert the tap and hammer lightly until the tap is snug in the hole. Hammering too hard will result in split wood, which takes longer to heal.

SUGARMAN STEVE

I've been sugaring for the better part of eight years. I started my first year with just twenty-five trees at a nature center I worked at, which grew to one hundred trees at the peak of production. For three years I worked on a sugaring operation of around five hundred taps, then for a few years tapped only a few in the backyard. At our farm now we tap about a hundred trees with buckets each year. One of my favorite aspects of working on all of these operations has been sharing this process with youth, which connects them to the powerful cycles of nature. There is no better way to get children excited about nature than to show them that sugar can come from a tree! Working with kids earned me one of my favorite nicknames, "Sugarman Steve."

My background in permaculture has affected my methods for sugaring, which I share below. Take them as opinion based on my experience, context, and reflections on my own goals. There is no single "right" way to do it, but my hope is that the following suggestions will be helpful for those trying to figure out their relationship to this wonderful process.

Small Scale (Noncommercial) Is Most Sustainable

In my opinion, tapping five to one hundred trees is the most sustainable in terms of personal health and well-being. It becomes really difficult to scale up to a commercial operation without compromising values. For instance, most of the larger producers utilize vacuum systems, which does result in a lot of waste, as tubing has to be changed out. It also takes an incredible amount of wood (or other fuel) to boil the sap from so many trees, which can compromise efforts to thin the woods in a healthy way. On our forest farm we try to differentiate between agricultural systems that are appropriate for scaling up to commercial production and those who are better on a scale to produce mostly for personal consumption (with a small surplus to share). For us maple sugaring has fallen into the latter category.

It's also a very taxing endeavor to sugar, and it affects not only your own personal health (and sanity) but that of your family as well. I found over the years that I am not a late-night person and that I enjoyed sugaring when it wasn't the sole activity of my January through March each year. Our current scale allows sugaring to mix with a number of other activities. I now fire up the boiler and then head off to do other chores. This isn't to say that I don't appreciate those who are sugaring at a larger scale, because I very much admire them. I am merely sharing my thought process as an example of what each person should consider. I am also keenly interested in seeing the number of backyard sugarmakers increase.

Practical Matters: Relative Location and Gravity Systems

Permaculture design (see chapter 10) led me to think a lot about the principle of relative location in setting up multiple sugarbush systems. This principle suggests placing elements to minimize energy expenditure and making use of gravity to save a lot of time and effort. When collecting buckets, it makes sense to start at the top of the hill and work down. It's also well worth putting the sugar shack (the place you boil) in close proximity to the woods you are tapping, ideally at the bottom of the hill. We debated about several locations on our farm before ultimately deciding that proximity was the most important factor in our planning.

Ethical Use of Materials

Outside materials that are not biological in nature always come from somewhere else and take energy inputs (fossil fuels) to create. It is thus important to invest in durable materials that are long lasting. While it's cheaper (at least in the short term) to use plastic buckets, they won't last nearly as long as their metal counterparts. Making use of used materials (old metal buckets) doesn't increase demand for new manufacturing.

Wood: Waste = Food

Sugaring takes a lot of fuel for boiling, period. So where wood is sourced can have great implications no matter what scale of sugaring. Since we've scaled our system to a smaller (one-hundred-tap) operation, we are able to make use of waste materials as our main source of firewood. Each season we head to two local sawmills and purchase a few trailer loads of black locust slabs (from one) and the ends and scrap of red oak (from the other). We also burn poorly performing mushroom logs in the fire. Pine and other softwoods are also good candidates for a sugar fire, since they are not appropriate for burning inside the house. It has been our decision that all sugaring will be done with wood that others consider waste, as it is an appropriate use of resources.

Further, good forest management can also arguably provide a good source of wood for sugaring. As much as many landowners would like to conduct timber stand improvements, which are a boon to forest health and necessary because of a long legacy of forest abuse, it's hard to find the incentive to conduct these thinnings. The promise of sweet syrup (and mushrooms) is enough to get anyone out of his or her chair and into the woods.

— Steve

Figure 4.45. Waste black locust and red oak from local sawmills along with spent mushroom logs are used for boiling sap at Wellspring Forest Farm.

Collection and Storage

When below freezing nighttime temperatures are followed by days of rapid warming above freezing (ideally around 40°F), a sap run will occur. Checking buckets or storage tanks becomes an exciting daily chore and teaches the sugarmaker a lot about the subtle dynamics of the awakening spring. No two runs, and no two seasons, are ever alike. There will be days when the operation is overwhelmed with sap and days when you are surprised by how little comes out of the trees.

After a run, sap should be collected and boiled as soon as possible. If temperatures drop below freezing at night, sap will be effectively refrigerated and will last many days until the sugarmaker is ready to boil. Sustained temperatures between 45 and 60°F can cause sap to spoil in as little as twenty-four to forty-eight hours. Spoiled sap will appear cloudy and taste bad. It is easy to keep a storage tank cool by piling snow around it, keeping a lid on it, and sheltering it from sunlight.

Boiling

The major time and energy sink of sugaring is in the boil, and efficiency can be maximized through several strategies. One key way is to outfit any backyard-sugaring rig with a stovepipe; a 6- to 8-foot rise will provide natural draft and keep the fire burning hot. Some folks install a small fan to blow air through the fire as well. Try to construct the evaporating rig to be as airtight as possible to direct the flow of air.

The reality is that sugaring takes time. Expect to get a fire rolling and be keeping watch over it for many hours. This provides a great excuse to have a party and share the fun of standing in the woods boiling with friends and neighbors who come around to warm by the fire, share stories (and a sap toddy), and welcome spring. Inoculating mushroom logs is a great task to engage in while doing a boil. There is also something to be said for sitting quietly in the forest, listening to the sounds around as the fire hums along. It's a truly wonderful time.

Finishing

No matter what size setup you use, it will be nearly impossible to bring sap to that magic 67.7 percent number on a fire-driven setup. Most sugarmakers boil as much as possible over the fire, then finish on a propane or electric stove, where the sap can be closely monitored and the heat source easily adjusted.

Syrup is finished when the boiling temperature reaches 219°F (the boiling point of syrup), which can be determined with a candy thermometer. For a more accurate reading, use a syrup hydrometer, which will measure the sugar content of your liquid. It is critical to get the syrup as close to the correct ratio of sugar (67.7 percent); if it's too low it will become moldy and if too high it will crystallize. Moldy syrup can always be revived by bringing it to a boil and skimming off the mold, so it's not that big a deal in the end. Remain vigilant when boiling on fire and stove; when the level becomes too

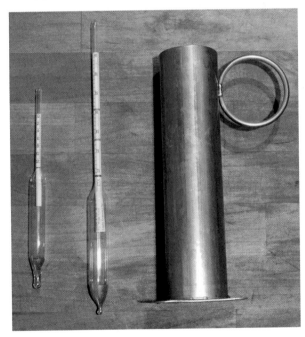

Figure 4.47. Hydrometers for measuring percent sap (left) and sugar content in finished syrup (middle). Purchasing the metal tube shown is helpful, as it makes it easy to place the hydrometer in and pour sap or syrup until full to get a reading.

Figure 4.46. Stoking the fire on a late-night boil at Sapsquatch Pure Maple Sugar. Photograph courtesy of Jen Gabriel

low or toward the end of finishing it is easy to scorch or burn the syrup—which is a sad fact, indeed—to have the product ruined after many hours of work!

Bottling

The best method for preserving backyard syrup is canning it in mason jars. Sterilize the jars and lids in boiling water, then pour the freshly boiled sap into the jars once the sap is between 180 and 219°F, which will be warm enough to seal the jars (no hot bath needed). Jars should be warm when you pour the sap in them, as the hot sap can break the jars if they are cold. Of course, a wide range of other containers can be used to store syrup, which if cooked to the correct percentage should be shelf stable. If you don't want "sugar sand" (the unfiltered sediment, which is harmless) to settle in the jars, use a cloth filter before bottling. Syrups should be stored in a cool, dark place away from direct light.

WALNUT SYRUP

As mentioned previously, walnut (*Juglans* spp.) trees are wonderful nut producers and also provide potential high-value wood products. If trees are not candidates for the latter, then tapping in addition to nut harvest can be a nice combination of yields for the forest farmer. The tapping of walnut also opens up the possibility of sugaring for forest farmers in warmer temperate climates found in the southern United States and parts of the Midwest.

While the basics of tapping, harvesting, and boiling walnut are the same as with maple and birch, the potential yields are much lower. Walnut trees have a similar sap-to-syrup ratio as maple syrup, but sap yields (volume per tree) from trees appear to be lower. Walnut sap flows in response to the freeze/thaw dynamics similar to those that make the maple sap run, so often the need to boil is concurrent with maple syrup. While there is still research to be done, it isn't highly likely that black walnut syrup will develop into its own market. Instead, it is recommended as a hobby pursuit or for the commercial grower to consider boiling walnut along with maple, and selling this combination for a higher price. Indeed, a small number of growers are doing this and getting as much as $60 a gallon for their walnut-maple

syrup (compared for $45 a gallon for pure maple). This is mostly because of the novelty.

On its own, walnut syrup is similar to maple syrup; it can be quite astringent but is infused with the nutty taste one might expect from it. Researchers from Kansas State University recently experimented with producing black walnut syrup, then did some consumer research on preferences for black walnut versus maple syrup. They found no significant differences on the likability scale between these two syrups and concluded that black walnut syrup could develop as a niche market in the Midwest.[43]

Usually it isn't recommended that trees destined for timber markets be tapped, whatever the species. Yet an interesting recent anecdote is that tapped maple wood has been successful when sold as a high-value wood in niche markets, where the staining is seen as a unique feature desirable to some for decorating homes and celebrating the rich tradition of sugaring. The same potential could hold true for black walnut, though it's hard to know. One major difference is that the discoloration left by a tap in maple wood is often a dark stain on white wood, but the black walnut develops a very dark brown heartwood, and it's unclear what effect tapping would have on this. So if timber is a clear goal for walnut trees, it's better to leave them untapped. Walnut trees with poor form and defects are good possible candidates for experimentation. Trees with a decent timber potential could also be tapped lower on the tree.

BIRCH SYRUP

Those who already tap maples may want to consider also tapping birch trees, should they be fortunate enough to have a stand in their woods. Birch sap doesn't usually begin flowing until the end of maple season, and since the same equipment is used for both, maple producers could simply switch over and continue to make syrup. Of course, there is a catch: While a gallon of finished maple syrup takes 40 to 50 gallons of sap, it's more like 100 to 200 for birch, because it has a much lower sugar content (1 to 1.5 percent).

The extra time and expense can pay off, however, as birch syrup is sold for $350 to $400 a gallon (maple is

BIRCH TWIG TEA

A short article by Benjamin Lord in *Northern Woodlands* magazine[44] offers a simple recipe for birch twig tea, which he describes as an "aromatic winter delight."

To make tea, harvest twigs of black or yellow birch in roughly 6-inch segments and stuff them as tightly as possible into a quart mason jar. Twigs must be fresh to get good flavor. Boil water in a kettle; once it boils let it cool for 15 to 20 minutes before pouring over the twigs (water that is too hot will evaporate the flavor). Seal the jar, and let it sit overnight. The next day enjoy a wintergreen, aromatic brew.

In fact, teas can be made from a few other trees, including white pine, spruce, and hemlock (from their needles), which are good sources of vitamin C, and sassafras (from the leaves and roots). Steep these in hot water for several minutes, tasting periodically until the flavor is desirable. All of these teas can be especially nice if made with maple sap as the tea water.

$45 to $60 for a gallon), and with demand far outstripping supply, it's a farmer's market. Most of the available birch syrup comes from Canada and Alaska, where birch forests are more common. But many northern states have the potential to "tap" into their birches as a source of syrup.

Birch syrup is not for pancakes. It's fruity, spicy, and sometimes reminiscent of molasses or licorice in flavor. The primary sugar in birch syrup is fructose, compared to maple, which contains mostly sucrose. The former is touted to be an easier sugar to digest and also contains the lowest glycemic index of all sugars, which makes it the most suitable sugar for use by diabetics. The syrup boasts a high vitamin C content and good amounts of potassium, manganese, thiamin, and calcium.

While maple and black walnut saps run in response to dramatic changes in temperature dynamics (also known as stem pressure), birch sap operates off root pressure, which requires that temperatures stay above freezing day and night. Thus, since collection and

boiling equipment is the same, birch could be seen as a form of season extension for sugarmakers.

In 2012 researchers at the Cornell Maple Program station in Lake Placid, New York, tapped around four hundred birch trees and produced about 30 gallons of syrup. Mike Farrell, director of the program, plans to expand production to six hundred to seven hundred trees and develop a significant research and Extension effort on "the biological, technological, processing and economic aspects of birch and walnut syrup production." Research objectives include determining the best times for tapping, sugar concentrations of trees, consumer preferences, and the impact on lumber quality, along with looking at the economics.[45]

Other Foods: Ramps, Spicebush, Sunchoke, and Groundnut

It's worth mentioning a few additional food crops in the context of forest farming. There are, of course, hundreds of plants that can be wildcrafted for yields, which falls out of the range of the criteria stated at the onset of this chapter. In this section the foods are leaves, bulbs, and tubers, all of which could be cultivated with reasonable yields for home or small commercial use.

RAMPS (WILD LEEK)

Anyone who has come across the ramp, or wild leek, *Allium tricoccum*, likely can't help but feel a sense of abundance; the leafy, bright green onion family plants often show up in clusters that can range from a few square feet to a solid quarter acre or more of green. It's a welcome gift of the forest in the early springtime of the year, as the forest wakes up from a long winter's nap. Ramps are spring ephemeral, which, like many early wildflowers such as hepatica, dogtooth violet, miterwort and others, "make hay while the sun shines" by taking advantage of the relatively high light levels in the deciduous forest before the trees have leafed out. The pungent leaves emerge in March and April, growing for around one month before dying back as temperatures warm and days lengthen. All of their photosynthesis and all of their food making (growth) take place during that short period of time. In June, a

Figure 4.48. A patch of wild ramps at Cornell's Arnot Research Forest. These populations have been the basis for propagation trials, which involve both seed collection and transplanting to other sites, including the MacDaniels Nut Grove.

flower stalk emerges and flowers bloom for a short time in summer, then form small, shiny black seeds that sometimes don't fall to the ground until the following winter. Not all of the increase in the size of a population of ramps is due to bulb division or splitting, like daffodil bulbs. Both bulb splitting and seed germination contribute to formation of a clump of twenty or more plants over several years.

Ramps have a long and storied history of wildcrafting in the United States and Canada. They were well established when colonists arrived, being so abundant in many places that locations were often named from them. Their range is vast, stretching to South Carolina and as far north as parts of Canada, such as Montreal and Ottawa, and as far west as Iowa. The term "ramp"

comes from Scottish and Irish settlers, who used the word to describe a plant back in the motherland, "ramson," *Allium ursinum*, a relative of chives native to Europe and Asia, known sometimes as the bear leek.

The plant is quite distinguishable with some basic identification and especially through the strong, garlic-chive odor it offers. The reddish stems lead to bright green leaves typically 5 to 9 inches long and 2 to 3 inches across. At some stages of growth, the plant could be mistaken for the foliage of lily-of-the-valley, a distant relative that is poisonous. The biggest distinction comes from the smell, but if unsure, one can dig up the bulbs to verify.

Ramps grow well in moist, rich soils with slight acidity, often the very type found in many hardwood

Figure 4.49. Large ramp bulb with leaves and flower stalk. This specimen, dug (then replanted) for educational purposes, should be left in the ground, as it will produce seed.

rule," which suggests that in harvesting one should gather no more than one-third of a population to ensure a stable community persists.

But what is this guideline based on? Research tells a different story, one that should be alarming to those interested in the long-term fate of ramps, which are considered a species of "special concern" in Maine, Rhode Island, and Tennessee.[46] Two prominent studies have looked at the effects of harvesting on long-term sustainability of wild leek populations. The first study in 2004 compared various levels of harvesting populations over a four-year period, and concluded that,

> harvesting wild leek is not sustainable except at very modest levels. Using the results of this study to predict recovery times, by assuming that growth rates and concomitant recovery times are affected in a consistent manner by levels of harvesting, the sustainable harvest level is predicted to be 10% or less, once every 10 years.[47]

Another study over five years in Quebec came to a similarly startling conclusion:

> In a particularly unproductive season like 1985–86, even a 5% harvest is deleterious, and in all other years a decline is predicted when a 15% harvest is stimulated.[48]

So as far as wild harvesting is concerned, harvesting over 10 percent is likely to be detrimental, but to be safe it's best to aim for a maximum harvest of 5 percent each year from a given population. This becomes trickier when harvesting from populations on public lands, where multiple people may come through hunting ramps. It demands that more time is taken to observe, catalog, and note the changes in populations from year to year. And when in doubt, it is best to err on the side of caution.

The studies cited above are based on the traditional harvesting method, which is to dig the leeks from the ground. Some evidence exists to suggest that managed wildcrafting, as mentioned previously in this chapter, could be beneficial. With ramps, taking only the leaves

forests. They often inhabit the same environments as sugar maple, beech, and hemlock. They share similar site needs to many interesting understory species such as mayapple (*Podophyllum peltatum*), trillium (*Trillium* spp.), bloodroot (*Sanguinaria canadensis*), and black cohosh (*Actaea racemosa*).

Ramps are traditionally wildcrafted, and there is almost no cultivation happening. One reason for this is that ramps appear to produce in the wild so prolifically. Why buy your lunch if you can get it for free? Many wildcrafters look forward to harvesting this tasty, tangy, and nutritious plant, and at first glance it appears that one could harvest a hearty share without inflicting any harm on the population. Often used as a guideline for wildcrafting is the common "two-thirds

A PLAN FOR CONSERVATION OF WILD LEEK

A recent report called "Plugging the Leak on Wild Leeks: The Threat of Over-harvesting Wild Leek Populations in Northern New York"[49] provides an in-depth account of the plight of wild leeks that provides useful background and information to those interested in leeks in any region where they are found. The publication notes,

> In New York State, the *Allium tricoccum var. burdickii* [a rare variation on the common ramp] species is listed as endangered and any harvesting of this plant is forbidden. The *Allium tricoccum* species is not far behind, meaning that conservation efforts to ensure survival of the species must be quickly developed and enforced.

The authors offer an analysis of five scenarios to consider for conservation of the plant:

1. Ban harvesting, as is currently done in the Great Smoky Mountains National Park in North Carolina and in Gatineau Park, Quebec.
2. Limit harvesting through some sort of permit system and through education to encourage harvesting of leaves only, not bulbs.
3. Encourage commercialization by giving harvest permits to a limited number of growers and potentially labeling and certifying legally collected material (official status is designated by government agencies), which would indicate to consumers that those ramps were harvested in a sustainable manner.
4. Reintroduce bulbs through a government program, similar to a successful effort in Canada, where 1,117 participating landowners planted and monitored bulbs for five years, with over 80 percent success in establishing plants.
5. Encourage cultivation by commercial and hobby growers, which could be supported through education efforts.

The paper concluded that likely a mixture of these options provides the best scenario.

> After considering the feasibility of all possible solutions to the problem, we have come to the conclusion that the harvesting of wild ramps should be limited through a harvesting permit program, cultivation should be encouraged, and educational programs must be put in place to make people aware of the issues created by over-harvesting and to expose them to the basics of plant conservation.

The full paper can be found at: http://web.stlawu.edu/sites/default/files/resource/wild_leek_conservation.pdf.

of a plant is a more appropriate way to harvest, a practice that native tribes such as the Cherokee practiced for centuries.[50] It may also be more sustainable if care is taken to harvest a mix of ages, with the majority of bulbs being near maturity yet a certain (unknown) portion left for reproduction.

Propagation of Ramps

Another prudent response to the threat of overharvesting wild ramps is to propagate them for cultivation. This is another form of the core tenet of forest farming, to engage in "productive conservation"—to act to conserve wild species through production. Planting sites should contain well-drained, nearly continuously moist soil. To prepare a planting bed, remove debris and unwanted weeds and tree sprouts. Loosen the soil and incorporate organic matter such as compost, shredded leaves. Sow seeds on top, and gently press into the soil. Cover with 2 to 4 inches of leaves. Studies have shown that moisture is critical to all states of growth, so mulching with leaves (especially sugar maple) is recommended. Shade is an important dynamic, with research indicating that seedlings emerged best in at least 30 percent shade.[51] Denser shade will likely increase the leaf surface area, which is desired.

Figure 4.50. Immature seed heads in early summer at the MacDaniels Nut Grove; 2013 was the first season that transplanted bulbs from previous years set seed in cultivated beds at Cornell.

Ramp patches can be successfully grown from seed or through transplants. Bulbs are the faster route and can be purchased or dug from wild populations, keeping in mind the importance of sustainable harvesting. The best time for transplanting is between September and March, with February to mid-March being the best time. Plant bulbs 3 inches deep and roughly 4 to 6 inches apart.

Seeding is best done in prepared beds in the late summer to early fall from collected or purchased seed. The fresh seeds have an underdeveloped embryo that needs a warm, moist period followed by a cold period to break dormancy (see chapter 7). This process can take upward of eighteen months.

Do not harvest any plants until they have filled the site, have large bulbs, and have flowered. If whole plots are harvested at one time, it is recommended to have enough plots to allow for a five- to seven-year rotation; that is, to have continuous harvests year

Figure 4.51. Clumps from healthy leek patches can be successfully transplanted in fall and spring. Take care to minimize disturbance to bulbs by moving large clusters along with the soil they are in. Cultivated plantings will need to grow for many seasons before a harvest can commence.

Figure 4.52. Dug ramps should be washed before eating or stored to preserve healthy bulbs.

Figure 4.53. Spicebush growing wild in a healthy forest in central New York. The plant is relatively common, especially in maple–beech–birch forest types.

after year, harvest only one-fifth or one-seventh of your production area each year. When harvesting a portion of a plot, no more than 15 percent of the ramps should be removed. If the thinning method is used, great care should be taken not to damage plants that are not harvested.

Tools for harvesting ramps vary with the person using them. A ramp "digger" tool can be purchased or made. This hand tool is the size of a hammer, with a long, narrow head similar to a mattock. Other suitable tools include a garden hoe, a pick, and a soil knife. For commercial operations having a tool that can be used comfortably all day is essential.

Digging methods are the same as those described in the transplanting discussion above. Again, great care should be taken not to damage the bulbs. While harvesting, keep the dug ramps cool and moist. When harvesting is complete, wash ramps thoroughly, and trim off the rootlets. Pack in waxed cardboard produce

boxes, and store in a cool place, preferably a walk-in cooler. Do not store in airtight containers.

SPICEBUSH

A small shrub that thrives in the understory, spicebush (*Lindera benzoin*) is found in moist woods, often among stands of poplar, maple, and beach. Easy to identify, a suspected plant can be confirmed by crushing the twigs or leaves, which releases a tangy lemon fragrance.

Its bright showy flowers in the spring and yellowing leaves in the fall make it a pleasant addition to the aesthetic of the forest. Since the leaves, which are used for tea or as a spice in cooking, are the main interest in this species, forests with denser canopies often encourage the leaves to get up to 5 inches long.

The fruits, called drupes, can also be dried and used as an allspice or pepperlike flavor in cooking. Used fresh, they are a perfect companion to apples, which their fruit season overlaps with. They are a food enjoyed also by robins, flycatchers, and catbirds. Spicebush is also an important plant to butterflies in the swallowtail family, including, naturally, the spicebush swallowtail.

Twigs, dried berries, and leaves also make a delicious tea. Pioneers referred to the plant as "fever bush," as a decoction (boiled leaves or roots) was used frequently to induce a sweat. Tinctured leaves are also a good fever medicine.

While spicebush is a minor crop, there is definitely potential for adding this to a forest farming enterprise. The folks at Integration Acres sell dried berries and leaves as "Appalachian Allspice" and get around $1 an ounce.

SUNCHOKES

A native tuber and a member of the sunflower family, sunchokes (*Helianthus tuberosus*) are a highly adaptable species that could easily be grown within the forest farm. The vegetation grows rapidly and can be used as an annual windbreak, reaching heights of as much as 10 feet in a season. Roots are dug in early fall after the flowers die back. They are an abundant source of food with a taste similar to potato with a bit of a nutty flavor.

The name certainly implies that the plant enjoys a full-sun environment, but production can still be adequate in shade conditions. Sunchokes are juglone tolerant and thrive in the dappled light of a walnut canopy. They can be observed growing in old fields, forest edges, road ditches, and gaps in the forest. They are one of the more adaptable species in this book.

The tubers offer some significant nutritional benefits, boasting a high amount of protein, iron, and potassium, as well as inulin, a carbohydrate associated with good intestinal health and a good food for diabetics. The probiotic qualities of this food are a mixed bag, as new eaters can sometimes experience a significant increase in flatulence from consumption. This can be easily remedied in two ways. The first is to ease your diet onto sunchokes by grating small amounts of the raw root onto salads and into other dishes. Another successful method is to first boil the roots and skim off any foam, similar to the processing of dry beans.

Sunchokes are remarkably easy to plant, grow, harvest, and propagate. To plant, obtain roots from a supplier or a neighbor with an established patch. Tubers are best planted in the fall but will do just fine transplanted in spring. Be sure to plant in a space where it is okay to have these roots reside permanently; it is very difficult to eradicate them entirely. To harvest, wait until flowers bloom in late summer, then die back completely; the roots can be harvested soon after or really anytime, as long as the ground isn't too frozen; some growers even mulch and leave tubers in the ground all winter, harvesting as needed. Storage of the roots is equally easy, if they are packed into sawdust and kept in a cool, root cellar–type place.

Propagation is accomplished by planting tubers in new locations. When harvesting a patch, dig the biggest plants and leave the others; harvesting around half of the patch will mean a good harvest is virtually guaranteed to be ample for the following season. The only real planting situation to avoid are areas that are consistently wet, where the roots may rot. While some small farmers' markets and CSAs have begun to offer sunchokes as an option, it is somewhat baffling that they aren't more widely available. Because of its ease of planting and harvest, this is a great plant to experiment with in the wide-ranging conditions found in the forest farm.

GROUNDNUT

Not to be confused with peanut, which is sometimes called by the same common name, the American groundnut (*Apios americana*) is another prolific tuber that originally served as a staple food for Native Americans. Other common names for the plant include potato bean, hopniss, and Indian potato. A most curious species, it is a nitrogen-fixing vine with beautiful purple pealike flowers. Although it's harder to establish than sunchoke, once it gets a foothold removal is equally impossible. The vines can be trained up a structure or simply allowed to grow horizontally as a groundcover. In the wild, groundnut is often found in damp bottomland soils or riparian areas, though it will grow in many soil types.

Groundnut is high in both starch and protein and on a dry weight basis has roughly three times the protein of potatoes.[52] A limitation for serious cultivation is partly due to the fact that tubers need two to three growing seasons for maturity, though modern breeding has resulted in selections that have superior growth characteristics, including one strain from Louisiana that yielded as much as 7 pounds per plant in a single growing season.[53] Plant stock remains limited in availability, and efforts to propagate more would likely be the more profitable venture for a forest farmer compared to selling the tubers.

Groundnut is a member of the legume family and is a moderate nitrogen fixer. It produces a seedpod, but seeds can take from one to three months to germinate. Propagation is by digging tubers from one location and transplanting them in another.

Getting Food from the Forest

Much of the interest in forest farming comes from the ability to cultivate and enjoy a wide range of food products that the forest offers. Chapter 5 will detail the most common food people envision when they think of forest farming—mushrooms—yet as this chapter has described there are many fruits, nuts, shoots, and

roots to incorporate into a forest farm. For good food production, the forest farmer will need to be willing to play with light dynamics in the forest and find appropriate niches to have crops thrive. A good strategy for starting out would be to establish plantings in high-, medium-, and low-light conditions within the forest farm, then evaluate the plants' response. In the end, each forest is unique and dynamic in its own right, which is what necessitates this type of experimentation.

Those seeking commercial options should note that fruits, nuts, and other foods have a long way to go when compared to medicinals, mushrooms, and nursery production, which are clear in terms of the potential profitability, at least for some crops within those categories. This is not to discourage experimentation, as it is important for practitioners to experiment to find new pathways. Shiitake mushrooms, for example, were not seen as a viable crop in the United States two decades ago, but today they are quickly gaining attention, and more growers are getting into the business each season. Without early adopters willing to try, make mistakes, and share their experience with others, forest farming will remain a marginal practice.

CASE STUDY: BADGERSETT RESEARCH CORPORATION
CANTON, MINNESOTA

A visit down the windy dirt roads of rural southern Minnesota brings you to one of the most extraordinary (genetic) libraries in the world. Here American, European, and Asian nut varieties comingle, hybridize, and adapt to the harsh climate of the area, playing out the dynamics of natural selection on their own terms. The Rutters, Philip, wife Meg, and son Brandon Rutter-Daywater, are merely orchestrators and monitors of the activity. And by "merely" I mean they have fully devoted their lives to the tedious work of research and development, monitoring tens of thousands of plantings. While the site itself is not technically a forest farm, it is very much a part of the overall forest farming and agroforestry vision. The Rutters are carrying out the research necessary to make these things work for the rest of us.

Figure 4.54. Philip Rutter and Brandon Rutter-Daywater are the father and son team committed to a multigenerational research project at Badgersett.

I met Phil on a weekday morning during my long jaunt across the Midwest; he graciously forgave my tardiness and proceeded to engage me on a six-hour tour of his outdoor living laboratory, telling endless stories of his journey of forty-plus years down the path of nut cultivation. At heart Phil is an ecologist, and while his PhD work was in the field of zoology, the underlying interest was always evolution and ecology, with a strong interest in plants and plant communities. His past work includes serving as president of the Northern Nut Growers Association; cofounding the American Chestnut Foundation in 1982, and developing the Wagner Research Farm in Meadowview, Virginia, a 20-acre site devoted to American chestnut research, with over five thousand seedling trees planted. His work in chestnut, hazelnut, and hickory ecology has resulted in dozens of professional papers that define the field of "woody agriculture," which Phil describes as,

> The intensive production of agricultural staple commodities from highly domesticated woody perennial plants. Permanent stands of the woody crop are established and seeds are harvested annually. Once every 5–10 years the wood is harvested for biomass by coppicing, whereupon the plants regenerate from the roots and resume production of the food crop one year later.

Badgersett is the home and field office of his work, which was started with the goal of "pursuing the intensive domestication of woody perennial plants for agriculture." The sheer number of trees on-site is impressive; more than eight thousand hybrid chestnuts have been screened, with about ten thousand currently under evaluation. In addition, hazelnut plantings exist in the area of sixty thousand bushes of various ages. To keep up with research, several thousand trees are planted annually, assisted by the construction of a greenhouse in 1992 that enables production of seventy thousand seedlings per year. The Rutter family, along with one other scientist, one employee, and several investors, supply the people power to barely keep up with the research.

Figure 4.55. The hawk-like kite that the Rutters fly is an indication of their commitment to working with the natural ecology.

The main ecological pattern utilized at Badgersett is based on the ecological concept of a "hybrid swarm," which is a phenomenon usually in "edge" ecosystems or in changes in climactic conditions where the range of multiple compatible species overlaps (for instance, wild hybrids of pecan with shagbark and bitternut and shellbark hickory). Over several years a wide genetic mixing occurs, with species segregating, recombining, backcrossing, and crossing again. It's the way nature "solves" problems with disease and pests, by essentially mixing up the gene pool and creating new combinations.

At Badgersett concept is human driven; for example, chestnut species don't normally cross paths in the wild, but at this farm one can find American, Chinese, Seguin, European, and Japanese chestnuts

all planted together. The seeds (nuts) from those next-generation hybrids are noted for the desired qualities and if persistent are grown into the next generation of trees. Badgersett does this same process with hazelnut, bringing together European, American, and beaked hazel, as well as with hybrid hickories, where a mixture of four species of pecans and hickories have matured to the point where over 95 percent of the hybrid hickory pecans are very thin shelled and extremely cold hardy—and some bear *annual* crops (a major problem in the pecan industry). Says Phil,

Our breeding process is so different from "traditional" that we've been forced to come up with a new name for it. And it's complex. The breakthrough process is Accelerated Guided

Evolution, AGE. The breakthrough outcome is a NeoHybrid crop. NeoHybrids are utterly different, at the genomic level, from normal hybrids. Both kinds of hybrids have huge advantages, but they're as different, technically speaking, as AGE is absolutely NOT standard breeding; it entails complex, specific processes developed over 30 years based on hybrid swarms. As the acronym suggests, it takes time, patience, and very large numbers of hybrids being tested. Nothing else will work.

HAZELNUT

The most developed crop at the farm is hazelnut, which is mainly a mix of native Wisconsin and Iowa wild hazels with the commercial European varieties (varying generations of *Corylus avellana* × *C. americana* × *C. cornuta*). Research and breeding have focused on three rounds of breeding, with each round focusing on a few variables (such as cold hardiness, yield, ease of cracking, ripening time, and so forth). Phil mentioned to me that it's important to take this work in stride, as seeking to breed for all characteristics at once becomes exponentially more complicated. Each succession of plantings contained about five thousand individual numbers of plantings, from which the following traits were observed starting at four to five years of age and continuing for upward of twenty to thirty years. The project is now on its fourth cycle. Each cycle narrows the range of hybrids based on the following criteria:

Cycle 1: Eastern Filbert blight resistance and cold hardiness
Cycle 2: Heavy crop and annual crop
Cycle 3: Nut size and flavor characteristics
Cycle 4: Machine harvestability and big bud mite resistance/tolerance

The results of this work are quite remarkable. The hazelnuts are stated to offer a wide range of desirable traits, including nuts that are on average 100 to 300 percent larger than wild hazels, are unquestionably hardy to zone 4, and have established resistance to eastern filbert blight, which affects the higher yield-ing/large nut cultivars of European chestnut. They also demonstrate drought resistance, respond well to coppicing, and, once established, eliminate any need for plowing, tilling, or fertilizing. The root systems are extensive and fibrous, making it an excellent crop for carbon sequestration. In addition to a food crop, hazelnut offers a potential as an energy crop, in both the harvest of biomass (every five to ten years) and for oil pressed from the nuts, which is also excellent as an edible oil, comparable to olive oil. Hazels, like most nuts, are very high in oil—60 to 70 percent, of which 70 percent is monounsaturated, the most heart-healthy kind—one ounce of hazelnuts contains 4.24 grams of protein, 178 calories, and 2.7 grams of dietary fiber, and nine different minerals.

The hazelnuts have been modified to a degree from their normal commercial form, which is often more treelike. The shrub form developed by Badgersett allows for machine harvesting, which has just begun after they acquired a rare antique blueberry harvester, a BEI Inc., which is so old it has no serial number. This machine is able to harvest a decent amount of nuts with, so far, never more than two passes each season. Breeding work now focuses heavily on getting varieties that will respond to machine harvesting, as the ultimate vision at Badgersett is to replace corn and soy fields with rows of nut trees, providing row-crop farmers a genuine alternative they can adopt. At this point, too, genetics for hazels are being split for selecting both for machinability and for hand harvest.

CHESTNUT

The next crop Badgersett has focused on is chestnut, which nutritionally resembles grains, with 8 to 20 percent protein, 2 to 5 percent oil, and 2 to 4 percent minerals. Badgersett began its breeding with hybrids from several amateur breeding projects connected through the Northern Nut Growers Association, including Douglass, Gordon, Gellatly, Szego, Clapper, and Perry hybrids, along with varying amounts of Japanese, European, Chinese, and Seguin chestnut. Whereas the American Chestnut Foundation (described in detail on page 111) is entirely focused on the idea of restoring a "pure" form of the wild American chestnut, this strategy focuses on domestication for production. The results are a mixture of species that begin bearing in the three-to-five-year range, exhibit cold tolerance to zone 4, and are showing resistance to chestnut blight, though only recently did it show up at the farm. Prior to the arrival of the blight, testing of Badgersett genetics in China and at Auburn University have shown that about 80 percent of their seedlings are "as resistant as Chinese."

Compared to the hazels, which have had the time to "restabilize" genetically after such intensive crossing, the chestnuts have proved to take a longer time and as a result have a high variability; the best are excellent, but some are always weak. There are also several challenges to overcome with respect to cold tolerance; one is that the bark is susceptible to freezing and thawing, which can open up sections to fungal disease.

When we walked into one of the chestnut groves, recent grazing by the farm horses had left the ground-cover short in preparation for nut harvest, which was coming in just a few weeks. It was easy to see the abundant amount of ripening chestnuts, as their yellow spiky husks contrasted against the dark shiny leaves of the trees, and the branches were weighed down with the harvest.

Walking through this grove, I had two revelations. The first was that, being in the fifth hour of the tour, I'd walked through probably 30 to 40 acres of highly productive nut crops. Yes, there was a wide range in size and readiness for commercial production, and yet there was complete abundance here. I was amazed that I'd become normalized in such a short time to walking through such an incredible amount of food.

The second revelation was how important it was to have people like the folks at Badgersett, who were committed and devoted to just a few crops and their development. Writing this book had been keeping me in the "big picture" mind-set, looking at the potential of a wide range of crops. But here was the opposite: a narrow yet critical focus on just a few species and their development over several lifetimes. This sense of a narrow focus was only in one direction, as while there were only a few families of tree crops in production, the genetic pool was more diverse for these species than anywhere else in the world.

HICKORY

The third and newest species is a hybrid hickory, which combines genetics from pecan, shagbark hickory, and bitternut hickory. This development is in fairly early stages, mostly because regeneration of trees is slow. Yet preliminary results have demonstrated good cold hardiness and large, tasty nuts that have thin, easy-to-crack shells. The plantings I saw were relatively young and were at the time hosting the resident flock of Icelandic sheep, which provided good mowing services and made harvesting the fallen nuts a lot easier.

THE FUTURE

The foundational belief and paradigm at Badgersett is multifaceted; there is first recognition that nature is the guiding force in this process. Mimicking the hybrid swarm was a strategy to respond to the environmental conditions: cold temperatures, floods, droughts, and fungal diseases. These were accepted variables considered while working toward production of sizable yields. This is domestication within nature's framework, very different from many other forms of hybridization that occur by scientists around the world, which in many ways ignore or avoid some of the basic principles of ecology.

But an even deeper fascination and attention to ecology was evident as I toured the land with Phil. Our first stop was in fact in the field of some of the original hazel plantings, where he worked to untangle a large kite shaped like a hawk, the latest strategy he was trying to deal with mounting pressure from birds—especially blue jays and crows. The kites, along with bird perches and nesting trees for predator birds, were attempts by Phil to thwart the pests, though he didn't talk of them as pests but with a surprising appreciation, noting especially how intelligent crows were. Perhaps he wasn't enraged because his primary business was research, not production, so yields were less of a concern. It appeared to me, however, that at one level he simply saw these animals as inevitable parts of working with ecosystems, as important to explore as the deep genetic work with his trees.

Another key characteristic of this project comes as an extension of the first: that economically profitable, farm-scale production of nut crops is feasible, even in

Figure 4.56. Hybrid hickories are the third nut crop to be added to the research work at Badgersett, with initial results showing promise.

the coldest of climates. Yet it could happen only if one accepts that it will take time, extensive documentation, and a lot of failures. Toward the end of the tour, as we stood overlooking a field of hazelnuts, I asked if they ever planned to name varieties. The answer was, "Maybe in another forty years"—there was just simply more work to be done before any consistency could be confirmed in a hybrid.

I also asked if they thought someday they would begin cloning varieties once they became exceptional. While they have worked on some cloning with the University of Nebraska, the sale of clones looks to be many years away. What they recommend is for farmers to plant the hybrid seedlings and when holes in their planting appear (inevitable) they should clone their own best plants, which both pushes their planting into higher production and develops locally adapted genetics further.

Driving away that day, I realized I had experienced something profound in this landscape—both in the actual work done for breeding of nut trees and for the philosophy and perspective that enabled this process to happen in the first place. The stewards of this project are rooted in the belief that, "Nature knows *more* than we do, much more; she corrects us constantly, and we learn every year" (Philip). This site was a prime example of a system in which people could work with nature to produce high-value food crops without compromising environmental health.

Forest farmers need people with this much devotion and focus paid to individual crops, if we are ever going to make production happen on a reasonable scale. In fact, potential farms might consider specializing a bit in one crop for production and breeding, in service to doing the work that needs to be done for the greater good, which ultimately comes down to observation, good record keeping, and selection, all with a sense of patience. When buying a hybrid hazelnut or chestnut seedling from Badgersett or any other quality nursery, the cost may appear to be much more on the surface versus a cheaper field-grown seed someone tossed in a container. The difference, in the end, is genetics. You get what you pay for.

— Steve

Badgersett is releasing a book from Chelsea Green Publishing in 2014 called *Growing Hybrid Hazelnuts*.

FURTHER READING

Rutter, P. A. 1987. "Badgersett Research Farm; projects, goals, and plantings." 78th Annual Report of the Northern Nut Growers Assoc.: 173–186.

Rutter, P. A. 1989. "Reducing Earth's 'greenhouse' CO_2 through shifting staples production to woody plants." Proc. of the Second North American Conference on Preparing for Climate Change: 208–213. The Climate Institute, 316 Pennsylvania Ave., SE, Suite 403, Washington, DC 20003.

Rutter, P. A. 1990. "Woody agriculture: increased carbon fixation and co-production of food and fuel." Paper presented to the World Conference on Preparing for Climate Change, Cairo, Egypt, December 1989. The Climate Institute, Washington, DC. Reprinted in 80th Annual Report of the Northern Nut Growers Assoc.

Rutter, P. A., G. Miller, and J. Payne. 1991. Chestnuts (*Castanea*). Pp. 761–788 in *Genetic Resources of Temperate Fruits and Nut Crops*. J. N. Moore and J. R. Ballington, Jr., eds. International Society for Horticultural Science. *Acta Horticultura* 290.

5 Forest Cultivation of Mushrooms

There is nothing more pleasing to a new forest farmer than her first flush of log-grown mushrooms. Shiitake (*Lentinus edodes*), native to Asia, are the most widely grown forest-cultivated mushroom in the cool temperate climate, either by the total weight sold or the total dollar value, as well as with respect to the number of forest farmers involved in their cultivation. Several North American species, including lion's mane (*Hericium* spp.), oyster (*Pleurotus* spp.), and Stropharia (*Stropharia rugoso-annulata*), are forest cultivated only

Figures 5.1, 5.2, 5.3, 5.4. Part of the reason for shiitake's (upper left) popularity among forest farmers is its relative ease of cultivation on hardwood logs. Other promising species for the forest farm with at least some successful cultivation include, clockwise from upper right, oyster, lion's mane, and wine cap.

in small, noncommercial quantities but have considerable potential since preliminary research at Cornell has been successful, and these are considered among the best of wild-collected species by the amateur mycophiles (mushroom collectors).

In addition to the exciting food and medicinal value these species offer to forest farmers, production can be linked to forest management and other production systems, as mentioned in chapter 3. Of all the forest products covered in this book, mushrooms present one of the better entry-level practices that also can yield promising results in a relatively short amount of time (one year) for both the home and commercial grower.

Mushroom Production in the United States

During 2012 and 2013 Americans consumed 877,097,000 pounds of white (button) mushrooms and their close relatives, portabella and cremini (all come from one species, *Agaricus bisporus*). Despite the American consumer's demand for these mushrooms, they do not lend themselves to forest cultivation. They are grown on compost in large climate-controlled production facilities, which bear no resemblance to forest farming.

By comparison production of specialty mushrooms was only 2 percent of *Agaricus* mushroom production. Shiitake amounted to about half of total specialty (non-Agaric) mushrooms produced, and most of that was indoor production on sawdust blocks. Only about

Figure 5.5. Indoor production of button mushrooms on compost.

10 percent of total shiitake production was forest cultivated on hardwood logs. Statistically speaking, forest production of oyster and other specialty mushrooms did not register at all (see table 5.1).

Conventional (indoor on sawdust) mushroom producers have nothing to fear from competition by forest farmers. The price per pound reported for shiitake applies to indoor sawdust-grown mushrooms, whereas research from Cornell and the University of Vermont has shown that in the Northeast the average price per pound for log-grown shiitake is about $15/lb.[1] Further, based on work with growers and spawn producers, some regions of the country report that prices can be as high as $25 a pound.

Table 5.1. Sales and Production of Specialty Mushrooms

Variety	Production (pounds × 1000)	Sales (× 1000)	Natural wood logs, outdoors (× 1000)	Indoor, sawdust (sq ft × 1000)	No. growers > 200 logs	Price per pound
Agaricus	877,097	$1,045,565	N/A	N/A	N/A	$1.19
Shiitake 2012–2013	8,617	$26,873	137	1,506	179	$3.33
Oyster 2012–2013	7,411	$21,045	Not reported	785	195	$3.02
Other 2012–2013	3,654	$16,802	Not reported	362		$4.99

Source: USDA National Agricultural Statistical Service, 2012–2013.[2] Production, sales, and price per pound include both forest farming production (natural wood logs) and indoor sawdust production.

IF YOU GO HUNTING FOR WILD MUSHROOMS . . .

Advice from the mushroomforager.com bloggers Ari Rockland-Miller and Jenna Antonino DiMare

Many Americans are mycophobic, or intimidated by the prospect of hunting wild mushrooms. Yet wildcrafting can be accessible, safe, and abundant if you approach it with patience and intention. Even though this book is not focused on wildcrafting topics, we thought it would still be important to include some brief tips on foraging from our friends at The Mushroom Forager. This is their guide to a safe, fruitful, and renewable harvest:

1. **Start by learning your region's most deadly species.**
   Wildcrafting can be extremely safe if you are responsible, but the stakes are too high to take risks. For example, the destroying angel (*Amanita bisporigera*) is one of the first mushrooms you should learn. This ubiquitous deadly mushroom is so white that it almost glows, bearing an uncanny and unsettling resemblance to the common button mushroom.

2. **Next, begin learning the edible species one at a time, starting with the most foolproof.**
   Instead of trying to learn every species in the forest, which is a worthwhile but daunting task, start by learning distinctive and delicious species such as the giant puffball, lion's mane, and chicken of the woods. Each season, learn additional species that you can confidently identify.

3. **Know the ForageCast!**
   The Mushroom Forager (www.themushroom forager.com) publishes a list of the most distinctive and delicious species in season in the Northeastern United States, updating the list often during the wild mushroom season based on reader feedback and reports from the region's fields and forests. Random, haphazard foraging is fun but often fruitless. Make your foray targeted, using the ForageCast, so you know when, how, and where to look for your favorite edibles.

4. **The first time you try a new species, have a mushroom or plant expert verify your ID.**

Always be 100 percent confident of your ID before taking a bite. Be sure to identify in the field, so you don't miss crucial ID characteristics, such as the presence of a volva—the often subsoil swollen base that is common in the infamous *Amanita* genus of fungi. Even if you are positive about the ID, try a small portion the first time.

5. **Never mix known edibles with unknowns in your basket.**
   When you arrive home, eager to cook up your bounty, a poisonous species may fall into the frying pan along with the edibles.

6. **Check back on known producing spots every year.**
   Soon you will have more spots than you have time to check on and may have to freeze, dehydrate, can, or pickle the surplus. Many gourmet and medicinal mushrooms will fruit in the same spots every year, but even the most reliable, hen-of-the-woods (*Grifola frondosa*), will rest every few seasons. Eventually, you will develop an intuitive sense for when your favorite species fruit based on temperature and rainfall.

Figure 5.6. A Chicken of the woods (*Laetiporus sulphureus*) mushrooms fruiting prolifically on an old oak log at the MacDaniels Nut Grove. The shelf form and bright orange and yellow cover make this one of the easier wild mushrooms to identify.

Figure 5.7. Decay of a dead tree stump by a saprophytic (white rot) fungus.

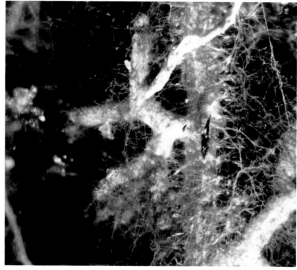

Figure 5.8. Mycorrhizal symbiosis is an intimate association between a fungus and a root. The ectomycorrhizal fungus ensheathes the pine root, and fungal hyphae project out into the surrounding soil.

In contrast to these relatively few species of mushrooms that are deliberately forest cultivated, there are many more edible species that are collected from the wild. Wildcrafting of mushrooms is a tradition handed down from generation to generation in some parts of the country and is becoming popular with a new audience of mushroom hunters as well. Young or old, most foragers collect wild mushrooms for personal consumption and the sheer enjoyment of the "chase." Since the 1980s, commercial sale of wild mushrooms in the Northwest has become an important "industry," with an estimated seven hundred to nine hundred collectors in Washington and Oregon. Most of the wild mushrooms collected in the Northwest are sold in Europe and Japan, but they can also be found in grocery stores throughout the United States. There are nearly twenty species commercially harvested in the Northwest. Much of the harvest is on federal land (US Forest Service, Bureau of Land Management). Some of the most prevalent of these wildcrafted mushrooms are chanterelles, boletes (e.g., porcini), morels, hedgehog, and the much-sought-after matsutake.

The commercial availability of wild mushrooms in the United States is limited by their relative scarcity in the wild, difficulty in shipping (poor shelf life), and, perhaps most importantly, by the unfamiliarity of most of the public with anything but the button mushrooms (*Agaricus bisporus*). Together these limiting factors might seem to make a case for deliberate cultivation of some of the more valuable of the wildcrafted species, except for the fact that for most, cultivation is virtually impossible, because of their mode of nutrition. Practically all of the mushroom species suitable for cultivation are saprophytic (decomposers), in contrast to mycorrhizal species such as boletes, chanterelles, matsutake, and other native forest mushrooms.

The terms "saprophytic" and "mycorrhizal" refer to how the fungus "makes a living"; that is, what the fungus "eats." Saprophytic fungi are scavengers in the sense that they grow by breaking down (decaying) and assimilating (consuming) dead organic matter, which is wood in the case of mushrooms that are suitable for forest farming. On the other hand, mycorrhizal fungi are "cooperators" rather than scavengers. They enter into a mutually beneficial association with the roots of living trees and other plants, as mentioned in chapter 3. The green plants provide the fungi with sugars, obtained photosynthetically, while the fungi provide

Figure 5.9. The black truffle (this specimen from Northern Italy) is the extremely valuable underground fruiting body (mushroom) of a mycorrhizal fungus, *Tuber melanosporum*. Photo courtesy of Blue Moon In Her Eyes (Flickr)

the trees with minerals and water obtained through an extensive network of threadlike mycelium that extend out into the soil. Both the plant and the mycorrhizal fungus benefit from this complex symbiotic relationship. In fact, it is the very complexity of mycorrhizal symbiosis that makes these fungi so difficult to cultivate deliberately. Cultivation of a mycorrhizal fungus would involve managing the exacting demands not just of the fungus but also its associated partner tree, whereas cultivating a saprophytic fungus, such as shiitake, involves managing just the fungus alone.

The assertion that mycorrhizal fungi cannot be cultivated is a useful generalization, but there is at least one notable exception: Truffles have been cultivated in a few instances, though failed attempts far outnumber the successes. Actually, it could be said that truffle cultivation involves not just two organ-

isms—the tree and the fungus—but also a third—the pigs or dogs that are trained to smell out the below-ground truffles for digging by the truffle hunter. For those who are stouthearted and not averse to failure, more information about cultivation and other aspects of the amazing truffle can be found in a book called *Taming the Truffle, the History, Lore, and Science of the Ultimate Mushroom*.[3]

Consumer Demand for Mushrooms

Aside from supply and demand, there are a number of other factors that affect the purchase price of both wild and forest-cultivated mushrooms. In 2010 Jim Ochterski, a Cornell Cooperative Extension educator in Ontario County, asked sixty wholesale mushroom buyers (chefs) from upstate New York to rank wild and cultivated mushrooms in order of their familiarity and preference (the chef's) and with respect to the "characteristics most important if you buy locally harvested mushrooms."[4] The results presented in figures 5.10 and 5.11 show predictable results with species familiarity as well as "Flavor" as the top-ranked choice. The second most popular choice was "a local sustainable source," which reflects the increasing demand for local production and chefs' willingness to pay for it. Chefs have no issue with the difference in price, once they see, taste, and work with the log-grown varieties.

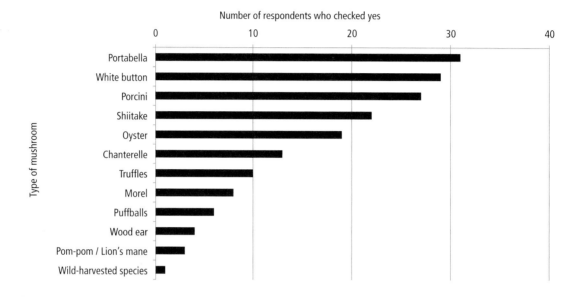

Figure 5.10. Familiarity of chefs with mushroom species in Jim Ochterski's 2010 survey of 60 chefs in Upstate New York.

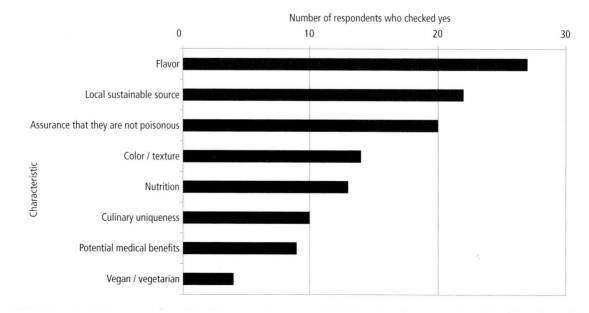

Figure 5.11. Characteristics of mushrooms bought by chefs. Courtesy of Jim Ochterski

As mentioned, locally grown shiitake in the Northeast retails for about $15 per pound and wholesales to restaurants and distributors for $10 to 12 per pound, in contrast to distant industrially grown shiitake, which are sold for $5 to $8 per pound. A three-year Northeast Sustainable Agriculture Research & Education (SARE)-funded project[5] worked with new growers in the Northeast and found that shiitakes are easily absorbed into the market without seeming to saturate demand and depress prices, although researchers at the University of Maryland reported that increasing numbers of growers in that area have begun to do just that.[6]

The third most important characteristic ranked in the Ochterski survey was "Assurance that they were not poisonous." This must be taken seriously by anyone selling mushrooms, but mostly as a point of education, as the level of concern about mushroom poisoning implied by figure 5.11 is about perception and overstates the actual risk. The fact is that fatal poisoning by wild mushrooms is rare, and poisonings associated with misidentification of toxic mushrooms among forest-cultivated mushrooms are unknown, according to several expert mycologists. Fatal poisoning from eating wild mushrooms under any circumstances has only rarely happened. A very small percentage of people have minor allergic reactions when eating "safe" mushrooms such as shiitake, notably if they are eaten raw[7] or in combination with alcohol.

According to the US Food and Drug Administration's *Food Code for 2009*, of about five thousand species of wild mushrooms in North America, sixty are toxic to humans but only fifteen are potentially deadly.[8] Regardless, concern about the safety of wild mushrooms has resulted in safety-related regulations in many states regarding sales to restaurants and other food vendors. These include, for example, California, Colorado, Iowa, Minnesota, New York, Oregon, Pennsylvania, Washington, and others. The FDA has a mandate to *recommend* food safety regulations to the states. The 2009 *Food Code* suggests that states require wild mushroom sellers to provide the following information:

- The Latin binomial name [of the fungus], the authority (author) of that name, and the common name of the mushroom species
- That the mushroom was identified while in the fresh state
- The name of the person who identified the mushroom
- A statement as to the qualifications and training of the identifier, specifically related to mushroom identification

States are at different stages of implementation of these regulations, and the appropriate regulatory agencies should be consulted on a state-by-state basis.

The next section describes the four-stage process for forest mushroom cultivation. After a brief generalized description of these stages as applied to any forest cultivated mushroom species, the emphasis will turn to a more specific discussion of the four stages as they apply to shiitake mushrooms in particular. The focus will be on shiitake cultivation, not only because of its proven track record and income potential but also because our understanding of how to grow shiitake in the woods is better established than for other species—shiitake has been cultivated in Asia for centuries and in North America for over three decades. Finally, we will address the issues involved in forest cultivation of other prospective specialty mushroom species, including oyster, lion's mane, and wine cap Stropharia.

The Four Stages of Forest Mushroom Cultivation

Forest cultivation of mushrooms is a rotten business. It begins with wood, which is the food source, or *substrate*, for the saprophytic fungi considered here, and ends with mushrooms and rotten wood (decomposed substrate). From a practical perspective the process can be seen as consisting of four stages. The four stages are substrate acquisition, inoculation, spawn run, and fruiting, as summarized in table 5.3.

Forest farming of specialty mushrooms is uniquely challenging, and not at all like growing fruits or vegetables. There are four more or less distinct stages that must

COMPARING SHIITAKE WITH GINSENG: THE ECONOMICS

At the present time there are only two mushroom species that are forest cultivated on a commercial basis: shiitake and oyster mushrooms. Cultivation practices for both are quite similar, but cultivation of shiitake is far more common. For most forest-cultivated mushroom producers who grow both, most of their operation is devoted to shiitake.

According to researchers at the University of Missouri's Center for Agroforestry, as of 2006 there were 383,000 shiitake logs under production in the United States.[9] The retail price of forest-grown shiitake mushrooms varies regionally but was consistently at least $10 or more per pound. The high price of forest-cultivated mushrooms (mainly shiitake and oyster) makes them attractive starter crops for forest owners who are just considering embarking on forest farming.

But how do income opportunities for forest-cultivated mushrooms stack up against other nontimber forest products? American ginseng is the most compelling example. Beginners tend to gravitate to cultivation of American ginseng because it sells for a mouthwatering several hundred dollars per pound dry weight, although the price of ginseng varies year by year and from one dealer to another. The price differential between forest-cultivated mushrooms and American ginseng may seem overwhelmingly compelling to a beginning forest farmer, but income potential is not the whole story. The forest farmer should carefully consider the costs, time, labor, income potential, and the environmental impact of both.

There are several different systems for growing ginseng (see chapter 6), but for the purpose of comparing ginseng with shiitake mushroom production in table 5.2, the wild-simulated method of producing is used because it requires the least upfront investment of labor and money compared to the woods-cultivated ginseng production method.

Wild-simulated ginseng takes eight to ten years from sowing of seed until the roots are ready for harvest. On the other hand, harvest of log-grown mushrooms begins about one year after inoculation. When a ginseng root is dug, the entire plant (above and below ground) is harvested. No regrowth of a second crop from a belowground root or rhizome is possible, as is the case of some other perennial plants, such as rhubarb, asparagus, and iris (for cut flowers). A shiitake mushroom log, on the other hand, produces several flushes of mushrooms (½ to 2 pounds per log or more each year) and continues to do so for three to four years.

In the end the hope is that forest farmers will consider both for production. Shiitake offers a quick yield, while ginseng can be considered more of a long-term investment.

Table 5.2. Comparison of Income and Cultural Factors between Forest Farming of Shiitake Mushrooms and of American Ginseng

	Shiitake Mushrooms	Ginseng
Unit price	Retail: $10–$20/lb (fresh wt) Wholesale: $10–$12/lb	$200–$500/lb (dry weight) Wholesale (only option)
Years until (first) harvest	1	8–10
Successive harvests	2 or 3 flushes/yr for 3–4 yrs	Only one harvest
Environmental considerations (site selection)	Shade (source and quantity) Substrate tree species Access to water for irrigation and soaking	Shade Soil pH, calcium, drainage Aspect Associated tree species and understory vegetation
Annual maintenance	Moderately high (toting logs, soaking, weekly harvest)	Very little until harvest at 8 to 10 years
Pests	Slugs, bugs	Deer, poachers

Table 5.3. Summary of Substrate Tree Species, Inoculation Method, and Stage of Production for Several Specialty Mushroom Species

Mushroom	Best Tree Species for Several Specialty Forest Mushrooms	1. Substrate Acquisition	2. Inoculation	3. Spawn Run	4. Fruiting
Shiitake	Oak, sugar maple, beech, hop hornbeam, ironwood	Hardwood "bolts" usually 36" long and 4–8" in diameter	Bolts are drilled and filled with plug (dowel), sawdust, or thimble spawn	12–18 months	Naturally or by soaking in water for 24 hours, early spring–late fall, depending on strain
Oyster	Poplar, sugar maple, beech	Bolts as above, or totem logs 8–12" in diameter and 24" long	Bolts as above or totems cut into two or more 12" sections with sawdust spawn layered in between	18+ months	Naturally, mostly in spring or fall
Lion's Mane	Beech	Totems, as above	Totems, as above	18+ months	Naturally, mostly in spring or fall
Stropharia	Mixed species of hardwood woodchips (no more than 25% conifer)	Layers of woodchips and sawdust	Remove OM to mineral soil, 2" layer of sawdust, spawn, and cover with 2–4" wood chips	6–12 months	Naturally, often after heavy rain or temp change

be undertaken to cultivate mushrooms successfully in the forest. The details of each stage vary among different mushroom species, but before considering the requirements for specific forest-cultivated mushroom species, each stage will be introduced here in general terms.

STAGE 1: SUBSTRATE ACQUISITION

In most cases, the substrate, or food for growing mushrooms, comes from intact logs or wood chips that were sourced from recently cut trees. There are several key issues to consider:

Will material be harvested on-site or purchased from another landowner, logger, or firewood dealer?
A basic starting point is that two trees 8 to 12 inches in diameter will need to be felled per ten logs for mushroom inoculation (each tree yields five-plus logs). If purchasing logs from someone else, the going rate is about $1.00 per log.

How will logs be moved to (and within) the laying yard?
A typical log for mushrooms weighs 30 pounds or more. Consider access to the woods where logs are harvested as well as to the final location where logs will be inoculated and set up for fruiting (the laying yard). Trucks, tractors, carts, sleds, and wagons can be used. If buying logs from someone else, he will usually pile them on the side of the road or may even deliver directly to your laying yard.

When should you cut the trees to be turned into mushroom logs?
The short answer: almost anytime, but the best time is in the middle of winter (January–February), with inoculations in the months of March through May. That said, almost any time of year works, except when they are budding out in late spring through early summer.

What kind of trees (species) should be cut?
This depends on the mushroom species. Table 5.3 provides the basic information, but more nuances will be covered for each species below.

STAGE 2: INOCULATION

For growth and *fruiting* to occur, the fungus must be brought into intimate contact with the substrate before it can begin colonization and commence the rotting/ decaying/decomposition process. The actual process of introducing the spawn into the substrate is called *inoculation*. The method of inoculation varies among different mushroom species and different systems for cultivating them.

Inoculation brings the fungus together with the wood substrate, but the fungus in this case is not the mushroom itself but the "hidden" part of the fungal body called the mycelium. Just like people, the fungus has both reproductive and somatic (nonreproductive) parts. In the case of basidiomycete fungi (shiitake, etc.) the reproductive part is the mushroom and its spores. The rest is the somatic (nonreproductive) part, which is mostly unseen. This is the mycelium, which is made up of tiny threads called hyphae. The mycelium of saprophytic (wood rotting) mushroom species is hidden within the substrate, which is the log or wood chips in the case of the fungi covered in this chapter. A good way to see the mycelium is to dig into a bed of wood chips in a well-established Stropharia bed to expose the white, fluffy mycelium. Pulling the bark off a well-colonized shiitake log will show you the same thing.

Biologically speaking, the fungus's "mission" is to produce and distribute spores, which are often thought of as the seeds of the mushroom. That seed metaphor is a little misleading. More accurately, fungal spores are analogous to pollen/egg (male/female gametes), which fuse to form the embryo that develops into a seed. As shown in figure 5.13, fungal spores germinate on a suitable substrate and

Figure 5.12. The somatic (nonreproductive) mycelium of the saprophytic fungus, such as Stropharia, grows by colonizing wood chips (or logs in the case of shiitake) and eventually produces mushrooms, which are the reproductive structures of the fungus.

send out a haploid (1) hyphal thread. When the hyphae of two compatible spore comes into contact, they fuse to form a (1n + 1n = 2n, diploid) vegetative hypha that grows and branches to become the *mycelium* that spreads throughout the substrate. Eventually, under appropriate conditions (discussed later) the mycelium gives rise to the mushrooms we know so well.

It is a common misunderstanding that the spawn used by growers consists of spores, but that is not the case. Rather than spores, the spawn is mycelium mixed with a carrier (food source) such as sawdust, grain, or wooden dowels. Once the fungal mycelium comes in contact with the substrate (wood) it begins to grow into the wood, causing it to decompose. This "invasion" of the wood by the mycelium is called *colonization* or the *spawn run*.

But that is just a bit of biological background. As it turns out in the case of cultivated mushrooms, reproduction from sexual spores, the sexual process, is rarely practiced by mushroom growers. As with any sexual process, reproduction from spores gives rise to offspring that are genetically variable (different) from both parents and from each other, just as when a boy and his sister have different-color eyes. When the goal is uniformity, as it is in cultivated mushroom production, asexual reproduction from the mycelium produces mushrooms that are genetically identical to the original spawn and mushroom type. That genetically unique mushroom type of spawn is called a *strain*, which is equivalent to the term "variety" as applied to plants (e.g., Macintosh or Red Delicious apples).

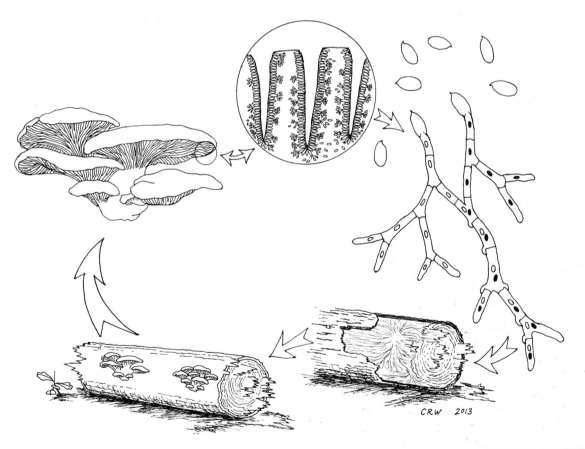

Figure 5.13. Life cycle of a saprophytic fungus (oyster, illustrated here; same for shiitake, lion's mane, and others). Basidia, which mature within the mushroom, release haploid spores, which germinate to form hyphae with one nucleus per cell (1n). Two compatible haploid hyphae fuse to become a diploid (2n) hyphae that develops into the mycelial network that colonizes the substrate (log, wood chips, etc.) and eventually produces more mushrooms. Illustration by Carl Whittaker

Figure 5.14. Grey Dove oyster mushroom.

Figure 5.15. Golden oyster mushroom.

Any particular strain is genetically different from all other strains, and the mushrooms it produces are all uniformly true to type, maintaining the characteristics of the original strain.

For example, Grey Dove and Pohu are two genetically different strains of the same species of oyster mushroom (*Pleurotus ostreatus*) whereas golden oysters are a different species (figure 5.15, table 5.7).

In the case of shiitake, strains differ less dramatically in color, but more importantly in the temperature range in which mushroom formation is triggered. The careful selection of fungal strains as a management technique will be discussed below.

To prepare a batch of spawn for inoculating a substrate, the mycelium must be introduced to a carrier, such as sawdust or wooden dowels. Sawdust, the most common carrier, is an intermediate substrate on which the mycelium grows until it is introduced onto the log or wood chips. Wooden dowels infused with the fungal mycelium are another type of spawn that can be introduced into the substrate log.

Spawn, which is produced in a laboratory (see case study, Spawn Production at Field and Forest), is ready when the rapidly growing mycelium thoroughly permeates the carrier (sawdust, etc.). Although it is possible for an ambitious, well-trained, and properly equipped mushroom grower to produce her own spawn, the effort is rarely justified because high-quality spawn is readily available from a number of commercial suppliers at very reasonable prices. For example, the amount of spawn required for inoculating a single shiitake log costs about $1.

Stage 3: Substrate Colonization (Spawn Run)

Colonization refers to the growth of the mycelium, within and throughout the substrate. Successful colonization leads to the decay of the log, which provides the energy and substance for mushroom production. After inoculation the period during which colonization takes place is called the *spawn run*. Mushroom production cannot begin until the spawn run (colonization of the log) is more or less complete, and in the case of shiitake, this can take a year or more.

During the initial spawn-run year, where the logs or wood chips are located is very important, as is how they are managed. Stacking methods will be introduced below along with the mushroom species they are most

CASE STUDY: SPAWN PRODUCTION AT FIELD AND FOREST PESHTIGO, WISCONSIN

Many people think of log-grown mushroom growing as "all natural"; that is, that all the cultivation occurs in the woodlot. While this is true once the logs are inoculated, the production of spawn takes place in a laboratory setting, where sterility and controlled conditions are critical to success. This past fall I was fortunate to visit Field and Forest Products in Wisconsin and walk with Joe Krawczyk and Mary Ellen Kozak about their business, which they stumbled across partially from necessity, as they couldn't find good spawn producers when they decided to move back to the family farm and homestead in the early 1980s.

From the desire to produce quality spawn came a thriving business, which now supplies growers around North America and offers one of the widest ranges of mushroom products, including ten strains of shiitake, eight kinds of oyster, and many others, including lion's mane, Stropharia, reishi, nameko, and even blewit (not for beginners!). The owners and staff are

Figure 5.16. Sawdust pile "curing" at Field and Forest Products before being made into spawn.

friendly and willing to spend time answering questions on the phone, and their catalog is essentially a beginner's guide to cultivation. We've used Field and Forest for much of our Cornell research spawn, so I was eager to have a look at the facility.

The process starts with a massive pile of oak sawdust out back. This is purchased from a furniture factory and given time to age before being used in spawn production. Field and Forest also makes a significant amount of grain spawn, for which they use organic rye and barley bought by the bag.

When they are making a new batch, the material (sawdust) is brought into their facility and put through an autoclave to sterilize it. This is no doubt the more energy-intensive part of the process, which heats steam to 121°C (251°F) several hours. The new autoclave at Field and Forest is built into the wall, so it can be loaded from the unsterile side, then unloaded on the other side, in a sterile room where active mycelium can be mixed in.

A frozen stock culture of an isolate (clone) is thawed and grown out on a petri dish containing potato dextrose agar (PDA). Next they are transferred to sterile grain and finally onto sterile sawdust. The entire process to make a new batch of spawn takes about three weeks. It's critical to regrow spawn continuously from the (frozen) source, to ensure that the finished product is strong and healthy, which equates to better success in the field. The inoculated sawdust is then bagged in plastic that has a filter patch, which allows oxygen in (necessary for mycelium survival) while keeping contaminants out. The bags can then leave the sterile environment and grow out further in the grow room for several weeks before they're ready to ship out to customers.

Spawn that gets too "old" to be sold is taken to the room next door, which is humidified to stimulate fruiting, allowing for a small amount of mushroom production and some testing to occur. Field and Forest offers ready-to-grow tabletop kits as well, which are essentially myceliated blocks of sawdust that are at the peak of growth. Over time Joe and Mary Ellen have found that rather than adding a bunch of nutrients to the mix (thought to stimulate better growth in indoor cultivation) they've selected certain strains, which perform better under block conditions. This focus on "selective breeding" is a particular strength of Field and Forest as a company.

Considering that spawn is relatively cheap (it costs less than a dollar to inoculate one shiitake

Fig 5.17. Inoculated spawn incubates in the grow room before being shipped to locations all over North America.

log), in most cases it makes sense to pay someone to produce spawn. It's not practical for each mushroom grower to maintain a spawn production facility, though more spawn producers would mean more local isolates so that mushrooms could be better adapted to local environmental conditions across the cool temperate regions. Just as with seed crops, diversity is key. With only half a dozen producers on the continent, as demand grows spawn production may become a bottleneck. As it is, Field and Forest can barely keep up with the demand, and more new customers call each year.

So while high-quality spawn ends up in the forest, it begins in the lab. Yet considering that out of the three- to five-year life span of a productive shiitake log just one month is spent indoors with high-energy inputs, the entire process is still much more sustainable than straight indoor cultivation, which requires constant energy inputs throughout the growing cycle. Field and Forest does a wonderful job sourcing materials and minimizing energy inputs to the process, focusing on high-quality and diverse strains as ways to improve mushroom production. Joe, Mary Ellen, and their staff work hard and take their business seriously, while having fun in the process. As their motto says, "Proud to be part of this rotting world!"

For more information: http://www.fieldforest.com.

— Steve

suited for. Proper management applies not only to the spawn run but also to the remaining productive life of the log. The location where all of this happens is called the *laying yard*. Selecting an appropriate site for your laying yard is very important and is discussed below.

The terms *rot*, *decay*, and *decomposition* all apply to the process of colonization that goes on during the spawn run. All of the edible fungi described in this chapter are white rot fungi, which have evolved the ability to break down complex organic molecules, most notably the tough lignins found in woody material. The growth of the mycelium is responsive to a number of dynamics in the laying yard, including temperature, relative humidity, light, wind, and log moisture

content (LMC). A forest farmer needs to understand how to manage each of these factors in the laying yard through canopy management, wind abatement, log configuration, and irrigation to maximize mushroom production. Despite all these variables, the good news is that fruiting success is high.

Step 4: Fruiting, Harvesting, Storage, and Marketing

Mushrooms are not fruit, at least not in the sense of an apple or a tomato, but the process of mushrooms growing from a log or other substrate is called "fruiting." Mushroom production from a colonized substrate begins when the following have all occurred:

- When the substrate is fully colonized (or nearly so)
- When moisture content is sufficiently high (~ >35%)
- When the temperature is in the permissive range (~ 55–85°F, depending on strain)

When these conditions are met, fruiting either can occur naturally (passively) or, in the case of shiitake grown on logs, can be induced by the grower ("forced fruiting") as described below. Other substrate configurations in the laying yard, including totems for oyster and lion's mane and wood chips for mushrooms, are immobile, so forced fruiting is not an option. In these cases natural fruiting prevails. This is a disadvantage for the grower, since natural fruiting is sporadic and to some extent unpredictable, which is not particularly conducive to marketing on a regular schedule. The reliability of shiitake is what gives commercial growers the upper hand.

One of the most important considerations regarding the harvesting and marketing of mushrooms is maximizing shelf life and minimizing spoilage. Forest-cultivated and wild mushrooms have a relatively short shelf life—rarely more than a week. Unless fresh mushrooms are taken directly to market and sold within hours, they must be refrigerated. Dried mushrooms, on the other hand, can be stored almost indefinitely.

Shiitake

Shiitake (*Lentinula edodes*) is the third most common cultivated mushroom in the world and certainly the best-known forest-grown mushroom in North America. It is a bit ironic that shiitake mushrooms, one of the most important nontimber forest crops used in North American forest farming, is not native to the continent. It is indigenous to Asia and has been cultivated in China since about AD 1100 and later in Japan, where the value has primarily been placed on its medicinal properties. A paper published in 1982 by Gary Leatham[10] was instrumental in popularizing shiitake cultivation in the United States. Today a half dozen spawn producers exist, but the number of growers is quickly on the rise.

Figure 5.18. A beech log that has been "forced" to produce a uniform flush of shiitake mushrooms.

In addition to their delicious taste, shiitake boast a number of positive nutritional and medicinal qualities.[11,12] The proteins contained in shiitake have an amino acid profile similar to the "ideal protein" for humans. These mushrooms are one of the best sources of protein, especially for vegetarians/vegans looking to eliminate animal proteins in their diet. Multiple studies conducted over the last ten years have demonstrated that an active component in shiitake called eritadenine "significantly decreased the plasma total cholesterol concentration, irrespective of dietary fat sources . . ." Shiitake mushrooms are considered a good source of three B vitamins (B2, B5, and B6); a very good source of six trace minerals (manganese, phosphorus, potassium, selenium, copper, and zinc); and a notable source of magnesium and vitamin D (see sidebar, Enhancing Vitamin D Content in Shiitake, page 187).

Figure 5.19. Freshly harvested log-grown shiitake mushrooms.

Figure 5.20. Shiitake mushrooms grown indoors on artificial sawdust "logs."

Another documented health benefit of shiitake mushrooms is as a supporter of the immune system. Shiitake also contains the polysaccharide lentinan, a (1-3) ß-D-glucan, which is associated with cancer prevention properties of this mushroom.

Producers of forest-cultivated shiitake mushrooms range in size from "hobby" scale (<200 logs), part-time commercial (500 to 1,000 logs), all the way up to full-time commercial (>5,000 logs). Few of these producers bother to produce their own spawn for log inoculation; instead they rely on companies that specialize in spawn production and also sell a wide range of mushroom-related supplies. According to the USDA National Agricultural Statistics Service (NASS, 2013), in 2011–2012 there were 151,000 natural wood logs in outdoor shiitake production, whereas in 2012–2013 the number of logs dropped to 137,000. It should also be noted that several experienced growers, and a major supplier of spawn to mushroom growers all over the United States, regard these shiitake-related NASS statistics to be unreliable because they believe most growers do not fill out the annual surveys sent to them by NASS, which may be the reason for the drop in the numbers. By all indications "on the ground" (attendance at Extension events, sales at spawn producers, involvement in grower groups), log-base shiitake production is well on the rise.

For beginning farmers shiitake is perhaps one of the best candidates as a niche crop, at least in the northeastern United States. The markets are more or less wide open, with consumers and chefs eager to get their hands on this tasty and nutritious food. Shiitake can be easily sold at farmers' markets, to restaurants, and through CSA models for $11 (wholesale) to $16 (retail) a pound. A beginner can start with 300 logs, which yield roughly 10 pounds a week, or $120 to $160 of sales, and add more logs until he is satisfied. The cost to inoculate each log is $1.50 to $3.00, which pales in comparison to the $50 to $60 of sales per log that will be gained over its lifetime. And other than drying out, the crop is forgiving of changing weather conditions, floods and droughts, and even the farmer's desire to take a vacation.

In the previous section, the four stages of mushroom cultivation were described that apply in general to any mushroom species suitable for forest farming. In this

Table 5.4. Number of Natural/Outdoor Shiitake Logs in Production (NASS data, endnote 1).

Year	Outdoor Logs in Cultivation
2010–2011	126,000
2011–2012	151,000
2012–2013	137,000

Figure 5.21. Midscale shiitake farming in the Blue Ridge Mountains of North Carolina. This grower derives a significant part of his income by selling shiitake mushrooms fresh at the local farmers' market.

BEST MANAGEMENT PRACTICES FOR OBTAINING LOGS FOR MUSHROOM CULTIVATION

1. Conduct logging in early spring before budbreak, two to three weeks before inoculation.
2. Cut deciduous hardwoods as part of a long-term forest management plan.
3. Cut logs into 3- to 4-foot sections, 4 to 6 inches in diameter.
4. Keep logs shaded and protected from wind before inoculation.

time of logging; selection of shiitake strains to maximize production; management of light, temperature, humidity to keep log moisture content optimal for fruiting; shocking of logs to induce fruiting; and harvest and marketing of fresh and value-added shiitake mushrooms. At the beginning of the section describing each stage of shiitake mushroom cultivation is a set of best management practices.

STAGE 1: SUBSTRATE ACQUISITION

As described in the sidebar on obtaining logs for mushroom cultivation, "substrate" refers to the food source for the mushroom, which in the case of shiitake may be either logs, used for outdoor production (forest farming), or sawdust blocks (artificial logs). Hence the term "log" will be synonymous with "substrate" from this point on.

section, each of these stages will be reconsidered in light of cultivating shiitake mushrooms in particular. This will include selection of appropriate tree species;

Log Dimensions

Typically, shiitake logs are 4 to 8 inches in diameter. Larger-diameter logs can be used, limited only by the strength of the grower. Smaller-diameter logs will begin to produce mushrooms sooner, but the logs tend to dry out faster because they have a higher surface-to-volume ratio. Larger logs (those greater than 6 to 8 inches in diameter) can produce mushrooms, but their greater weight makes them more difficult to manage (carrying, stacking, soaking) in the laying yard.

Aside from the risk of back injury, larger logs may produce fewer mushrooms (by weight) per pound of log than those produced by smaller-diameter logs (discussed further under Stage 3: Substrate Colonization [Spawn Run]). Larger logs can also be inoculated with cool weather (CW) strains, which fruit in response to temperature swings rather than soaking in water, which eliminates the need to move them to a soak tank for forced fruiting.

Tree Species

There are a number of issues to consider when deciding which trees to cut for shiitake mushroom cultivation. One of the best management practices for sustainable forest management and forest health is *to not cut trees merely to obtain logs for shiitake cultivation. Obtaining those logs should be part of a comprehensive forest management plan.* For example, timber stand improvement (TSI) (see chapter 10) is a forest management strategy that involves improving overall forest quality by removing low-quality trees to favor more desirable ones. These culls can often be used for shiitake cultivation. Suitably sized branches of large trees that are taken down for timber or other reasons are also often well suited for shiitake cultivation.

Among experienced shiitake growers and perhaps to a greater extent among beginners a frequent topic of conversation is "what tree species is best." If a vote was taken among experienced growers, oaks would probably get the most votes, but those who voted for some other species are not necessarily wrong. The "best" tree for growing shiitake mushrooms depends not only on the inherent characteristics of that species, such as wood density and bark thickness, but also on a number

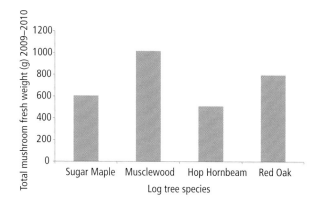

Figure 5.22. Effect of four tree species on shiitake mushroom production (average fresh weight per log, 2009–2011): sugar maple, musclewood, hop hornbeam, red oak.

of external factors, such as how vigorously the tree is growing on a particular site in response to exposure, soil moisture, and other site-specific conditions, as well as the strain of the shiitake fungus used to inoculate the log. Figure 5.22 is just an example of the relative performance of four species of trees cut, inoculated with shiitake spawn WR 46, and incubated at Cornell's Arnot Forest. Note that the oak, which is often cited as the "gold standard," was not as good at producing mushrooms as a nonoak species, musclewood. At a different location with different circumstances, this might not be the case. In a separate experiment involving red oak and three different species, once again red oak was not the most productive but rather American beech was.

The fact is that there is no single species of tree that makes the best substrate for growing shiitake mushrooms. Not surprisingly, a grower with a limited palette of available tree species may decide that one of those tree species is the "best" and the others are to be avoided or to be used as a last resort. Another grower with a different set of species to choose from might come to a different conclusion. Factors that affect tree species performance include both internal and external factors, including those that follow.

Wood Density

Tree species in the Fagaceae family (oak, beech, hornbeam, and chestnut) are generally good substrates for shiitake colonization and mushroom production, mainly due to the fact that they have high-density wood, which means more substrate (lignin) for the fungus to feed on. Conversely, lower-density "soft hardwoods" such as willow and poplar will support mushroom production for a year or so but "run out of food" and do not produce mushrooms for as many years as tree species with higher-density wood.

Sapwood-to-Heartwood Ratio

Another factor with regard to the amount of substrate available to support mushroom production over time is the matter of the ratio of sapwood to heartwood. Sapwood refers to the living outer ring of light-colored wood, which is more favorable for colonization by the fungus. The heartwood is the darker-colored, nonfunctional inner core of the log. Not only is it nonfunctional with respect to water transport, it is also indigestible for the shiitake fungus, so it is not available for mushroom production.

All other things being equal (although they rarely are), a log with a greater volume of sapwood is likely to produce more shiitake mushrooms. Some species that have relatively thicker sapwood (higher sapwood to heartwood ratio) include maple, ash, hickory, hackberry, and beech, whereas other species have relatively thin sapwood, including chestnut, black locust, mulberry, Osage orange, and sassafras. This is another tree species–related factor that may influence differences in the relative performance between or among species.

Differences in sapwood-to-heartwood ratio may account for differences in the productivity between two or more logs of the same species. Figure 5.23 shows two "cookies" (cross sections) from logs of similar diameter from two different red oak trees cut from the same stand. The one on the top has a considerably wider sapwood width than the one on the bottom, so it is likely that the one on the top is growing more vigorously and will support production of shiitake mushrooms for a longer period of time. External factors can also affect the sapwood-to-heartwood ratio. A tree that is growing vigorously will have a higher ratio of sapwood to heartwood

Figure 5.23. The outer ring of light-colored wood is the sapwood, which is still functioning to transport water upward. The darker core is heartwood, which is no longer functional in water transport. The shiitake fungus colonizes and decomposes the sapwood (mostly), so the greater the sapwood-to-heartwood ratio the longer mushroom production will be sustained. Both the sapwood layer and the bark are thicker in the log on the top than the one on the bottom. The logs shown here are about 5 inches in diameter. Photograph courtesy of Steve Sierigk

than a less vigorous tree of the same species. Vigor is influenced by how favorable the site conditions are, including soil quality, soil moisture, nutrient status, and light availability.

While it is true that logs with greater sapwood volume will produce mushrooms longer and have more usable substrate than logs with less sapwood, it would be inadvisable to cut more trees than necessary to choose the ones with greater sapwood volume, even if the lesser ones are intended for firewood.

Bark

One of the most critical factors in successful mushroom production is moisture content of the substrate (in this case, logs). One important way moisture loss from a log occurs is when liquid water in the log evaporates through the bark. Moisture loss by evaporation through the bark is influenced by the thickness of the bark and the tendency of the bark to peel away from the underlying wood. In figure 5.23, not only does the tree on the top have a greater volume of substrate (sapwood) for mushroom production than the one on the bottom but it also has significantly thicker bark. This should result in less drying of the log over time, which will also contribute to the overall productivity of the log.

Species with thin bark tend to lose moisture faster than thicker-barked species, and when the logs are not managed correctly, the bark is more likely to peel on thin-barked species, figure 5.25. Poor management, in this case, includes cutting the beech tree after it has emerged from dormancy in the late spring. Prior to this, while the tree is fully dormant in winter or early spring, the bark is held tightly to the underlying wood, and the bark remains intact during all or most of the time the log is in the laying yard. When cut later in the season when the tree is emerging from dormancy (late spring), the bark loosens up (described as "bark slipping"), such that after the tree is cut and the bark begins to dry in the laying yard, it can crack and pull away from the underlying wood, causing the log to dry out more rapidly than if it had been cut while dormant.

Availability

For some forest farmers who are skilled with a chain saw, access to trees on their own land is preferable, as long as harvesting trees is part of a deliberate forest management plan. Prospective forest farmers who have

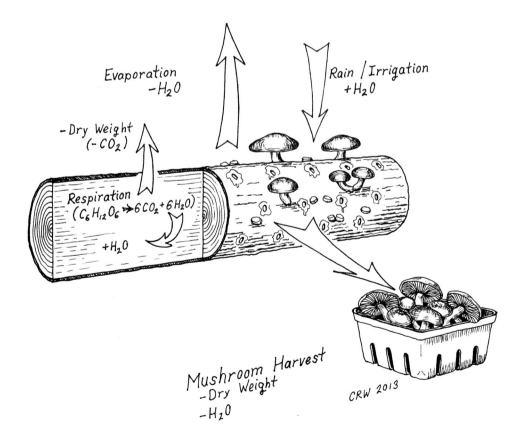

Figure 5.24. Illustration of the factors affecting the moisture status (gain and loss) of a shiitake log. Illustration by Carl Whittaker

Figure 5.25. American beech is a species with thin bark, which tends to crack and peel off, especially if the tree is cut when it has begun actively growing in the spring.

only limited or no access to appropriate tree species on their own land should consider purchasing freshly cut logs from someone else who is selectively thinning her own forest. Another source of shiitake logs is loggers or arborists engaged in cutting down large trees for timber. Only the main trunk is likely to be of commercial timber value, and the larger branches will go to the slash pile (coarse woody debris, as foresters call it) or be cut up for firewood. Suitable-size branches can be bucked (cut into 3- to 4-foot logs with a chain saw). They are typically sold for $1 to $2 each if the seller does the cutting. If the forest farmer does the cutting they may be less. State forests in some states will sell U-cut licenses.

Season

What time of year should trees be cut and inoculated? Conventional wisdom is that trees should be cut and inoculated when they are dormant in the spring, before the buds have begun to swell. As long as there is not too much snow on the ground, this is often a convenient

HOW MANY LOGS?

One question enthusiastic new growers want to know is how many logs they should be inoculating to get a reasonable yield. Their response is often met with the one familiar to many gardeners and farmers: "It depends." The trick is matching the quantity of logs with the goals for yield and the willingness to invest time in proper management.

Assuming that the logs will flush ¼ to ½ pound per log on average, then need a rest period of seven to eight weeks after soaking, here are a few estimates:

Enough to cook for one meal a week = 8 logs
This means you soak one log each week and get ¼ to ½ pound with each flush. That's enough for a decent meal (or two). You could easily stash this number of logs under a porch or a single tree and soak each one in a trash can or even an old bathtub.

Enough to feed family and friends = 32 logs
Soaking four logs a week should yield between 1 and

2 pounds per week, which is plenty for eating and dehydrating some for the off-season or to give as gifts. An old kiddie pool would suffice for soaking.

Enough to make a little side income = 320 logs
If you soaked forty logs a week, you could gross between $60 and $160 a week. That's a yield of 5 to 10 pounds that you sell for $12 a pound wholesale or $16 a pound retail. We aren't talking about a huge investment of time here; a well-managed system could be maintained in five hours or less per week.

Enough to make it a career = 10,000 logs
Now we are getting serious! Soaking 1,250 logs a week would yield 300 to 600 pounds of mushrooms, which at $10 a pound would gross $60,000 to $120,000 over a twenty-week period, June through October. Expenses are considerable; at this scale mechanization and hired hands would be necessary. It's possible to make 40 to 60 percent of this gross as profit.

time for forest farmers to get out into the woods, when other responsibilities are not so pressing as at other times of the year, but snow cover can sometimes overwrite convenience. Despite this belief that spring is best, some mushroom farmers inoculate at other times of the year with acceptable results. In terms of time management, some growers find it more convenient to inoculate at least some of their logs in the fall. The question, "Does it matter what season?" was one that research at Cornell's Arnot Forest wanted to answer, with a study that compared logging and inoculation at each of the four seasons with both red oak and beech.

Shiitake production from winter- and spring-inoculated logs was higher for both red oak and beech, compared to logs cut and inoculated in summer and fall. This is as would be expected from the conventional wisdom that trees should be in a dormant state when cut, but the differences were not as much as might be expected. The graph shows that mushroom production decreased somewhat when inoculated in the summer and fall, especially for the red oak. Yet the decline was relatively small, and perhaps not enough to discourage a shiitake grower from inoculating in later summer or in fall if there were offsetting advantages to be gained

by more efficient scheduling. The results shown in figure 5.26 should be interpreted with caution because only two tree species were involved, and there was only a single set of external conditions (spawn type, laying yard, tree vigor, etc.).

Figure 5.26 suggests that season may affect different tree species differently, but in this case the differences between beech and red oak were not very great. On the other hand, there is a very interesting report out from the University of Missouri's Center for Agroforestry,[13] which showed that when trees of sugar maple, white oak, and red oak were cut for shiitake logs in February, sugar maple logs produced significantly greater mushroom fresh weight than either of the oaks.

When the same three species were logged in May, there was no significant difference in mushroom yield among the three. The reason for this seasonal difference is not well understood but may have been associated with the relatively high sugar content of the sugar maple sapwood at about the time when sugar maple trees are tapped to make maple syrup. These findings are consistent with the conventional wisdom that logging when trees are fully dormant (winter) is better than later in the spring, but the results also

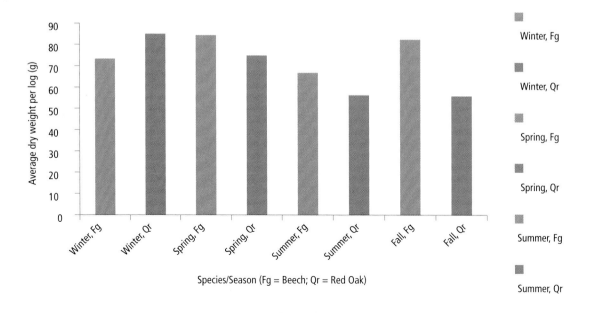

Figure 5.26. Effect of season and tree species on shiitake mushroom production over three years.

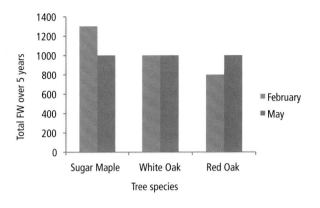

Figure 5.27. Interaction between time of cutting of trees and tree species on shiitake mushroom production. The results shown are combined from two separate experiments. Graph modified from Bruhn et al.

illustrate that the productivity of one species compared to another depends on the time of year when the trees are cut.

Although there is wide latitude over when trees can be cut for shiitake production, there is at least one time of year when cutting should be avoided. From late spring when buds begin to swell until early summer when leaves are fully expanded, trees should not be cut for shiitake logs.

Recommended Species for Substrate

With all the above factors at play, it is clear that there is no simple answer as to the best species for shiitake cultivation. Table 5.5 offers some levels of good, better, and best, along with species that are not recommended. The only "ranking" should be among tiers, and even

that should not be considered absolute. In other words, tier 1 species are likely to perform better than tier 2 species (but not under all conditions). The reason oak (tier 1) is often considered to be the "gold standard" is that much of the time, under some conditions but not all, it tends to come out ahead when compared to other tree species. As for tier 3, there have been many who have tried these species and come away disappointed, although there are a few growers who swear by them.

It would be convenient if a beginner could consider this list to be the final word in choosing trees for shiitake cultivation on logs, but unfortunately it is not that simple. The best advice for beginners is to *compare several readily available tree species under your laying yard conditions.* About ten logs of each species would be adequate. It is useful to continue observations of experimental logs for several growing seasons, because the relative performance of logs in their second and subsequent production years may differ from the first year. Measuring weights of harvest and observing general deterioration of logs are good ways to collect some simple data to help with a long-term decision. It is also recommended to contact and visit with experienced growers in the local area. The Northeast Forest Mushroom Growers Network (http://blogs.cornell.edu/mushrooms), which offers an online directory of growers, is one resource to find such people.

Log Management prior to Inoculation

Once species have been chosen and bolts cut, the prime directive of successful mushroom cultivation goes into effect: *moisture management.* Reduction in log moisture content (drying) is the worst enemy;

Table 5.5. Recommended Species for Shiitake Cultivation

Tier 1 Excellent	Red oak, white oak, musclewood (*Carpinus caroliniana*), sugar maple
Tier 2 Very Good	Hop hornbeam (*Ostrya virginiana*), American beech, bitternut hickory, sweetgum (*Liquidambar strictifolia*)
Tier 3 Good–Fair	Alder, black birch, yellow birch, red maple, yellow poplar
Tier 4 (not recommended)	Aspen, willow, poplar, white ash, fruitwood (apple, cherry, peach, etc.), any conifer (pine, hemlock, spruce, fir, etc.), black locust, walnut

Note: The above is a generalization based on research and grower feedback of the best species for shiitake cultivation in the northeastern United States, arranged by tiers.

TRYING LOGS ON FOR SIZE

As I got into mushroom production, I started with sugar maple because at the time I was working at a nature center and the logs were a byproduct of a timber stand improvement. For a workshop we hosted a local grower who swore by oak, only he wasn't optimistic about the performance of the maple, which turned out to do just fine.

A few years later, as I decided to get into commercial-scale products, I was involved with a number of land management projects around the area, many of which had been "high graded" of the best trees and in several cases only red maple (*Acer rubrum*) was left, which is a low-quality tree. "How perfect!" I thought, for it would be great to have a use for all this wood I was thinning, since it makes poor firewood, building material, and so forth.

That was in 2011. This year (2013), I finally gave up on red maple. I found that much of the bark would start to flake off after a few soakings. Some logs fruited, but not nearly as well as the sugar maple and oaks I had. Now these logs act as pathways in the forest, and only occasionally do I see a mushroom fruiting from one.

This is the reality of "trying logs on for size." Some come by chance, some are deliberately harvested, some are, we hope, going to work. Ultimately, I rely on red oak and sugar maple as my mainstays now, but I continue to try new species. We are planting red alder and European (sweet) chestnut as windbreak species, and mushroom growers in England claim they are good substrates. We will see.

— Steve

in fact it is one of the major causes of failure. This issue will come up again and again as the process moves from tree cutting to log inoculation and so on. Excessive moisture loss at any stage of the forest mushroom cultivation process can and must be avoided at all cost. From the standpoint of moisture management it is best to cut appropriate-diameter trunks or branches into 3- to 4-foot lengths, then move them soon to a protected (shady) location. Some operators find it convenient to leave the tree trunk or large branches lying on the forest floor right where they fell, for days or even weeks. However, this is less desirable than cutting the trunk into bolts and moving them to a protected location with shade and protection from wind. A recently cut log on the forest floor may be exposed to hot, drying sunlight. To

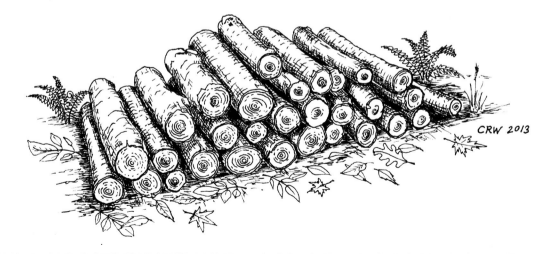

Figure 5.28. Dead stacking (like firewood) of fresh-cut logs minimizes moisture loss that would otherwise occur with a more open stacking configuration. Illustration by Carl Whittaker

begin with, they should be *dead stacked* like firewood (no space between logs) to conserve moisture, figure 5.28. Dead stacking, however, is not a suitable way to stack logs in the laying yard after they have been inoculated, when they need more aeration.

Another frequent question concerns how long to wait before inoculation—the so-called *curing* period. A common assertion (conventional wisdom) is that it is necessary to wait two or more weeks between cutting and inoculation so that (alleged) fungal inhibitors present in the living tree and in fresh-cut logs have time to dissipate before inoculation. This may be the case for some tree species, but if so, the phenomenon is not well documented. A disadvantage of protracted curing beyond a few weeks is undesirable drying of the log. The authors recommend that logs be inoculated as soon as possible after felling—within weeks rather than months.

STAGE 2: SUBSTRATE INOCULATION

Inoculation is a labor-intensive aspect of cultivating shiitake mushrooms, second only to cutting down the trees and carrying the fresh-cut logs out of the forest to the inoculation site. The task of inoculation is certainly the most exciting activity for beginners. Many growers have found that folks who want to learn how to grow shiitake will either volunteer to work for free, in exchange for the hands-on opportunity to learn, or even pay for the privilege by way of a workshop registration fee, especially if they get to take home a log they have inoculated. Some nonprofit organizations use training events not only as a way to educate the public but also as a way to get some "free labor" that contributes logs to their own educational laying yard or to raise "donations" for their organization via registration fees. Sometimes growers will conduct no-cost "inoculation parties" at their farm, open to the public. This helps them with free labor during peak inoculation season in exchange for invaluable hands-on learning for beginners. It is wise for beginners not only to read this chapter carefully but also to attend a formal or informal educational event as described above. There is no substitute for hands-on learning.

BEST MANAGEMENT PRACTICES FOR INOCULATING LOGS WITH SHIITAKE SPAWN

1. Using a high-speed angle grinder (~10,000 rpm) with a drill bit adapter and a 7/16" bit, for each 1 inch of log diameter, drill one row of holes along the length of the log, evenly spaced around the circumference. Space holes 4" apart, within rows. Offset rows by 2" every other row, creating a diamond-shaped pattern.
2. Fill the barrel of an inoculation tool with sawdust spawn of shiitake strain WR 46. Place the tip of the barrel opening against a hole, and depress the plunger to deposit the spawn into the hole. Repeat for all holes.
3. With a paintbrush or dauber apply molten food-grade cheese wax to each hole to seal it completely. The molten wax should be barely smoking.
4. Label each log as to date, spawn type, and any other relevant information, such as your initials.
5. Avoid direct sunlight on the logs throughout this procedure, and as you transfer the logs to the laying yard.

Spawn Selection

Just as the tree species have to be carefully chosen before proceeding with inoculation, the shiitake strain must be carefully chosen as well. A fungal strain is a genetically unique clone selected for one or more desirable properties. A strain is used to produce spawn, and spawn is used to inoculate logs. Spawn consists of mycelium of a particular strain, mixed with sawdust or other substrate for the mycelium to colonize (grow into the mass of sawdust).

Spawn consists of two things: the fungus (mycelium) and a carrier. There are two commonly used spawn formulations for shiitake cultivation that differ only by carrier. These are *sawdust spawn* and *wooden dowel spawn*. Grain spawn and thumb spawn are less widely used and will not be covered here.

Sawdust spawn is generally preferred among growers inoculating more than just a few logs. The

Figure 5.29. Birch dowels (wooden pegs) that have been impregnated with shiitake mycelium in a commercial lab are set in a ⁵⁄₁₆-inch hole drilled into the log; a hammer is used to pound them down flush with the log's surface. Photo courtesy of Jason Grauer & Matthew Goodman and HaveYouEverPickedACarrot.com

reason for this preference is that actual delivery of the spawn into the hole is faster than hammering a dowel into a hole. On the other hand, the investment in equipment is greater for sawdust, which requires an inoculation tool. The consensus among growers, and the result of research at the University of Missouri,[14] is that overall mushroom production is higher with sawdust compared to dowel spawn. A slight disadvantage is that sawdust dries out faster when exposed to air, but this is not a problem if waxing proceeds within a few minutes after filling the holes with spawn, as it should. In general it is recommended that hobby-scale growers (less than one hundred logs inoculated) use the dowel spawn, while commercial growers should make the additional investment and use sawdust.

Another consideration besides the substrate composition is strain selection. Genetically unique strains, available from commercial suppliers, differ from each other with respect to one or more deliberately selected characteristics from which the buyer can choose. While some differences are more about aesthetics, the most important consideration is the temperature range at which they fruit. Strains fall into one of three broad temperature-range categories. These are cool weather (CW 45–70°F), wide range (WR 55–75°F), and warm weather (WW 70–85°F) strains.

By inoculating some logs in the laying yard with one strain of a certain temperature range and other logs with a strain of a different temperature range, an experienced grower can extend the growing season to produce and sell mushrooms for a longer period of time

MAKING SPAWN

Isolation of new fungal strains and production of spawn for inoculating logs involves growing a small piece of fresh mushroom tissue under laboratory conditions, as shown in figure 5.30, then using the new strain to produce sawdust spawn, which can then be used to inoculate logs. It is not necessary or desirable to start from a new mushroom each time. Spawn of a given strain can be multiplied indefinitely. Proceeding counterclockwise, the process of isolating a new strain begins with a fresh whole mushroom (top right). The cap is split open, and a small piece of tissue (explant) is taken from the inner mushroom flesh (not spores). This is transferred under sterile laboratory conditions to a petri dish containing a nutrient medium (potato dextrose agar). Mycelium grows from the mushroom explant, which is genetically identical to the mushroom from which it originated.

This is the new strain or isolate. To make spawn from the new strain, the mycelium is transferred first to sterile grain, then sterile sawdust, where it grows to colonize the sawdust in the bag. The spawn can be used by growers to inoculate logs or other substrates. Normally growers do not produce their own spawn. Spawn production is usually done in commercial labs that specialize in the process and sell high-quality spawn for reasonable prices (about $1 per log). Typically, a commercial lab will mass-produce spawn of a particular strain by adding a small amount of well-colonized sawdust spawn to fresh sterile sawdust under laboratory conditions to make more spawn of the same type. Only rarely are new strains (isolates) started from mushrooms. Cornell has done this to establish new strains of lion's mane as described below in the section Strains Matter.

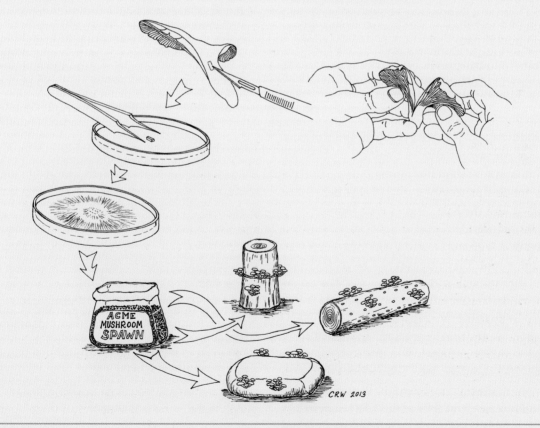

Figure 5.30. Process of making spawn from isolates. Illustration by Carl Whittaker

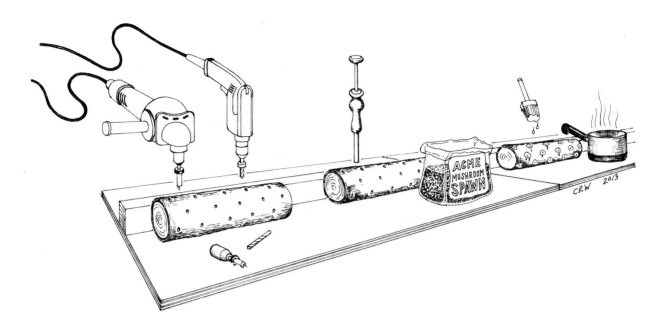

Figure 5.31. Sequence for inoculating a log with shiitake spawn. From left to right: drilling, inoculation with spawn (from bag) into drilled holes, followed by waxing holes. Illustration by Carl Whittaker

each year. This strategy will be described more in the section on fruiting in this chapter (see Step 4).

Inoculating the Logs

Once substrate logs have been acquired, and spawn of an appropriate strain has been obtained, it's time for inoculation. It consists of three steps:

1. Drilling
2. Spawn insertion
3. Waxing

The sequence of log-inoculation activities is as shown in figure 5.31. It begins with drilling holes, followed by inserting the spawn into the holes (inoculation per se), and finally waxing. The inoculation station is usually arranged with each of the three stations (drill, inoculate, wax) along a long table, with one person assigned to each station. With lots of logs to inoculate and a large crew of inoculators, as well as enough tools, two such inoculation tables can be set up, or a second line on the opposite side of the same table (six people).

Figure 5.32. A high-speed angle grinder can be used to drill holes more rapidly than a typical shop drill.

Figure 5.33. Holes in each row are offset by 2 inches from adjacent rows, creating a diamond pattern for a total number of about thirty holes in a 3-foot log. Holes are 1¼ inches deep and ⁵⁄₁₆ inch in diamater for a dowel inoculum and ⁷⁄₁₆ inch deep for sawdust spawn.

Drilling

The power tool chosen depends on how much time and money a grower wants to invest. A standard ~2,500 rpm shop drill is adequate if only a few logs are being inoculated, but for the inoculation of more than a hundred logs, "time is money," as they say, so you will want to use a high-speed angle grinder (~10,000 rpm) with a ⁷⁄₁₆-inch bit. This tool enables drilling of three times as many holes in the same period of time as with a slower shop drill. Angle grinders are designed to take paint off the side of your house rather than to drill holes, so it is necessary to purchase a custom adapter for the angle grinder that will accommodate an appropriately sized drill bit. A ⁷⁄₁₆-inch-diameter, high-speed drill bit manufactured specifically for use with an angle grinder adapted for mushroom log inoculation has a built-in "stop" that makes every hole exactly 1¼ inches deep.

Safety goggles are a must with both tools, but particularly with the angle grinder, which spews out "sawdust" (small wood chips) in all directions.

The holes should be more or less evenly spaced to facilitate even distribution of spawn and subsequent colonization of the log by the fungal mycelium (threads that digest the wood and turn it into mushrooms). In a straight row down the length of the log, drill holes about 4 inches apart. In the case of a 4-inch-diameter log, roll it about 90 degrees to make the next row of holes (2 to 3 inches from the first row). To make a diamond pattern of well-distributed holes, as shown in figure 5.33, the first row of holes should begin ½ to 1 inch from the end of the log, and the next row should be offset by starting the first hole of the second row about 2 inches from the (same) end of the log. The first hole of the third row begins back at ½ to 1 inch from the end of the log. A rule of thumb to follow is to aim for one row of holes for each inch of log diameter. The diameter of the holes drilled into the logs differs for the two main types of spawn. It is not advisable to drill all the holes for many logs at once before beginning insertion of the spawn because the log will tend to lose moisture by evaporation from empty holes.

Spawn Insertion (Filling the Holes)

Long before she is ready to put the spawn in the holes, a grower should have decided if she is going to use sawdust spawn or dowel spawn.

For sawdust spawn an inoculation tool is needed to put the sawdust spawn into the holes drilled in the log. This tool is used to pick up a measured amount of sawdust spawn and "inject" it into the holes that have been drilled into the log. Consisting of a hollow barrel and a plunger that delivers a measured amount of spawn to each hole, the inoculator is filled with spawn by plunging the tip into the bag of sawdust spawn and moving it over a hole drilled into the log, where it is injected by depressing the plunger. For dowel spawn a hammer is needed. Simply place the wooden dowel over the hole and tap it into the hole with the hammer, flush with the top of the hole.

Waxing

Finally, the holes must be "waterproofed" by applying molten food-grade cheese wax with a small paintbrush or dauber, to completely seal off the hole and the spawn

Figure 5.34. The inoculation tool is filled with spawn by plunging it once or twice into the bag of sawdust spawn. Photograph courtesy of Allen Matthews

Figure 5.35. The tip of the inoculation barrel is placed over the hole, in contact with the bark, and the plunger is depressed to force the spawn into the hole. Of all the tasks that go into inoculation, this is the most time consuming and can easily create a "bottle jam" if you don't carefully assign the people involved in the day's activities.

Figure 5.36. After spawn is inserted into the hole, a paintbrush or dauber is used to apply molten cheese wax to seal the hole.

within. Wax is melted in a pot on top of a burner or hot plate. A word of caution: Hot wax can cause very painful, serious burns. Always wear eye protection and long pants.

After the spawn itself, food-grade cheese wax is the most expensive supply involved in the inoculation process. Cheese wax has a relatively low melting temperature, so heat damage to the spawn is minimal. Waxing prevents moisture loss from the spawn and prevents the entry of spores of potentially competitive fungi. When waxing the holes the bark should be dry. Otherwise the wax will not adhere to the bark, and it may peel away, exposing the spawn, which is easily dried out when exposed.

Completion of inoculation is the point at which the countdown begins for the spawn run and eventu-

ally the arrival of the first mushrooms. Some growers apply wax to both ends of each log, but our research at Cornell has found no significant difference in mushroom production for waxed or unwaxed ends, and therefore we do not recommend waxing the ends, which just wastes expensive wax.

The following is a list of equipment and supplies and how to use them. See table 5.6 for more detail.

- Wax. Only food-grade wax should be used. Beeswax is acceptable but is more expensive than food-grade cheese wax.
- A pot to melt the wax in. A standard cooking pot is fine, but don't expect to use the pot for food ever again.

Figure 5.37. The wax should completely cover the hole to prevent moisture loss and ingress of competitive fungi.

- Heat source. A propane camping stove is often used, but the combination of an open flame and flammable (molten) wax can be dangerous, especially if it is overheated. We prefer to place the wax pot into an electric fry pan, which melts the wax and has plenty of room to contain a spill.
- Paintbrush or daubers to apply the wax to the holes in the log.

BEST MANAGEMENT PRACTICES FOR SPAWN RUN IN THE LAYING YARD

1. Locate the laying yard on level ground that is well shaded year-round, is protected from wind, and has access to water.
2. Stack logs in a crib-stack configuration.
3. Irrigate logs periodically if necessary to avoid excessive drying.

- Labels. There can be many variations on this theme, from soft aluminum labels to colored flagging.

STAGE 3: SUBSTRATE COLONIZATION (SPAWN RUN)

Colonization or *spawn run* begins in the laying yard as the mycelium in the spawn holes begins to grow out into the surrounding wood. The spawn run is complete when the fungal mycelium has more or less completely colonized the sapwood of the entire log and the log is ready to fruit.

The shiitake fungus is a primary decomposer, meaning that it most efficiently invades "clean" substrate uncontaminated with other fungi competing for substrate; that is, undecayed wood. Colonization of a log by the shiitake fungus will not occur as rapidly (if at all) or as completely as it would in the absence of competitors. In other words, the use of fresh logs with

Table 5.6. Time, Material, and Supplies, and Approximate Costs Required for Log Cultivation of Shiitake Mushrooms

Task	Labor	Materials and Supplies	Alternatives
What will it take to acquire 100 logs?	17 hours	Chain saw Chain saw oil Gasoline Tarps Transport (pickup, etc.)	Bolts could be purchased from outside for ~$1–$2 each
What will it take to inoculate 100 logs?	29 hours	Sawdust spawn ($100) Inoculation tool ($30) Wax ($31) Angle grinder ($100) Drill bit adapter for angle grinder ($35) 7/16" dia. drill bit ($13) Stove to melt wax Wax applicator ($2)	Plug spawn rather than sawdust spawn (no inoculator required). 5/16" diameter drill bit Drill (~2,500 rpm) rather than angle grinder (10,000 rpm)
What will it take to force fruit (shock) 100 logs?	14 hours	200+-gallon soaking tank Agricultural cloth Tarp to keep rain off mushrooms	Stream, pond, etc.
What will it take to harvest 100 logs?	9 hours	Balance for weighing mushrooms Knife for removing mushrooms from log / killing slugs	

Note: Estimates based on data collected from seventeen new shiitake growers during 2011–12 participating in a Northeast SARE project, Cultivation of Shiitake Mushrooms as an Agroforestry Crop for New England.

bark intact will help prevent other fungi from colonizing the log and outcompeting the shiitake.

The location within the forest farm where the spawn run occurs is called the *laying yard*. Logs remain in the laying yard not just through the spawn run (six to eighteen months) but throughout their productive life, which may be anywhere from three to four years.

Laying Yard Management

The goal of laying yard management is to promote rapid colonization and subsequent fruiting of the log. Moisture management refers to maintaining log moisture content within a range that permits abundant mushroom production. It is the single most important consideration in laying yard site selection and management. The five key site selection criteria that pertain to moisture management in the laying yard include:

1. Shade (direct sun dries the logs)
2. Slope and aspect (north- to northeast-facing slopes are cooler and less drying than south- to southwest-facing slopes)
3. Air circulation and wind (wind causes drying of the logs)
4. Access to water (necessary for periodic irrigation and forcing)
5. Vehicle access (pickup truck, four-wheeler, etc., for transport of logs; the laying yard should be laid out to facilitate vehicle access)

Log Placement in the Laying Yard

Ken once visited a small-scale shiitake grower in Georgia who had a hundred or so logs in his laying yard. The logs were scattered about, as if tossed randomly. Some logs were lying flat on the ground, and others were leaning up against each other at no particular angle, like a bunch of Lincoln Logs or Tinkertoys dumped by an angry child all over the floor. There were no rows or adequate spacing to allow easy access for harvesting mushrooms other than climbing and perhaps tripping over the logs. Obviously, this was not an optimal configuration for arranging logs in a laying yard. In fact, how logs are stacked and configured within the laying yard are important considerations,

Figure 5.38. Three common log-stacking methods for the laying yard: (left) crib or rick stack, (middle) high A-frame, and (right) Japanese hillside stack. Illustration by Carl Whittaker

not just from the standpoint of the efficient use of space and materials handling but also with respect to air circulation and humidity (microclimate) that may affect drying. Stacking for any given log may change over time to facilitate picking and even overwintering. Besides, an orderly laying yard, like evenly planted rows of vegetable crops, will impress even the most uninformed visitors.

Three of the most common stacking methods for shiitake are shown in figures 5-39.

Crib Stack or Rick Stack

The crib stack or rick stack consists of alternating rows of four to five logs, stacked four or five rows high. The crib stack has a small footprint, making it the most space efficient. A disadvantage to the crib stack is that it is not an efficient arrangement for picking mushrooms. It can be difficult if not impossible to get one's hands into the interior of a crib stack where mushrooms are liable to be hiding.

High A-frame

With the high A-frame the mushrooms are well exposed and easier to harvest than other stacking methods. Because logs are exposed and more prone to drying, some growers will only use the A-frame stack immediately after shocking until the mushrooms are harvested, then will place the logs back in the crib stack. Other growers with humid, well-shaded laying yards will use the high A-frame throughout the year.

Japanese Hillside Stack

A more novel stacking method that is not generally covered in other shiitake mushroom cultivation guides is the Japanese hillside stack. A former Cornell student who graduated and went to worked on a shiitake farm in Japan returned to Cornell some years ago to show us the technique, which is used in hilly regions of Japan where bottomlands are used for rice cultivation and shiitake farming is practiced on the adjacent steep hillsides. This method is well suited to steep slopes that would not be suitable for other log-stacking configurations—or most other forest farming activities, for that matter. Illustrated directions for constructing a hillside stack are shown in figure 5.39.

In the section on fruiting of shiitake mushrooms (see Step 4), it will become apparent that a given set of logs may alternate between either the crib stack or the Japanese hillside stacking method and the high A-frame (or other stacking configurations), depending on their fruiting status. In other words, in the same laying yard there are likely to be two or more different stacking methods in use at the same time, each serving a purpose that is different and complementary to the others.

Moisture Management in the Laying Yard

Understanding factors affecting log moisture is essential to successful laying yard management. Excessive drying of logs during the spawn run is the single most common reason for failure, so it is critical to keep it

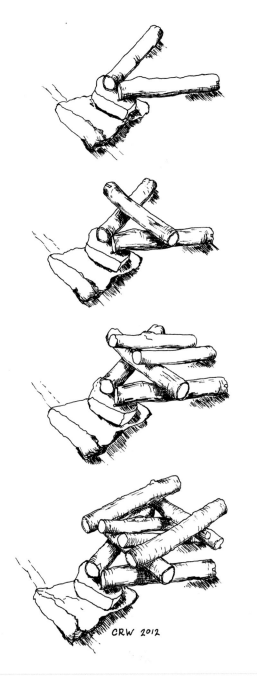

Figure 5.39. Construction of the Japanese hillside stack. Illustration by Carl Whittaker

is practiced, as it is by most growers seeking income by growing mushrooms. Beginning after the spawn run period, "forced fruiting" is used to trigger uniform mushroom production. Since it involves soaking logs in water for twelve to twenty-four hours and repeating this every seven to eight weeks, the logs tend to stay well hydrated from that point on, as long as they continue to be well managed. Forced fruiting will be discussed in detail below (see Forcing [Forced Fruiting]).

Laying yard site selection and management can go a long way toward passive moisture management. Active water management (irrigation) will be discussed below. The most important passive factors controlling the rate of evaporation and drying of the log are temperature, light, wind, and humidity. Evaporative water loss is lower from a log that is cooler than from a warmer log. Light, particularly direct sunlight, causes the logs to heat up and evaporation to increase. A hotter log is eventually a dryer log. The dark-colored bark of a red oak log, for example, can be several degrees warmer than the surrounding air if it is in direct sunlight. In shade there would be little if any difference in the temperature of the air and the surface of the log. The solution to avoiding log warming and evaporative moisture loss is a well-shaded laying yard. Most forest mushroom cultivation takes place under the natural shade of a forest canopy, although some outdoor operations use artificial shade such as lath, shade cloth, or even pine boughs suspended over the logs by a framework of poles. With respect to natural shade from the forest canopy, nearly complete canopy cover is desirable.

Some species of trees are better than others for shading the laying yard. Evergreens (pine, hemlock, and to a lesser extent spruce) make the best canopy for a laying yard. Hemlock is especially effective because it casts a deep shade year-round. A deciduous forest canopy provides shade for up to nine months each year, which may not be good enough to avoid excessive log drying. After leaf drop in the fall, and before leaf-out in the spring, a deciduous forest canopy is largely "transparent" to sunlight, and the relatively direct sunlight during that period can cause excessive drying even in the winter, when the midwinter air temperature is cool or even cold.

to a minimum. Drying of a log during the spawn run occurs mainly by evaporation from the surface of the log. Log moisture content after fruiting has begun is just as critical if not more so than during the spawn run, but it is usually less of a concern if forced fruiting

Figure 5.40. Laying yard with crib-stacked logs under evergreen canopy. Hemlock provides deep shade year-round. Note that the stacks sit on top of a wooden pallet to keep the lowest level off the soil. Uninoculated logs may be used for the same purpose.

Figure 5.41. Laying yard under deciduous canopy. This laying yard at the Wellspring Forest Farm is shaded by maple during the growing season but not during the winter as shown here. Black shade cloth has been draped over the stacks to protect them from winter sunlight.

In addition to shade, other factors influence the environment of the laying yard and the moisture status of logs therein. Excessive wind can effectively steal moisture from a log by promoting evaporation, so sites should be well protected from wind by vegetation or other barriers. The slope and aspect (see chapter 3) are two related factors that can affect moisture management in the laying yard. A north- to northeast-facing slope is cooler and likely to have higher soil moisture and higher relative humidity than a south- or southwest-facing slope. Ideally, a flat site should be chosen just for ease of moving logs and other materials, but if that is not available a well-shaded, gently sloping site (<10 degrees) will suffice. On the other hand, if a steeper slope is the only alternative in a laying yard that meets most of the other criteria, the Japanese hillside stacking method can be used.

Access to water on the site (hose, pond, or stream with an appropriate pump) is essential not only for forced fruiting of the logs when the time comes (see Step 4 below), but it is also indispensable if it becomes necessary to irrigate the logs under dry conditions when their moisture content may have become low enough to interfere with the spawn run and subsequent mushroom production. Of course, the need for the irrigation is minimized if all the other factors are optimal.

Some growers are fortunate enough, or more likely planned carefully enough, to have chosen their laying yard with these issues in mind, such that no irrigation may be necessary during the entire spawn run (~ twelve months). This is the case with the shiitake production by the Rockcastles at their Green Heron Growers farm in western New York described in the case study at the end of this chapter.

Even if careful attention is paid to the siting of a laying yard to take advantage of passive moisture management (shade, for instance), the logs may still dry to the point that the growers need to give them a "drink" by way of supplemental irrigation (active moisture management).

As described above, the goal of moisture management is to prevent the log from drying out, so let's consider the limits of drying out. The log moisture content (LMC) of a live standing tree or a fresh-cut log is in the range of 45 to 60 percent. Some moisture loss in the laying yard is unavoidable, so the goal is to keep LMC at least as high as 35 percent or more throughout the spawn run period and for the rest of the productive life of the log. If the LMC drops below about 25 percent, it is "game over," since at that point the fungus is dead. Just as a point of reference, the LMC of kiln-dried lumber is 8 to 10 percent. It is possible for the mushroom grower to measure

ESTIMATING LOG MOISTURE CONTENT

- With a chain saw cut a thin "cookie" about 1 inch thick from the end of a fresh-cut log. For an older log it is necessary to cut off and discard about 3 inches from the end of the log before cutting the cookie from the new end.
- Weigh the cookies as accurately as possible to obtain the fresh weight (FW).
- Dry the cookie in a cool oven (200°F) overnight.
- Reweigh the cookie to obtain the dry weight (DW).
- Calculate the % LMC = (FW – DW) / FW.

LMC at the beginning and during the spawn run period and beyond, but it is not practical because it is destructive and/or inaccurate (see sidebar, Estimating Log Moisture Content). Most growers simply do not bother to do it. Typically they rely on intuitive/subjective estimation of log moisture status by "hefting" a log (estimating log weight by picking it up) and by observation.

Logs should be observed carefully for signs of excessive drying, especially during the spawn run period (approximately eight to eighteen months). After the spawn run period, excessive drying is less of a concern if logs are being shocked in water for twenty-four hours every seven to eight weeks to bring about fruiting. Some growers choose to allow logs to fruit naturally (without shocking). In this case they must be monitored for signs of drying throughout the productive life of the log.

Even in the case of natural fruiting (that is, no soaking), practiced by some commercial growers (though not many), well-managed logs tend to maintain relative high LMC throughout their productive life. This was the case for an experiment performed from 2006 to 2010 at the Arnot Forest in which both mushroom production and LMC were measured on logs from four different tree species. The logs were fruited naturally (no irrigation or soaking) except for the final year (2010), when they were force fruited (soak, twenty-four hours). Figure 5.42 shows that the change in LMC from 2006 to 2009 was surprisingly little, and the LMC remained well above the critical 35 percent level. Mushroom

production during the 2007 season was low. It was higher in 2008, but in 2009 production was no higher than that in 2007. From these low levels of production for 2007 to 2009, we speculated that mushroom production from these naturally fruited logs was nearly exhausted by 2009, despite the fact that the LMC was relatively high (figure 5.42). Nevertheless, when the logs were force fruited in 2010, mushroom production was considerably higher than in any of the previous years.

Surprisingly, mushroom production was not any greater for aspen, which was the tree species with the highest LMC of all the species. In fact it was the poorest performing tree species. This suggests that cool, well-shaded conditions in the laying yard, along with adequate rainfall, can go a long way toward maintaining well-hydrated logs. But despite the best intentions and preemptive management, dappled sunlight, direct sun over part of the day, and/or wind may be a factor in excessive drying of the log. It is worth repeating that even if adequate shade under a leafy canopy prevails in the laying yard during the summer and early fall, later, during late fall and winter, the logs may be exposed to some direct sun during winter when the trees are defoliated. This explains why an evergreen canopy (hemlock, pine, etc.) is preferable to leafy hardwoods that defoliate each fall.

Some growers do perform regular or occasional maintenance soaks (irrigation) when they judge their logs to be at risk of excessive drying. So how do they know when to irrigate? Some have a system based on monitoring rainfall. One grower we know has a rain gauge in her laying yard. If it indicates that less than 1 inch of rainfall occurs during any given week, she makes up for the difference by irrigating with a rotating-type lawn sprinkler mounted upside down in a tree above the rick stacks in her laying yard. Other growers judge the need for irrigation based on a combination of temperature and rainfall, either by direct measurement or by a more general impression of weather conditions. However one decides when to irrigate, it can be done by sprinkling logs for several hours (often overnight), draping a soaker hose over a stack (less effective), or soaking them in a trough of water, a pond, or a stream for two hours or so—long enough to rehydrate the logs but not

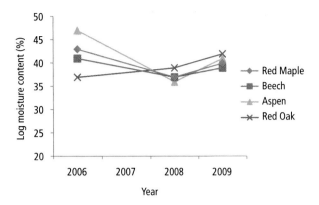

Figure 5.42. Change in log moisture content of shiitake logs over three years, for red maple, beech, aspen, and red oak. All except red oak lost moisture for the first two years but increased in moisture during the last year. None of the species ever approached the 25 percent that is fatal to the shiitake fungus.

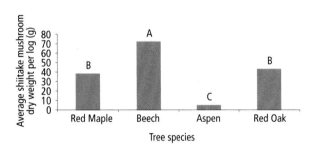

Figure 5.43. Shiitake mushroom production for four tree species over three years.

long enough to shock or force them into fruiting (twelve to twenty-four hours), as discussed below (see Step 4).

Keep in mind that not all logs in the laying yard dry at the same rate. For one thing, a smaller-diameter log dries more rapidly than a larger-diameter log because the surface-to-volume ratio is lower for larger-diameter logs. A log of a species with thin bark (red maple, for example) dries more rapidly than a log with a thicker bark (e.g., white oak). A stack on the edge of a laying yard may be more exposed to wind than one in the interior.

Step 4: Fruiting, Harvesting, and Marketing of Shiitake Mushrooms

Having completed the critical spawn run period in the laying yard, the next stage is fruiting. The

---

BEST MANAGEMENT PRACTICES FOR FRUITING SHIITAKE MUSHROOMS

1. After one year or when logs show signs of complete colonization (sporadic mushrooms and/or white patches at ends of logs), begin forced fruiting by shocking (soaking) logs for twenty-four hours in cold water.
2. Remove logs from water, and stack loosely in a high A-frame.
3. Harvest mushrooms in seven to ten days, and keep refrigerated.
4. Rest recently harvested logs for seven to eight weeks before shocking again.

---

process of actual mushroom formation arising from the mycelial network that has infiltrated the wood (colonization) is called *fruiting*, even though mushrooms are not technically fruit in the botanical sense. Understanding how to manage this final stage of the process is critical to success. Fruiting will occur spontaneously to a limited extent, which is a good sign that a "young" log is ready to be shocked (force fruited). Beyond that, however, this natural (spontaneous) fruiting is not an especially good thing in terms of predictability if the goal is income generation. It is better to force the logs to fruit when and as you need them to, rather than to wait until they want to fruit, although this is considered by some to be a philosophically debatable point.

When the spawn run (colonization of the substrate) is more or less complete, fruiting will begin naturally (spontaneously), but most growers force fruiting by shocking. The completion of the spawn run and the onset of fruiting occur anywhere from six months in the warmer climates of the South to twelve to eighteen months after inoculation in the cooler North. The earlier onset of fruiting and the extended length of the growing season in the South do not mean, however, that total mushroom production over the life of the log is any greater than that of a similar log in the North.

In both cases, the total mass of the mushrooms produced over the lifetime of the logs is similar, assuming

Figure 5.44. Crib (rick) stack is the most common laying yard configuration, because it contains the maximum number of logs on the smallest footprint. White patches at the ends of the logs indicate that the log is fully colonized and the spawn run is complete. The stack is perched on top of two uninoculated logs to keep production logs off the ground.

other factors such as sapwood volume are similar. In other words, total mushroom production from a log is limited by the amount of colonizable substrate in the log, not by how frequently or how many times it has been forced, which only affects the rate of mushroom production.

The first step toward forced fruiting of shiitake is to make sure that the logs are fully colonized; that is, the spawn run is completed. A sign that colonization is proceeding (but not necessarily complete) is the appearance of white patches on the ends of the logs, corresponding more or less to the position of each row (four or more) of inoculation holes (figure 5.44). This is an indication that the mycelium has progressed from the inoculation sites, through the xylem (water-conducting vessels in the wood), and out to the ends of the log. Eventually this white growth may cover the entire end of the log as colonization proceeds. White patches indicate that the log is completely colonized or nearly so, and the spawn run is complete. As indicated above, the most direct indicator is that a few mushrooms appear uninvited, without forcing (natural fruiting). At this point a regular schedule of forced fruiting can be initiated.

Figure 5.45. Forced fruiting of shiitake mushroom logs can be accomplished by soaking them in a tank of water for twelve to twenty-four hours. This small steel tank at the Wellspring Forest Farm will hold about twenty logs.

Forcing (Forced Fruiting)

Forcing is a strategy, unique to shiitake mushrooms, for triggering uniform mushroom production. Only shiitake can be induced to produce a coordinated flush of mushrooms that appear within a few days of each other. This facilitates the labor involved in picking compared to naturally flushing mushrooms that occur sporadically over weeks or even longer. It is also a great convenience when it comes to marketing, since it is much better to trigger flush when you want it to happen, so it coincides with market day, such as a farmers' market, or filling an order with a local restaurant.

Shocking or soaking is one of several practices used to force-fruit shiitake logs. It consists of soaking one or usually more logs that have completed spawn run in water at least overnight or preferably a full twenty-four hours (figure 5.45). The water should be as cold as possible, but this is limited by what is available. Researchers at the University of Missouri Center for Agroforestry reported that logs soaked in ice water (literally) yielded more mushrooms than logs soaked in water of a temperature more typical of a soaking tank or stream. But ice for shocking is not remotely practical in the laying yard. Cold water from a well or a cool stream water is better than tepid water that has equilibrated with the air temperature on a warm summer day.

This process of shocking begins with moving logs to the water. A set of logs can be moved from a compact crib stack or from the upper layers of the Japanese hillside stack (or other stacking method), and placed in a 200- to 500-gallon tank or larger with enough water to cover the logs. A cattle trough (metal or hard plastic) works well for this purpose. Fill the tank with logs and weight them down with something (stones, etc.) so that even the logs on the top of the tank are completely submerged if possible.

"Thumping" is a forcing-related practice that is usually combined with shocking in water as described above. Some growers swear by thumping, while others just scratch their heads in skeptical amazement. Thumping consists of banging one end of a log smartly with a hammer, or forcefully dropping it on its end, from a distance of a foot or so onto a hard surface. The authors of this book have no experience with thumping, although one of the most experienced and most successful growers we know is a big fan of it.

Once the logs have been removed from the tank (or stream) it is best not to stack them back directly into the crib stack, because when logs flush in a crib stack, they are difficult to harvest in those configurations, and some become deformed when they encounter another log. It is better to remove the logs from the soaking tank and stack them into a high A-frame stack, or lean them up against a tree, figure 5.38 middle.

Scheduling

It is important to know how to fruit logs either naturally or by forcing, but it is equally important, to commercial growers at least, to manage the process of forced fruiting so that:

1. Harvest day coincides as closely as possible with when you want to sell them
2. Long-term (seasonal) management of forcing cycles extends the season as much as possible

A key point here is that logs can be forced more than one time in a growing season (late spring to early fall). The time between the first forcing and the next is called the resting period. Depending on whom you listen to and what you read, the resting period can be anywhere from six to eight weeks or more. In our judgment seven weeks is just about right.

Figure 5.46 illustrates a seasonal forcing schedule that allows two or even three forcings during a typical Northeastern growing season. In a nutshell the strategy involves dividing the logs available for fruiting into seven different groups, which are shocked sequentially over a seven-week period. Then the cycle repeats with those same seven groups for a second round and maybe a third. The role of not only careful scheduling shocking but also selecting appropriate spawn strains to extend the production season as long as possible is described in Figure 5.46.

Before further discussion of season extension, there is another important consideration about forcing that must be understood. It sort of brings the "miracle" of forcing down to earth; it falls into the category of "no free lunch." Compared to natural (spontaneous fruiting), forcing does not increase total production per log over the productive life of a log. A log has a given amount of substrate (lignin and cellulose; i.e., wood) for mushroom production, and when it's gone, it's gone. All forcing can do is synchronize the fruiting and increase the pace of production. In other words, if a log has enough substrate to produce 3 pounds of mushrooms over its lifetime, it might do that in four years if allowed to fruit naturally, whereas the same log subjected to repeated forcings (every seven weeks or so)

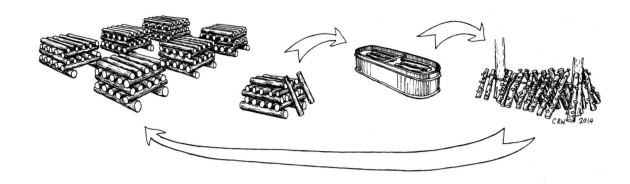

Strain	Jan.	Feb.	March	April	May	June	July	Aug.	Sept.	Oct.	Nov.	Dec.
CW												
WR												

Figure 5.46. Use of two shiitake strains — one cool weather (CW) and the other wide range (WR) to extend mushroom production season from about nineteen weeks (WR only) to about thirty-five weeks using the combination of CW in spring and fall, and WR from May through September. CW strains will begin fruiting in the spring when it is too cool for WR strains, which begin early summer and extend through September until lower fall temperatures shut them down, at about the time the CW strains begin fruiting again for several more week. Scheduling is the key to maximize fruiting of WR strains during the summer. It begins by dividing the forcing logs into seven groups (crib stacks in the figure), and shocking (twenty-four-hour soak) one stack each week for seven weeks. After seven weeks the first stack is sufficiently rested to be shocked for the second time, followed by the second stack the week after that, then progressing through the rest of the seven stacks for a second time, and then for a third time if late summer temperatures remain high enough. Note that after each weekly soaking the logs from that stack are fruited and harvested in a high A-configuration, and then returned to a crib stack until the next seven-week cycle. Illustration by Carl Whittaker

might only last for two or three years, while producing no more or no less yield than the naturally fruited log.

Season Extension

With respect to scheduling, for a commercial grower it may be desirable to extend the production season for as long as possible over a given season. One of the most important tools to accomplish this goal is to take advantage of different temperature range strains, as was suggested earlier in the section on inoculation. Choosing an appropriate combination of isolates for inoculation will set the stage for a scheduled progression of mushrooms harvested over the course of a season and for years to come.

There is another consideration regarding the management of cool weather (CW) strains. Because they are triggered by cool weather rather than forcing, they don't need to be moved around the laying yard as much, since they don't need to be soaked. Less toting of

Figure 5.47. Logs inoculated with a cold weather strain (CW) can be larger and heavier than logs inoculated with wide range strains (WR) because they do not have to be moved to soak them for forcing. Once inoculated and placed in the laying yard, they can stay in one location for years without moving.

BEST MANAGEMENT PRACTICES FOR HARVESTING SHIITAKE MUSHROOMS

1. Harvest mushroom when the rim of the cap is still slightly rolled under (see figure 5.48). Once the edge has flattened out, the mushroom is suboptimal for picking—its flesh is not as firm as a fresh mushroom's, and its shelf life is reduced. Nonetheless it is still edible.
2. Harvest mushrooms from logs using a sharp paring knife or by simply breaking them off by hand.
3. Be sure to pick off slugs, beetles, and so on.
4. Mushrooms with minor damage due to slugs can be saved for slicing and drying.
5. Refrigerate fresh mushrooms in paper bags (not plastic) for no longer than one week before marketing.

logs means less labor, and less labor is good. This means that it is possible to use larger-diameter (heavier) logs, since they don't need to be handled after inoculation and initial placement in the laying yard.

Harvesting

In the event of rain the mushrooms must be protected, for three reasons: (1) Caps become slimy, making them at least visually unappealing, so that they are either unsalable, or nearly so, and need to be heavily discounted; (2) A heavy rain can inadvertently disrupt the most careful planning, by triggering fruiting of all the logs at the same time. This can be a disaster if you only planned to harvest 20 pounds of mushrooms that week, but instead you are stuck with 80 pounds; (3) Even worse, those logs won't be ready to be harvested again for another seven to eight weeks.

To prevent an overabundance of mushrooms because of rain, it is necessary to protect the logs from excessive

Figure 5.48. The shiitake on the left is perfectly ripe for picking; the sides are just slightly curled under.

BEST MANAGEMENT PRACTICES FOR SUCCESSFUL MARKETING OF SHIITAKE

1. Approach restaurant chefs with a sample of your product, attractively packaged.
2. Branding: Print paper bags with your farm logo, as well as other useful information, such as the name of your farm, your name, contact information, quantity (weight) of mushrooms in the bag.
3. Stress the fact that your mushrooms are *log grown*.
4. Print a mushroom recipe or two on the back side of the bag.
5. Use paper bags that "breathe," not plastic bags, which will sweat the mushrooms.
6. If you are selling your product fresh, keep it refrigerated, but not for more than a week. Sooner is better to maintain high-quality mushrooms. If you don't expect to sell your mushrooms fresh within a week, consider drying them for long-term preservation.

ENHANCING VITAMIN D CONTENT IN SHIITAKE

Shiitake mushrooms have been long valued in many cultures for their health benefits, but the exceptional nutrition comes not only with fruiting, but can be "value added" as well. One of these is its ability to accumulate vitamin D when exposed to UV rays, whether synthetic or natural sunlight.

One study[15] looked at the use of pulsed UV light to increase vitamin D content in button, cremini, oyster, and shiitake. The results of this study demonstrated that "after a very short exposure time of about 1 sec (system generates 3 pulses per second) the Vitamin D2 content of these mushroom varieties can be increased from very little to upwards of 800% DV/serving."

Another study mentioned by Aloha Medicinals[16] noted that even drying shiitake in the sun (a less intense form of UV exposure) for at least three hours led to an increase of vitamin D by up to five times the normal amount. This means that through simple exposure we can increase the already impressive array of health benefits offered by shiitake.

rain with shade cloth or agricultural fabric or other means as necessary. Some growers build a rain shelter (e.g., poles and corrugated fiberglass roofing) to protect logs that are in the process of fruiting (after soaking), but not necessarily to cover the entire laying yard. After mushrooms are harvested the logs are moved back out into the laying yard.

Rain is not the only reason for covering logs on which mushroom production is under way. Newly developing mushrooms, known as pins, can easily desiccate in dry or windy weather, which results in mushrooms damaged or aborted. Lightweight agricultural cloth (Reemay, Agribon, garden fabric) does little or nothing to protect mushrooms from getting slimy in the rain, but it does help maintain a uniformly humid environment without damaging the mushrooms as a trap or heavier covering would.

Marketing Shiitake Mushrooms

If growers don't take marketing mushrooms as seriously as they do growing mushrooms, they may find that they've wasted a lot of time and energy growing

mushrooms that they can't sell or get an acceptable price for. Before deciding to grow any nontimber forest products, it is prudent to determine what market opportunities exist; that is, where products will be sold—roadside stands, farmers' markets, restaurants, or CSAs, for example. For small-scale producers, a CSA is a good arrangement. Clients buy a share of the farmer's production and are guaranteed a portion of the produce (mushrooms) every week.

To market mushrooms successfully, orders need to be filled reliably, particularly if the client is a restaurant chef who expects the grower to provide an agreed-upon supply of product, such as 10 pounds per week. Nothing disappoints a restaurant chef more than to learn that her order cannot be filled because of a production-related problem or any other reason. As described earlier, to ensure an even, reliable supply of shiitake from a laying yard, a grower should develop a forcing schedule using a judicious mix of spawn types (CW, WW, WR), so that logs will produce the quantity of mushrooms needed each week, throughout the entire growing season.

In the Northeast, unless a grower can get at least $10 to $12 per pound, it will be difficult to make a go of shiitake cultivation as part of a small farm enterprise. Based on recent shiitake enterprise research by Mudge, et al.,[17] shiitake sold for an average of $15 a pound at farmers' markets, when sold in quarter-pound units. To restaurants shiitake sold for about $12 per pound. The farmers' market price for shiitake in the South is $8 to $10 a pound, although data are limited. In the Midwest, farmers' market prices range from $10 to $16 a pound (Joe

Krawczyk, personal communication). Wholesaling your product to grocery store chains is unlikely to be profitable because grocery chains typically buy shiitake mushrooms wholesale at $3 to $4 per pound that are mass-produced on sawdust at large indoor commercial production facilities and retail them for $7 to $8 per pound. It is unlikely they will pay more for log-grown mushrooms despite their superior taste, appearance, and nutraceutical value.

Value-Added Products

Besides selling fresh mushrooms, there are some profitable value-added products to consider, including:

- Pâté: The recipe for shiitake hazelnut pâté shown in the sidebar (see Shiitake-Hazelnut Pâté Recipe) is a big hit at parties, some would say "to die for."

Figure 5.49. Value-added products from Green Heron Growers, where the Rockcastles process extra shiitake into sliced and dried mushrooms ($10/oz) and medicinal tincture ($10/oz).

SHIITAKE-HAZELNUT PATE RECIPE

Adapted from Green Heron Growers,
Panama, New York

4 oz of shiitake mushrooms
1 clove garlic
2 Tbsp butter
⅛ tsp thyme
¼ tsp salt
⅛ tsp pepper
1 tsp parsley
¼ cup toasted hazelnuts
3 oz cream cheese (for vegan, use avocado
   or blended tofu)
2 tsp dry sherry (for nonalcoholic use use
   plum or apple cider vinegar)

In food processor, blend mushrooms (cap) and garlic. In a skillet melt butter. Add garlic and mushroom mixture, and sauté for five minutes, stirring in spices. Blend parsley and hazelnuts in food processor, adding cream cheese (or substitute) until smooth. Add sherry and mushroom mixture, and process until uniformly mixed. Chill in a covered dish for at least one hour. Makes 1 cup (multiply recipe by four for a party-size serving).

Several commercial growers make and sell this item from mushrooms they have grown.

- Soup: We know a grower who makes an excellent mushroom barley soup and freezes it, so she can take it to their weekly farmers' market.
- Medicinal tinctures

Dried Shiitake Mushrooms

In Japan dried is practically the only way shiitake are sold. There are other value-added products as well. Steve and Julie Rockcastle in western New York (see case study at the end of this chapter) have found that it takes about 3 pounds of fresh mushrooms to make 1 pound of dry. They then sell a 2-ounce jar of dried shiitake for about the same price as for 14 ounces of fresh mushrooms before drying ($10).

Selling Preinoculated Logs

Whole or half logs (16 to 24 inches) can be inoculated as usual and either sold immediately or sold after most of the spawn run. The buyers of preinoculated logs are mostly hobbyists, chefs, and so on who wish to have a few fresh mushrooms on hand but who do not wish to tackle the process of growing shiitake from start to finish. From the grower's perspective, selling preinoculated logs shortly after inoculation avoids the effort and risks (drying, slugs, contaminating fungi) associated

Table 5.7. Oyster Mushroom Rainbow of Colors

Species	Strain
Pleurotus ostreatus	PoHu (off white)
Pleurotus ostreatus	Grey Dove (steely gray)
Pleurotus ostreatus	cite (white)
Pleurotus ostreatus	Blue Dolphin (pewter gray)
Pleurotus cornucopiae	Golden oyster ("luminous citrine yellow")
Pleurotus pulmonarius	Italian oyster (brown)
Pleurotus djamor	Pink oyster

with spawn run and later stages of production. The grower can charge more because most of the work has been done for the customer. An instruction sheet can be provided along with the log so consumers will have a better chance of succeeding.

Oyster

Oyster mushrooms include several species in the genus *Pleurotus*. They occur naturally throughout eastern North America, growing mostly on dead and occasionally living trees and other wood debris. Worldwide they are the second most commonly cultivated mushroom after the white buttons, but

RODNEY AND HEATHER WEBB: SALAMANDER SPRINGS FARM

In 2004, in the early days of the development of the MacDaniels Nut Grove, eleven trees were cut down over a 3-acre area to open up the area to more light. A few months later the Cornell Mushroom Club returned to the site with cordless drills, dowel plug inoculum (Grey Dove oyster), a hammer, and some wax and drilled ⁵⁄₁₆-inch-diameter holes vertically around the perimeter of these 10-inch-diameter stumps, just inside the bark. Holes were drilled about 1.5 inches apart. A dowel spawn plug was tapped into each hole, and the holes were sealed off with a soft grafting wax. Now, stump cultivation of mushrooms has never been a particularly high-priority activity at the nut grove,

so the club project was virtually forgotten for some time.

Eventually, only two of the eleven stumps ever produced any oyster mushrooms, and they did not fruit until about two years after inoculation. Even then, fruiting was sporadic for the next several years. That is how I came to the opinion that stump production of oyster mushrooms was not great shakes—that is, until I met Rodney Webb at Salamander Springs Gardens in the Blue Ridge Mountains of North Carolina. Most of his mushroom-related income at the farm was from conventional shiitake log cultivation, but there was this one time when a big 10-inch diameter breast height (DBH) tulip poplar tree had to be cut down for some reason or another. Instead of cutting the tree down low a foot or so off the ground, he cut it considerably higher, at about 6 feet.

With his chain saw he cut and removed about 20 evenly spaced watermelon-shaped wedges about 6 inches wide and several inches deep into the wood. The notches were then packed with oyster mushroom sawdust spawn, and the wedge was pounded back into the notch and secured with nails. The first picture (figure 5.50 left) shows Rodney beside the dormant stump in March. Later, he sent me the other picture (figure 5.50 right) of the stump about a year later, covered with masses of Golden (yellow) oyster mushrooms bursting from the spaces between wedges and the accompanying notches. This lovely golden monolith continued to flush several times a year, and over the course of three years, Rodney harvested about 60 pounds of mushrooms, which he sold for $10 a pound, so he made about $600. Not bad, but I wouldn't recommend running off to cut down all the 10-inch-diameter trees on your property. Remember our experience at the MacDaniels Nut Grove, where only two of eleven trees produced any mushrooms at all. Must be that Blue Ridge mountain air.

Figures 5.50. (Left) Rodney Webb with his tall poplar, which he "stump" inoculated by cutting out wedges, packing in spawn, and nailing the wedges back in place. (Right) The same stump the next season with Golden oyster mushrooms, which yielded over 60 pounds in three years.

among forest-cultivated mushrooms, they are a distant second to shiitake. Oysters are widely cultivated indoors on various easily digestible high-cellulosic substrates such as newspaper and straw, coffee grounds, and even crude oil[18] (mycoremediation). Cultivation outdoors on log substrates is less commonly done than log-grown shiitake. When oysters are grown on wood, in a forest-farming setting, they perform better on low-density woods such as poplar and willow, unlike shiitake, which performs best on high-density hardwood species such as oak, beech, and hard maple.

In our experience oyster mushrooms are most productive when grown on totem stacks (described below in the section on lion's mane), compared to the bolt-size logs that are used for shiitake production.

Another system that is occasionally used to grow oyster mushrooms is stump cultivation (see sidebar on Salamander Springs Farm). When a tree is cut down for whatever reason, the stump remains with its roots in the ground. Even as the stump dies these roots continue to passively absorb water, so the aboveground portion does not dry out as rapidly, and that is favorable for the growth of the fungus.

Lion's Mane

Lion's mane (*Hericium erinaceus* and *Hericium americanum*) is the common name for several species of fungi in the genus *Hericium*. Two of these *Hericium* species native to North America are considered choice edibles by mycophiles (wild mushroom collectors). Lion's mane mushrooms generate a lot of interest among wildcrafters, forest farmers, and gourmet chefs, but so far there have been few attempts to cultivate it on logs as a nontimber forest crop for forest farming. For one thing, its "mushroom" (spore-producing structure) doesn't look a bit like the classic mushroom. Some epicureans describe the taste of lion's mane as resembling seafood, especially lobster. Most mushrooms take to being sautéed in butter with a little garlic, and lion's mane is no exception. It is said to soak up flavors of whatever it's being cooked with, like a sponge. In traditional Chinese medicine, and more widely, it is regarded as having medicinal properties. Research has shown that it stimulated nerve growth, improved cognitive ability, and stimulated growth of damaged nerves.

Figure 5.51. Lion's mane (*Hericium americanum*) with branched spines. Photograph courtesy of Jonathan Landsman

Figure 5.52. Lion's mane, *H. erinaceus*, on bolt-size beech logs.

Lion's mane is pure white—strikingly so—with spiny teeth that look like tiny icicles. The two species of *Hericium* that are sought after for fine dining are *H. erinaceus* and *H. americanum*, although they both share the common name lion's mane. *H. erinaceus* is sometimes marketed by the name "pompom" because it looks like a cheerleader's pompom reduced to the size of a chubby golf ball. Other common names, applied to both species, are monkey's head, hedgehog mushroom, and others.

The fruiting body (mushroom) of *H. erinaceus* is globoid (more or less spherical) in form, several inches in diameter (figure 5.52). It has unbranched spines up to a centimeter in length that emanate from a more or less central core. *H. americanum*, on the other hand, is irregular in shape and can be six to eight inches across or larger. It has longer spines than *H. erinaceus,* which emanate from a branched structure, and overall it forms a less compact structure than *H. erinaceus* (figure 5.51).

Lion's mane is avidly hunted by wild mushroom collectors, but there is essentially no commercial forest production of either species, although they are cultivated noncommercially by a few enthusiasts. As with shiitake and some other specialty mushrooms, *H. erinaceus* (pompom) is grown on sawdust in some large factory

mushroom houses, but lion's mane makes up a very small portion of their sales. Indoor, sawdust-cultivated pompom is occasionally found in groceries for prices (~$13/lb) well above those of sawdust-grown shiitake (~$8/lb). Despite the fact that there is virtually no commercial forest cultivation of lion's mane, it is a mushroom with considerable potential as an income-generating NTFP for forest farming. There are three things that make it an attractive candidate for further development:

1. Its unusual and exquisite taste and unusual but attractive appearance give it an "exotic" allure for consumers.
2. Once established on logs using the totem production system, it requires little if any annual maintenance. Most importantly, it requires no shocking to induce fruiting, as is the case with shiitake. Because of the mass and the low surface-to-volume ratio of the large-diameter logs used in totem cultivation, moisture loss from the log by evaporation is less of a problem.
3. In our experience lion's mane totems (for both *H. erinaceus* and *H. americanum*) will continue fruiting for at least six years (as of 2013).

On the other hand, there are some drawbacks:

1. It takes about eighteen months to begin fruiting, compared to about a year for shiitake.
2. It fruits over a relatively short period of time (several weeks in the fall).
3. Storage life is somewhat shorter than shiitake's.
4. There is an almost complete lack of an existing market for, or consumer awareness of, log-grown lion's mane mushrooms.
5. It cannot be force fruited, to generate mushrooms "on demand" as shiitake can be; that is, to meet an externally imposed schedule.
6. Few strains are commercially available.

In our over six years at Cornell's Arnot Forest, lion's mane has performed better on totems than on bolts. A totem consists of a vertical stack of two or three logs that are larger in diameter (10 to 12 inches) but shorter

Figure 5.53. To prepare logs for inoculation, cut a 2-foot-long log in half, then cut a small 2- to 4-inch cookie from the top. Each section will have spawn layered in between the sections of logs.

in length (1 foot) compared to the smaller 4- to 6-inch-diameter, 3- to 4-foot-long shiitake logs.

Figures 5.53 through 5.55 show a sequence for constructing and inoculating a totem. Totems also can be used for mushrooms other than lion's mane; totems are the preferred configuration for cultivation of log-grown oyster mushrooms. Most growers use plastic garbage or paper leaf bags during spawn inoculation.

It has been noted by various passersby that bag covered totem stacks scattered about in the woods (laying yard) give the appearance of an invasion of R2D2s (from the *Star Wars* series) or, worse yet, black plastic bags full of garbage, neither of which leaves a particularly good impression on visitors. If the totem stacks are covered with white plastic bags, they can be mistaken for a

Figure 5.54. Construction of a totem stack for cultivating lion's mane or oyster mushrooms consists of cutting 10- to 12-inch-diameter logs. At the bottom of a log place a piece of cardboard, then an 8-ounce cup of sawdust spawn, and set one of the logs vertically on top of the spawn. Then place another 8-ounce cup of spawn on top of the first log and another log on top of that. Finally, top off with the cookie piece.

Figure 5.55. Cover the stack with a large paper bag. This should help keep moisture in and competition out. Some growers also cover with a plastic bag to ensure a greater degree of sanitation and moisture retention.

coven of ghosts. As an alternative some forest farmers choose to dispense with the bag entirely. In that case the pile starts with a 12 × 12-inch piece of cardboard placed directly on the ground, and the first portion of spawn is placed on that. The rest of the construction remains the same, except that some prefer to cut one more 2-inch section (cookie) off the upper 12-inch log and place this at the very top of the stack with a portion of spawn between it and the upper 10-inch-long log beneath it. This allows the fungal mycelium to colonize from the top down as well as from the bottom up.

Strains Matter

One of the more interesting (although not particularly surprising) experimental findings from our research at the Arnot Forest was conducted by then graduate student Jeanne Grace, see figure 5.3. She collected five different samples of wild *H. americanum* from various locations near Ithaca, New York, and used a sterile laboratory procedure to isolate each of the accessions (separate collections from the woods) in pure culture (five genetically different strains or clones), and from there each was used to make sawdust spawn. In addition to the four local strains she used one strain of *H. erinaceus* from a commercial source (Field and Forest Products, Peshtigo, Wisconsin). We used all five of the isolates to initiate totem cultures at the Arnot Forest.

Two interesting findings emerged. One is that the *H. americanum* isolates consistently outperformed the commercial *H. erinaceus* isolate. On the one hand, this difference could be due to the fact that all the *H. americanum* strains were from the vicinity of where the experiment was conducted and were therefore better adapted to local conditions than the "exotic" *H. erinaceus*. On the other hand, the difference might be due simply to the fact that the local isolates (*H. americanum*) were of a different species (genetically different) from the commercial isolate (*H. erinaceous*). One additional observation from this experiment, regarding cultivation of lion's mane, was that annual fruiting of this was very seasonal and of short duration. All three of the *H. americanum* isolates fruited only for about three weeks in the fall, during

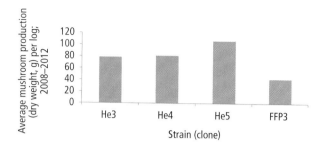

Figure 5.56. Comparison of lion's mane mushroom production from three local isolates (strains) (He3, He4, He5) and one commercial isolate (FFP3).

which mushroom production was considerable. Of interest is that the exotic isolate, *H. erinaceous*, fruited briefly, with relatively low productivity, during late spring and then again in the fall.

As of fall 2013, these results have held up for six successive seasons. This last speaks to the fact that totem culture may be productive for longer than the three to four years that are typical of shiitake using the bolt cultivation system. On the other hand, shiitake bolts usually fruit in the Northeast within about one year of inoculation, whereas lion's mane took eighteen months to begin fruiting with the totem system.

Since practically no one has experience with marketing forest-grown lion's mane to the public there are some important questions left to be answered. What is the overall cost of production of lion's mane, including labor, spawn, and other expenses, and what is the consumer demand for this mushroom? It can be said with some degree of certainty that the demand is not very much, since very few consumers have even heard of it. Nonetheless, if lion's mane can be marketed as a novelty—exotic and special, along with a more widely accepted product (shiitake), public demand could increase over time. The very fact that fruiting is confined to several weeks in the fall will necessitate that a grower will not be able to "rely" on this product exclusively, as is often the case with shiitake. Pairing the sale of lion's mane with that of shiitake would seem to make sense as a marketing strategy.

Stropharia

Also known as red wine cap or garden giant, Stropharia (*Stropharia rugoso-annulata*) is a mushroom native to North America and commonly found wild in garden beds, lawns, and forest edges. It is also the easiest to cultivate in terms of the time it takes from inoculation to fruiting compared to growing shiitake on bolts or lion's mane and oyster on totems. Research at Cornell is in the early stages: Comparison plots are being grown to look at the difference in light regimes (full shade, half shade, full sun) and substrate (wood chips vs. straw). Preliminary observations indicate that wood chips do the best, especially in wet years, and that Stropharia colonizes well in all light regimes,

though it appears to fruit slightly better in part shade and full sun.

Inoculation is rather straightforward, with several layers of organic materials layered with purchased spawn to create optimal conditions.

MATERIALS AND LOCATION

Stropharia grows best under partial shade. The following sequence is a quick and easy way to grow mushrooms in your back yard.

- Somewhat fresh (less than a year old) wood chips of mixed hardwood species); about two wheelbarrow loads *or* a fresh straw bale
- 5-gallon bucket of sawdust or shavings

Figure 5.57. Three phases of growth help to identify the Stropharia mushroom. Note the "crowns cap" ring (annulus) around the stem, a remnant from when the cap grew apart from the stem.

Figure 5.58. A patch of woodland cleared down to ground level, then covered in sawdust, with the sprinkled mycelium spawn on top. The next step is 2 to 4 inches of wood chips, then time!

- 5-gallon bucket of finished compost (optional)
- Sawdust spawn from a producer (see our website for a list)

The best locations for inoculation are already existing beds and places that are permanently installed, where there will never be any tilling, which would destroy the mycelium. Also consider establishing Stropharia with other plant cultivation (such as currants) where wood chips or straw are already being used as mulch and are likely to be watered if dry. Finally, consider a location that is well traveled, as the fruiting and maturation of the mushrooms can happen rather rapidly, and it's a shame to miss out on this tasty mushroom, which is similar in size and taste to a portabella.

INOCULATION PROCEDURE

First, measure out a spot that is approximately 16 square feet of bed space. This is approximately what a 5-pound bag of spawn will inoculate; you can inoculate one continuous section or multiple smaller areas. Make sure no inoculation is smaller than 4 square feet or a quarter bag of spawn.

Inoculation can occur as early as April or as late as September, with spring being the preferred time, as it often results in fruiting in the same season. To inoculate, remove organic matter down to bare soil. Add about ½ inch of sawdust or wood shavings and spread evenly. Layer the spawn on top of this, breaking it up into fine particles while also leaving some chunks in the bed. On top of this, layer about

4 inches of wood chips or straw. Soak the bed thoroughly with water.

MAINTENANCE

Stropharia requires little maintenance and can live and fruit for many years. In dry seasons water patches as with plants in a garden. It is best to add 2 to 4 inches of fresh wood chips or straw in the fall to provide fresh feedstock and protect the mycelium from damaging frosts. Once a patch has colonized an area for one full season, the mycelium can be divided into multiple-handful chunks and spread into other areas of the garden. A unique feature of Stropharia is that spawn does not have to be repurchased but will more or less "naturalize" to the site and can be propagated by dividing well-established patches and establishing new locations elsewhere. Growers who have been doing this for many years often say they don't ever know where the mushroom will show up, which makes for a fun surprise.

HARVESTING

It is important to properly identify Stropharia mushrooms before harvesting, as there are many mushrooms that can emerge from mulched garden beds. That said, Stropharia is rather easy to identify by the following characteristics:

- A reddish-brown cap that changes from dark to light as the mushroom matures
- Gills that begin as a light black and turn darker as the mushroom matures
- A "king crown" annulus ring around the stem
- A stem that is fibrous and full of air pockets
- No noticeable bulge where the mushroom meets the ground

Taking a spore print is an important tool in definitively identifying a mushroom. This is done by harvesting a cap, cutting off the stem, and placing it on a piece of white or black paper (depending on spore color) overnight. The Stropharia mushroom leaves a black-purple spore print. When in doubt, don't eat it!

The commercial potential for Stropharia is unknown, and like oyster and lion's mane the fruiting is sporadic, most often in the spring and fall. Those interested in commercial sales should establish markets with shiitake, then can likely offer Stropharia when available as an additional "surprise."

Mushrooms and Forest Farming

Mushrooms are certainly one of the more novel and exciting aspects of forest farming. The authors have provided an extensive amount of information for interested growers, but it's important to keep in mind that one of the best parts of mushroom cultivation is that the basics are easy and success with new inoculations is high. For those whose personalities are more geared toward reliable, predictable crops, shiitake is by far the best option to pursue. Others who possess the willingness to experiment, observe, and tweak the management of their mushrooms are encouraged to work with the other species—oyster, lion's mane, and Stropharia—where some success is likely but the growing system is less robust than with shiitake.

Keep in mind, though, that shiitake was not always this way. It has taken many years of on-farm experimentation and sharing of the results among growers to develop shiitake into the viable forest farming enterprise that it is today. This approach of on-farm experimentation and communication among growers to determine best management practices could be used as a template for other mushroom crops. Ultimately, if mushroom cultivation and forest farming as a whole are to succeed, there simply need to be more people farming the woods, in more places.

CASE STUDY: STEVE AND JULIE ROCKCASTLE, GREEN HERON GROWERS PANAMA, NEW YORK

Steve and Julie Rockcastle and two big birds make up Green Heron Growers, a highly diversified organic farm located in the rolling hills of southwestern New York near Panama, New York. The oldest heron—the Great Blue Heron Music Festival—has been going on for over twenty years. For three days each year the farm/forest is a melodious mass of seven thousand or so music lovers, most of whom camp out either on the forest side of the property or across the road on the pasture side. The other big bird is Green Heron Growers, Steve and Julie's organic farm that produces an eclectic mix of grass-fed beef, egg-laying chickens, meat chickens, shiitake mushrooms, and veggies. That's not all. These creative entrepreneurs host mushroom and other workshops and a unique event during the summer called Night Lights at the Heron that features "creative lighting installations" and live music in the woods.

Steve and Julie started growing shiitake about five years ago with 816 logs and increased that to about two thousand logs in rotation at the present time. This makes them one of the biggest shiitake farmers in the Northeast. Their venture into shiitake cultivation was inspired by a friend of the family and recent college graduate, Nick Laskovski, who began learning to grow shiitake mushrooms when the technology was young, from his mother, over thirty years ago. Later at Cornell, Nick was involved in forest farming research, teaching an Extension program run by Ken. Today at his farm in Vermont, Nick is a successful shiitake farmer in his own right.

The first year Steve and Julie ventured into mushroom farming Steve began by cutting down enough live red maple trees to provide exactly 816 logs from their own 100-acre woodlot. Their choice of red maple is something they regret to this day because red maple has performed as poorly for them as it has for many other growers in the Northeast. Red oak, which they paid to U-cut at a nearby state forest, has done well, as have sugar maple and beech cut from their own woods.

Of course, one of the major tasks involved in shiitake production is log inoculation, which requires power equipment to drill many holes into each log.

Some growers take the logs to the electricity, where they are inoculated, then transport them to the laying yard, where they spend the rest of their productive life. Not so for the Rockcastles. They take the electricity directly to the laying yard by way of a generator that powers the drills. The site even has plumbing (fresh water), thanks to infrastructure in place for the hundreds of folks who camp in their woods during the Great Blue Heron Music Festival. Inoculating all the logs they need every year is more than enough for two people. So, for four weekends in late April/early May they invite some friends and a few others who are anxious to learn about mushrooms to a party in the woods that includes plenty of drilling, inoculating and waxing, to be sure, but also a hearty meal and plenty of rock and roll (in the woods) powered by the same generator that runs the drills.

Inoculating five hundred logs not only takes happy volunteers but also a well-organized flow of materials. Steve keeps the process well stocked with logs via his front-end loader. In this case the inoculation "assembly line" consists of four stations, the first of which involves drilling about thirty holes in each log × 500 logs = 15,000 holes. It didn't really seem like that many, but then I was only there for one day. It wouldn't be possible with regular 2,500 rpm drills, but two or three high speed (10,000 rpm) angle grinders get the job done. The holes practically drill themselves. On YouTube, you can see Steve using an angle grinder to drill holes in an oak log (http://www.youtube.com/watch?v=kCiBt9foTFY). He's fast, and it's not just the tool. After drilling out each log, he pushes it down a roller track to the next workstation. Mushroom equipment suppliers don't sell them, but they sure make the job easier.

The second stop is the inoculation station, where Julie is waiting. We refer to this entire three-step process (drill, insert, wax) as "inoculation," but it's here at the inoculation station that the fungus meets the wood and the magic begins. In one hand Julie holds a plunger-style inoculator (a sawdust "syringe"), which she pokes twice into a 5-pound bag of spawn, filling the barrel to a depth of about an inch. Then she moves the tip of the inoculator barrel over one of the freshly drilled holes and depresses the plunger by smacking it with a flat rock, depositing a slug of spawn into the hole. Julie and her rock can be seen on YouTube (http://www.youtube.com/watch?v=twUUO7mq eLI&feature=plcp). If you have a chance to see the video you'll see how efficient Julie, her inoculator, and her rock really are at picking up spawn and depositing it in a hole thirty times per log and thousands of times over their eight-day/eight-hundred-log inoculation period. If you go out and purchase one of these plunger-type inoculators, be advised that the rock is not included.

Next the freshly inoculated log trundles along another section of roller track to the waxing station where a 20-pound propane tank and burner are used to melt thick chunks of food-grade cheese wax. The molten wax is just hot enough to smoke a little. The waxer volunteer feeds chunks of wax into the melting pot as necessary and uses a cotton dauber to apply wax to each of the spawn-filled holes in the log. Finally, at Green Heron both ends of the log are painted over with more molten wax using a 3-inch paintbrush.

After all those logs are inoculated, the task of moving them into the laying yard is greatly simplified by the fact that they are already at the laying yard, thanks to that generator. The laying yard at Green Heron, where the logs are stacked for the next four years, satisfies all three of the criteria for a good laying yard: (1) It has a leafy evergreen canopy, in this case of hemlock, that provides dense shade for the entire year; (2) It is on gently sloping land; and (3) It has a reliable source of water, which in this case is from the plumbing described above.

Immediately after inoculation, the logs are crib stacked. The logs sit quietly in the Rockcastles' laying yard for about a year until they are well enough colonized to start converting wood into mushrooms. They know it's time when white patches appear at the ends of the logs. Shocking at Green Heron begins in early July. Steve and Julie have found that shocking earlier in the spring or summer, during cooler weather, is ineffective. When the mushrooms begin to emerge from the log about a week after soaking, the A-frame configuration allows for easy picking, rather than the contortions that would be necessary to harvest mushrooms from a crib stack. At Green Heron,

unlike most other shiitake farms, they leave the logs in the A-frame configuration for the duration of the three- or four-year life of the log. This is possible at Green Heron, where they have plenty of laying yard space, but for a mushroom grower for whom space is more limited, logs can be more efficiently placed in a crib stack configuration, which has a much smaller footprint than the A-frame configuration.

During periods of high production, the laying yard is checked every day, and mushrooms are harvested as they are ready. This process may sound (relatively) simple, but unfortunately there is a fly in the ointment—slugs. It seems that every shiitake grower I know has his own favorite method for slug control, including beer, stacking logs on a gravel bed, or copper wire. Julie's is more direct than most. One time when I helped (mostly watched) her harvest mushrooms from the laying yard, she used a sharp pocket-knife, not only to cut the base of the mushroom stalk from the log, but also to deftly slice each slug in half. During peak season this takes as much as two hours a day. She tells me that sometimes she and Steve have a late-night slug-eradication "date," with headlamps and determination. Steve bragged that he offed over three hundred slugs in one night!

Julie and Steve sell their farm produce at two venues. About a third of total sales are made right out of their home. Their garage serves as the Green Heron Growers' Farm Store. The store is self-serve, and patrons can choose from a variety of items grown on the farm, including fresh-harvested shiitake mushrooms, 100 percent grass-fed beef, certified organic chicken, eggs, and veggies. There is an electronic scale for patrons to weigh out portions and pay accordingly. Steve and Julie are too busy to hang around the store, so sales are on the honor system.

Only about half of their shiitake are sold fresh. Value-added shiitake products make up the other half, including a to-die-for shiitake-hazelnut pâté and shiitake tincture for whatever ails you. According to Julie, it is an immune builder, blood purifier, and cholesterol-lowering agent. They also sell a frozen shiitake barley soup, a shiitake duxelles (a paste made from sautéed mushrooms, onions, and butter), and dried shiitake mushrooms. All of these add to the bottom line, not only by diversifying the product line but also by extending the sales season well beyond the five-month shiitake growing season. Value-added products make use of mushroom "seconds" that aren't perfect enough for fresh sale, including misshapen caps, caps that are too small, and ones with just a wee bit of slug damage.

Value-added products bring as much income as the equivalent amount of fresh mushrooms from which they were made —sometimes more. Dried mushrooms are a good example: A pound of fresh shiitake that sells for $16 dries down to about 2 ounces, which sells for $20. Before counting your profits from the dried mushrooms you have to take into consideration several dollars that were spent on the labor to slice the mushrooms, the mason jar container, and the custom-made label. Another consideration regarding value-added mushroom products is the regulations in New York State that require that value-added (i.e., modified from the fresh) must be prepared in a certi-

fied kitchen, which requires a 20-C food-processing license. Many other states have similar regulations. A certified kitchen is an expensive proposition, but in this case food preparation for sale at the Great Blue Heron Music Festival has already paid for the kitchen.

The remainder of the mushroom crop, along with Green Heron's other fresh and frozen farm products, are sold at a farmers' market near Buffalo, New York, starting in late July. It is safe to say that the Rockcastles have not gotten rich growing forest-cultivated shiitake mushroom (yet), but the mushrooms contribute a substantial share to a broadly diversified farm income and make productive use of their woods that would otherwise sit idle (economically speaking) for all but the three days of festival camping each year.

Steve and Julie's success is based on hard work, innovation, and the satisfaction that they obviously obtain from organic farming.

Table 5.8: Green Heron Growers

	Year 1	Year 2	Year 3	Year 4	Year 5
Logs inoculated	800	500	600	400	400
Logs fruited	0	660	1100	1400	1100
Pounds harvested	0	150	260	662	685
Average pounds/log	0	.23	.24	.47	.62
Income	$0	$987	$3,109	$6,969	$7,069
Expenses	$4,239	$2,954	$1,612	$1,963	$1,989
Labor	$1500	$2,330	$2,420	$3,650	$3,030
Sales	—	$987	$3,109	$6,969	$7,069
Total expenses (w/o labor)	$4,239	$2,954	$1,612	$1,963	$1,989
Net profit (w/o labor)	−$4,239	−$1,967	$1,497	$5,006	$5,080
Profit (labor subtracted)	−$5,739	−$4,297	−$923	$1,356	$2,050

6   Forest Medicinals

Long before there were drug stores and pharmaceuticals for whatever ails you, people of all societies have relied on traditional knowledge of native plants to ameliorate or ward off sickness. Some of these "folk remedies," which we consider medicinal herbs, are plants of the forest that have been collected (wild crafted) for generations, with little regard to their long-term conservation. This chapter attempts to indicate which forest medicinals lend themselves to cultivation on the forest farm, and which lend themselves to "managed wildcrafting."

Defining Medicinal Plants

In chapter 1 the term *productive conservation* was used to describe an important aspect of forest farming. Nowhere is this truer than in the case of the two most commonly and profitably cultivated medicinal nontimber forest products—the medicinal herbs ginseng and goldenseal. Both grow in the wild in eastern North America. Some wild populations of both are considered at risk, and their harvest from the wild is regulated for that reason. The nearly unlimited demand for American ginseng mostly comes from Asian countries, particularly Hong Kong and Korea, and results in high prices for wild and to a lesser extent for cultivated roots. Both species have fascinating life cycles that should be understood and managed if either is to be successfully integrated into forest farming.

There are other wild species of forest herbs, including black cohosh, blue cohosh, bloodroot, false unicorn, and others, shown in table 6.1, that are valued for their medicinal properties. *Medicinal herb* refers to any type

of plant product used traditionally for health-related reasons. These certainly may have a place in forest farming, but presently they have less potential for income generation. These herbs are referred to as "minor medicinal herbs." Most of these species have not undergone the rigorous clinical and safety testing required by the United States Food and Drug Administration[1] to be classified as pharmaceutical drugs. Nonetheless, they are well established in traditional folk medicine, and their annual trade volume (see table 6.1) attests to their popularity, which varies widely among species. While the traditional appeal of these plants is for their perceived medicinal value, some have additional valuable attributes for forest farming, including their production and marketing as ornamentals (see chapter 7).

One thing many forest farmers have in common is the need to generate some income to reward their efforts. The duo of American ginseng and goldenseal are by far the most valuable and potentially most profitable legal crops available to the forest farmer. The list of herbs in table 6.2 does not imply that the cultivating of the particular species as a forest farming crop is likely to be profitable, which depends on a number of variables that will be discussed below. Nonetheless, enterprise budgets for ginseng, goldenseal, and other medicinal herbs are presented below and may be considered as predictors of potential profitability.

Regulation of Medicinal Plants

Many of these plant species have not undergone the rigorous and expensive laboratory and clinical testing necessary to meet the criteria of "safe and effective" as

Figure 6.1. A large American ginseng plant at Dave Carman's forest farm. This is one of the specimens that he has been harvesting seed from for many year. The deer barrier is an essential part to his seed production system.

defined by the US Food and Drug Administration. In 1994 a special category was carved out by the Dietary Supplement Health and Education Act (DSHEA)[2] to allow for minimal regulation for commerce in traditional natural products, including medicinal herbs, that don't qualify as "safe and effective" by FDA standards. According to DSHEA, traditional herbal medicinals may be labeled as "dietary supplements" rather than drugs. All dietary supplements must fall into one of the following categories: minerals, vitamins, amino acids, herbs or other botanicals (excluding tobacco), combinations of the above, or substances historically used by humans to supplement the diet.

According to the National Center for Complementary and Alternative Medicine[3] (NCCAM), a dietary supplement may not claim to cure, mitigate, or treat a disease; otherwise it would be regarded as an unauthorized drug. The law does not exclude all health-related claims for dietary supplements. Marketers of dietary supplements are permitted to make structure/function claims such as *reduction of nutrient deficiency, support-*

ing health, or linkage to a particular body function, although all such claims must be accompanied by the following disclaimer printed on the label:

> This statement has not been evaluated by the US Food and Drug Administration. This product is not intended to diagnose, treat, cure or prevent any disease.

It should be noted that there are many scientists and others in the mainstream medical establishment who consider the rigorous clinical and other testing of pharmaceutical drugs versus the less rigorous, after-the-fact testing of "dietary supplements," including medicinal herbs, to be an unjustified double standard from a health perspective.[4] They often cite the adverse effects of some herbal medicinals that have been demonstrated in clinical trials and other studies. Saw palmetto, ephedra, and bloodroot (when taken internally) are a few examples. Medicinal herbs should be used with caution and with full awareness of adverse effects that may be present.

Conservation Status of Minor Medicinal Herbs

It is often assumed that one of the advantages of forest cultivation is that it should offset some of the demand for wildcrafted herbs, resulting in environmental benefits, but alas, forest cultivation isn't making much of a dent in the demand for medicinal herbs at the present time. Only about 5 percent of the amount of black cohosh purchased is from cultivated sources. One particular conservation-minded organization, United Plant Savers (UPS), is particularly focused on this issue. Its mission is "to protect native medicinal plants of the United States and Canada and their native habitat while ensuring an abundant renewable supply of medicinal plants for generations to come." United Plant Savers recognizes two different categories of wild medicinal plants that they feel are currently in decline and that are most sensitive to the negative impact of human activity. These are "At Risk" and "To Watch." The UPS classification is shown for each of the plants listed in table 6.2 (on page 215).[5]

Undoubtedly UPS's goal of "ensuring an abundant renewable supply of medicinal plants for generations to come" is a worthy one, and forest farmers can help to achieve it. There are three ways the UPS goal can be achieved:

1. Forest cultivation of medicinal herbs to reduce collection pressure on wild populations
2. Education to increase public understanding of the role that cultivated medicinal herbs can contribute to the conservation and sustainability of wild populations
3. Managed wildcrafting involving the use of deliberate cultural practices, such as timber stand improvement and elimination of invasive weeds, to enhance the productivity of wild populations of forest medicinal herbs

American Ginseng

Ginseng is a slow-growing perennial herb, with a valuable fleshy storage root, which is the part harvested for medicine. The English word "ginseng" translated into

Figure 6.2. American ginseng plant in midsummer, with three "prongs" (palmately compound leaves) and an unripe berry cluster.

Figure 6.3. American ginseng plant showing aboveground shoot with compound leaves, berries, and belowground storage root, fine roots, and rhizome with a dormant bud.

Figure 6.4. An exceptionally large, anatomically correct "man root" that was grown in a forest garden for over ten years.

Chinese is something akin to *man root*. The Chinese character pronounced "renshen" translates as either *man* or *ginseng*. An anatomically correct resemblance can be seen in the exceptionally large ginseng "man root" shown in figure 6.4.

Typical claims that can be found on product packaging for ginseng include reduction or tolerance of stress; increased alertness and mental clarity; reduction of fatigue; improved memory; and, last but not least, especially for the guys, enhanced sexual performance. Unlike some of the other claims, there is scientific evidence for ginseng's enhancement of sexual performance, albeit for male rats, not people, as described in a 2009 publication.[6] American ginseng, like many other medicinals, is often used in combination with other types and sources of ginseng (e.g., Asian ginseng, *Panax ginseng*), as well as with other herbs.

While many health-related claims *are* made for American ginseng, the most consistent claim regarding its effect on the human body is that it functions as an "adaptogen," which means that it increases the body's resistance to stress.

The "American" in American ginseng is not just a reference to where it grows in the wild but is also an indicator of a clear distinction between the two different commercially important species in the genus *Panax*, American ginseng (*P. quinquefolius*), which is native to North America, and Asian (Korean) ginseng (*P. ginseng*), which grew wild in China and Korea. Describing its geographical range in the past tense relates to the fact that Asian ginseng has been so extensively harvested from the wild for so long that it is nearly extinct in Asia. This is because Asian ginseng has been so highly valued for its medicinal properties for at least two thousand years. According to tradi-

Figure 6.5. A pile of 'sang roots recently wildcrafted (collected from the wild). Dry weight is approximately 1 to 2 pounds, worth $500 to $1,000 or more. Bob Beyfuss

tional Chinese medicine, American ginseng is said to promote yin energy and have a calming effect. This is in contrast to Asian ginseng, which is said to promote qi and yang energy. Because of considerable demand, ginseng is by far the most valuable of medicinal herbs native to eastern North America.

Today Asian buyers will pay hundreds of dollars per pound (dry weight) for top-quality wild (forest-gathered) American ginseng root and nearly the same amount for forest-farmed ginseng grown by the wild-simulated method. In Asia, American ginseng is considered a companion to the Asian variety, not a replacement. Occasionally a single exceptionally large or old (one hundred years) American ginseng root is sold for more than a thousand dollars. Over 90 percent of ginseng grown or wildcrafted in the United States is exported to Asia via Hong Kong. According to Scott Persons, as of 1994, Wisconsin produced 90 percent of cultivated ginseng exported from the United States. Canada is an even larger exporter than the United States.

According to the 2007 National Health Interview Survey[7] (NHIS), four out of ten American adults used complementary and alternative medicine (CAM), "natural products" in the past twelve months. Of these natural products, 14.1 percent was ginseng in some form. This relative popularity of ginseng with Americans is somewhat ironic, considering that most

over-the-counter or online ginseng preparations, such as the ever-popular ginseng-fortified teas or soft drinks sold in cans or bottles or the capsules sold in health food stores, are not American ginseng (*Panax quinquefolius*) at all but rather cultivated Asian ginseng (*P. ginseng*). In other words, most ginseng sold in the United States is Asian ginseng imported from Korea, while most (> 90 percent) American ginseng is exported to Asia. There is little or no cultivation of Asian ginseng in the United States or Canada, but a great deal of American ginseng is cultivated in China and Korea under artificial shade.

HISTORY OF AMERICAN GINSENG

American ginseng has been highly valued in traditional Chinese and Korean medicine ever since it was exported from North America in the early seventeenth century. Most commercial demand today is from these countries, especially China, and to a lesser extent other Pacific Rim countries. The history of the "discovery" of American ginseng and the origins of its export to Asian countries is rather fascinating. It was used by numerous Native American tribes for a variety of medicinal or health-related purposes. The first European to take note of it in North America was Jesuit missionary Joseph-François Lafitau, near Montreal, Canada, in 1716. From the work of Father Pierre Jartoux published in the *Memoirs of the Royal Academy* in Paris,[8] Father Lafitau was aware of the use and value of Asian ginseng in China. Lafitau had an interest in botanical pursuits, so he began deliberately looking for a similar plant in North America, "discovering" American ginseng. Before long a rapid and highly profitable trade began from North America to China involving the fur magnate John Jacob Astor and other speculators. In fact, it was noted at the time that trade in American ginseng was more lucrative than furs. In 1788 even the legendary frontiersman Daniel Boone was involved in export of ginseng to China. One version of the story has him paddling a canoe full of ginseng down the Ohio River on his way to Philadelphia. The canoe tipped over, and the entire cargo load was lost. According to this version of the story, he went back up the river and collected another canoe full of ginseng. This

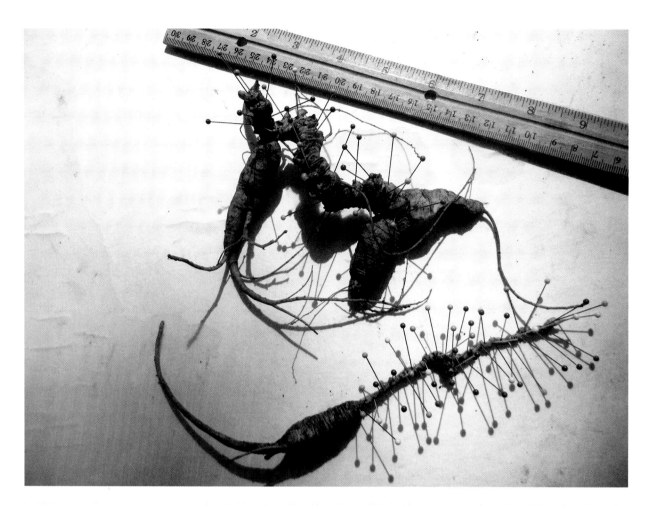

Figure 6.6. Ginseng rhizome (neck) with two roots attached. The age of a ginseng plant can be estimated by counting the bud scale scars on the "neck" (rhizome). The age of this twenty-five-year-old plant is determined by counting annual bud scars. The terminal bud just beyond 25 on the ruler is the growing point for next year's aboveground shoot. Bob Beyfuss

time his trip down the river was successful, and he made a great deal of money. In other versions of the story, he transported twelve "tuns" (small barrels) of ginseng down river in a barge or pole boat. The most historically accurate version of the story, based on letters from a number of sources, including his son, has it that Boone and companions were transporting a considerable load of ginseng by packhorse. As they were fording the Monongahela River the horses were spooked by the scream of a panther. The horses reared and bolted, and the cargo was dumped into the river. They collected about half of it, took it ashore, and dried it on a bed of coals, then continued on their way to Philadelphia.

Natural History of American Ginseng

The ginseng plant has an unusual growth habit and life cycle that contribute to the relative difficulty and long cycling of cultivating it as a crop. More typical plants such as echinacea, black cohosh, or tomato have a root system, stem(s) with nodes, internodes, and leaves. And at the junction of each leaf with the stem is a bud that may grow into a branch. This is the normal pattern of development for most plants, but not ginseng. A mature ginseng plant consists of a single stalk that looks like a stem but isn't. Actually, it's a sympodium, which is essentially several leaf stalks that fused to perform the role of stem. From this come true leaves, which bear three to seven leaflets but most commonly five, which is the basis for its botanical species name, *quinquefolius*.

All this growth dies back to the ground in the fall, and the plant overwinters as a narrow underground rhizome (belowground shoot) about the thickness of an earthworm with a bud at one end and one or more thick tuberous roots at the other end. These structures are dormant until the following spring, when the bud elongates and a new sympodium emerges from the ground. The leaves unfurl as the sympodium continues to elongate to its full height of a foot or less. Several weeks later the small white flowers are fully developed and ready for insect pollination, after which they develop into green berries each with one seed. By late summer or early fall they ripen into bright red berries that are easily recognizable in the forest understory. Also by late summer the belowground rhizome develops a single bud, which will remain dormant until it emerges next spring.

What distinguishes ginseng from many other forest herbs is its determinate growth habit. In other words, once it completes its full development from bud to full-size plant by late spring, there is absolutely no additional increase in height or number of leaves and no secondary branching. The only growth during summer and early fall is subsurface—root enlargement—which increases its size, weight, and economic value. The plant's slow growth rate is an adaption to its low-light environment, which unfortunately for forest farmers accounts for the fact that forest-cultivated ginseng takes eight to ten years to reach harvestable size.

The storage (tuberous) root, of course, is what all the excitement is about. It looks something like a branched carrot. The more it looks like a person (arms and legs and even a male organ on more valuable specimens), the more valuable it is to traditional Asian buyers (figure 6.4). The root (and other parts of the plant) contains the pharmacologically active compounds known as ginsenosides, although other compounds in the root may be pharmacologically active as well.

Wild American ginseng has very specific site requirements, including shade, aspect, soil, slope,

and associated vegetation. To be able to locate wild ginseng for wildcrafting or to find a suitable site for wild-simulated cultivation requires an understanding of those ecological preferences. One can make several generalizations about the ideal conditions where ginseng is likely to be found, including:

- Approximately 70 percent shade
- North- to northeast-facing slope (aspect)
- Well-drained soil that is high in organic matter and calcium
- The presence of certain indicator tree and herb species, such as sugar maple and maidenhair fern

This list represents a good start, but any particular site is not likely to conform to all of these recommendations. Luckily, there is a systematic way to assess the likelihood of finding wild ginseng. It is called the Visual Site Assessment (VSA) tool. This is covered in more detail at the end of this chapter.

WILDCRAFTING OF AMERICAN GINSENG

Before getting into cultivation, it is worth understanding the constraints associated with the wildcrafting of American ginseng, as its status in the wild directly affects the demand for cultivated ginseng. As previously mentioned, American ginseng is a slow-growing, shade-loving, perennial forest herb that produces a valuable belowground storage root. It occurs typically in hardwood forests over most of eastern North America, where some wild populations are considered to be in decline over much of its range. According to the conservation organization NatureServe (NatureServe Explorer, http://explorer.nature.serve.org), out of the thirty-two states and two Canadian provinces where it is found, wild American ginseng is ranked *Critically Imperiled* in six states, *Imperiled* in five states plus the two Canadian provinces, *Vulnerable* in fourteen states, and *Apparently Secure* in seven states.

Scientific research by forest ecologists and other experts, as well as anecdotal reports by ginseng hunters and conservation groups, differ with respect to the extent of its decline.[9] Internationally, it is one of many plant species considered at risk for extinction. As such

Figure 6.7. Distribution of wild American ginseng. Illustration courtesy of Halava

it is listed in the Convention on International Trade in Endangered Species of Wild Flora and Fauna (CITES), Appendix II ("may become threatened unless trade is closely controlled").[10] Since the United States is a signatory to this treaty, the US Fish and Wildlife Service[11] (USFWS) is charged with verifying that export of ginseng will not be detrimental to the survival of the species. Responsibility for monitoring its export was delegated by the USFWS to individual state environmental conservation departments. To this end, each state establishes regulations regarding harvest of wild plants that are intended to ensure that continued harvest will not be detrimental to its survival in the wild. These regulations vary to a small degree from state to state, but all address four key issues:

1. Establishment of a legal collecting season
2. A minimum harvestable plant age or developmental stage
3. A requirement that ripe seed must be planted in the vicinity of the harvested plant
4. If harvest from state-owned land is permitted or not

Since ginseng cultivated by the wild-simulated method is essentially the same as wildcrafted ginseng as far as sale value is concerned, the forest farmer should understand the CITES-motivated regulations that affect its sale. Wild and wildcrafted ginseng roots can be sold only to a state-certified dealer, who must report the number and weight of roots purchased back to the state regulatory agency, which then passes this information on to the USFWS. Based on reporting from all exporting states, USFWS issues a "finding" that certifies whether or not export will be detrimental to survival of the species. Since the CITES treaty went into effect (1975), the annual USFWS finding has always indicated that export has not had a negative effect on populations.

This is not to say that all individual wild populations are stable but only that continuation of wild harvest is permissible for one more year. The 2012 finding[12] indicated that 14,683,604 plants were harvested from the wild, yielding 62,831 dry pounds of ginseng root. If we assume that the price per dry pound of wild American ginseng is (very conservatively) $400, then total income to wildcrafters was at least $25,132,400. It's no wonder Scott Persons gave his earlier book (1994) the title *American Ginseng: Green Gold*.[13]

Cultivation of American Ginseng

In terms of income potential and popularity with most beginners, ginseng is the rock star of forest farming. It gets all the attention because its price to the grower is vastly greater than any other forest commodity because of an insatiable demand for it in China, fueled by a lore that is as fantastic as it is improbable. Whereas the market demand for goldenseal, which is mainly centered in European countries, is high, the demand for American ginseng, which is almost entirely from Asian countries, is well beyond that for goldenseal and can only be characterized as extremely high. Hence, forest farming (as an alternative to wildcrafting) of these herbs is rewarded not only by an attractive income opportunity but also by the satisfaction of contributing to the conservation of these increasingly scarce species in the wild. This is another example of the recurrent theme of forest farming as productive conservation.

There are three different methods for cultivating American ginseng, all varying in intensity of cultivation; that is, how much time and other expenses are required to grow a crop. The following sections are listed in order of least to most intensive.

Wild-Simulated Production

The wild-simulated method refers to minimalist cultivation under a natural forest canopy involving little more than scattering seed and waiting eight to ten years for harvest, with little if any maintenance in the meantime. We regard this as the preferred method for forest farming. More information about wild-simulated ginseng can be found later in this chapter.

Woods-Cultivated Production

The woods-cultivated production system is somewhat more intensive than the wild simulated. It requires six to eight years and involves growing in raised beds beneath a natural forest canopy. Management is more intensive than wild simulated, involving site clearing, tilling, soil amendment with gypsum and organic matter, mulching, fungicide application, and more.

Figure 6.8. Bruce Phetteplace in his woods-cultivated ginseng garden. His plants are exceptionally large (knee high).

Artificial Shade Production

The artificial shade method involves constructing a costly artificial shade structure rather than growing the crop beneath a natural forest canopy. High-density planting and rapid growth produce a crop in three to four years with much higher yield of roots per acre than forest-based production but requires weeding and regular fungicide application because of the high planting density. Despite this intensive chemical control of fungal diseases (mostly *Alternaria* and *Phytophthora*), pathogens build up in the soil to the extent that ginseng cannot be replanted on the same land once the first crop is harvested.

Comparing Natural Shade vs. Artificial Methods

The trade-off between artificial shade cultivation and forest cultivation (either woods cultivated or wild simulated) is that, while artificial shade production is much more capital and labor intensive than forest cultivated, it produces a crop in three to four years, while forest-cultivated ginseng requires eight years or more before harvest. On the other hand, the price per pound is inversely related to production intensity; that is, price for one pound (dry): wild simulated > wood cultivated > artificial shade. In other words, you get the most per pound for wild-simulated ginseng. The price for artificial-shade ginseng is $30 per pound or less, whereas wild-simulated ginseng is usually worth ten times as much.

Regardless of whatever advantages or drawbacks are involved in artificial-shade production of ginseng, it is really not compatible with forest farming. It is a treeless monocropping approach to production and is antithetical to forest farming not only because there are no trees involved but also because, unlike real forest farming, it requires elaborate artificial-shade structures and other expensive infrastructure, fungicides, fertilizers, and higher overall resource inputs. Woods-cultivated and wild-simulated production of ginseng, however, are both legitimate forest farming practices. From the standpoint of the permaculture approach to forest farming, wild-simulated ginseng most completely combines production with conservation.

Woods-Cultivated vs. Wild-Simulated—
A Contrast in Sustainability

Wild-simulated and woods-cultivated production of American ginseng is a contrast in sustainability and self-renewal. Woods-cultivated ginseng is usually grown in raised beds at a fairly high density. Seeds are planted, and as the seedlings grow they may require applications of fungicides to prevent fungal diseases and herbicides or hand cultivation to suppress weeds. A bed of woods-cultivated ginseng is essentially an even-aged stand, which is harvested all at the same time. Intensively cultivated ginseng like this can only be planted once in the same place because of the buildup of pathogens.

Wild-simulated ginseng, on the other hand, is modeled after a natural forest ecosystem. It takes a few years longer to begin harvesting roots, but harvest may be continual for years. This is because it mimics the way ginseng grows in the wild. This is a good example of what permaculturists mean by designing agricultural systems modeled from natural ecosystems. Seeds are sown on the forest floor and left to

germinate and grow on their own for eight or more years. Alternatively, with the wild-simulated method, plants are more widely spaced, and three to five years after seeding the more precocious seedlings will have matured to the point at which they are reproductive and are producing seeds of their own. These will germinate where they fall to the ground, beginning a second generation of wild-simulated ginseng on that same site. This cycle will happen yearly for several more years as the size of the population increases in numbers and size/age. Eventually, after eight to ten years the largest plants from the original seeding (first generation) will be harvested. Each year, additional plants (generation 2, etc.) will be large enough to harvest, and their numbers will be replenished as younger plants take their place.

Theoretically, this cycle can go on indefinitely as long as the forest farmer (or poacher or white-tailed deer) does not harvest too much at one time. Hence, a wild-simulated population, like a multiaged stand of trees, is much more sustainable both biologically and economically than a woods-cultivated system,

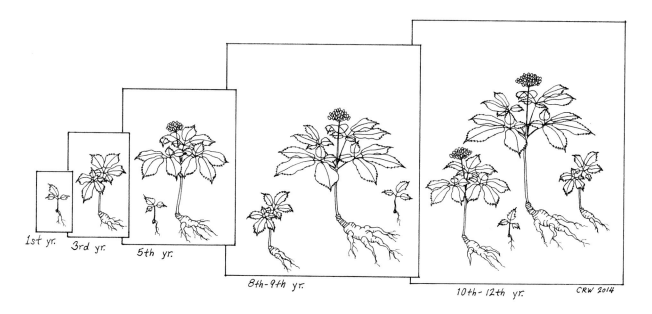

Figure 6.9. Graphic illustrating the indefinitely sustainable nature of wild-simulated production of American ginseng. As plants grow from the original seeding, they begin to produce seeds when three to four years old, which continues yearly until they are harvested at ten years old. Meanwhile younger generations give rise to the next generation until they too are harvested. This scenario does not, however, factor in the depredations of poachers or deer. Illustration by Carl Whittaker

which must be replanted year after year. In New York, a wild-simulated population of several acres was planted more than ten years ago at the Arnot Forest. Despite the fact that this population has been selectively harvested every year, the population size continues to increase annually. This wild-simulated planting was initiated by Bob Beyfuss, now a retired natural resource educator from Cornell Cooperative Extension of Greene County, New York. He has authored numerous ginseng-related publications, has advised many beginners in ginseng cultivation, and is an avid "'sang hunter" (wildcrafter).

SITE SELECTION

Site selection for cultivating ginseng, especially by the wild-simulated method, begins with locating an area that is as close as possible to ginseng's natural habitat, regardless of whether there is wild ginseng on the site or not. A very useful tool for site assessment for growing ginseng has been developed by Bob Beyfuss, who was instrumental in establishing the Agroforestry Resource Center in Greene County.[14] Beyfuss developed the Visual Site Assessment (VSA) based on years of observing wild populations and noting the ecological characteristics of those sites. From this fact it follows that the VSA is also a good predictor of where wild ginseng is likely to be found (see the sidebar at the end of this chapter).

The assessment is based on six categories that have a direct bearing on the performance of ginseng, including (1) dominant tree species, (2) exposure, (3) slope, (4) soil characteristics, (5) understory vegetation, and (6) security. Evaluation of a particular site involves assigning points to each of these categories, elaborated on the form. A final score is an indication of how well suited a given site is for growing wild-simulated ginseng. Development of the Northern VSA was based on his observation and understanding of the ecology and distribution of wild American ginseng in the Catskill Mountains of New York and farther south throughout the Appalachian Mountains, while the VSA for the Midwest is a straightforward modification of the Northern version to reflect different forest types found in these two regions.

Another useful tool for evaluating the suitability of a particular forest site for forest cultivation of American ginseng was described in a Virginia Cooperative Extension bulletin,[15] in which the authors list seven different categories that should be considered for planting:

- Operating conditions (Briars? Poison ivy? Steep slopes?)
- Signs of wildlife (Heavily used game trails? Groundhog holes? Lack of understory vegetation?)
- Site quality (Forest Soil Quality Index [FSQI] and Site Index [SI], soil type, moisture regime)
- Stage of stand development (initiation, stem exclusion, understory reinitiation)
- Dominant overstory species
- Presence of ground vegetation: if not, why is it not growing here?
- Presence of indicator species: Jack-in-the-pulpit; trillium; bloodroot; Solomon's seal; lady's slipper; mayapple; baneberry; spicebush; jewelweed; galax; ferns; wild yam; black cohosh; wild ginger; pea vine; Indian turnips; ginseng; goldenseal

Using these tools should give a forest farmer the indication of whether or not a given site is appropriate for ginseng cultivation. Unlike some of the other crops in this book, it is not recommended that ginseng be grown in less-than-ideal circumstances, given the long growing time from planting to harvest. Woods-cultivated plantings at the MacDaniels Nut Grove, which is a poor site for ginseng by all accounts, have been only moderately successful over the last ten years. Students transplant rhizomes from the wild-simulated Arnot Research Forest patch mentioned above, but less than half actually make it alive through their first year. Nonetheless, we consider this acceptable because the goal of the nut grove is to demonstrate, not sell, these crops.

PLANTING AND MANAGEMENT

After determining an appropriate site for ginseng cultivation, the following steps are taken to establish and manage ginseng plantings:

Wild Simulated (from Persons and Davis)

Once you have used the VSA or otherwise decided where to grow ginseng here is some advice from Scott Persons about how to proceed.

1. *Site Preparation:* Remove litter, branches, loose stones, saplings, and so forth. Rake off leaves and duff carefully so you can reapply as a mulch layer after seeding.
2. *Seeding:* Scatter seeds to achieve a final spacing of one to two plants per square foot, by sowing four to five seeds per square foot (50 percent attrition).
3. *Mulching:* Apply a layer of leaves so that all bare ground is covered. If the mulch is likely to blow away, apply sticks and branches to hold the leaves in place.
4. Inspect your planting occasionally for diseased plants, and remove them if found.
5. Harvest the original generation of seedling in eight to ten years, being careful not to damage younger seedlings as they develop from the first generation. After the first harvest you should be able to harvest more or less yearly, more or less indefinitely.

Woods Cultivated (Raised Bed)

This method is adopted from recommendations from Bob Beyfuss.[16]

1. Choose a loamy soil at pH6 to 6.5, or amend with lime if necessary. If you do a soil test (local Cooperative Extension) adjust phosphorous to the recommended level with bonemeal and potassium with greensand as recommended. If the soil test indicates that calcium is less than 2,000 pounds per acre, add gypsum (crushed wallboard) at 50 pounds per 1,000 sq feet.
2. Construct a three-sided wooden frame with black locust lumber or logs, leaving the uphill side open to runoff from upslope. The frame should be 4 to 5 feet wide by at least 10 feet long. Rebar can be used to hold logs in place if necessary. The frame can be recessed in a shallow pit, especially if you need soil to fill the frame later.

Figure 6.10. Bob Beyfuss ("Mr. Ginseng"), an avid 'sang hunter and ginseng educator, holding an exceptionally large ginseng root (stem and leaves still attached) that he has just dug from somewhere in the Catskill Mountains of New York. He'll never tell you where. Photograph courtesy of Robert Beyfuss

3. Fill the frame with 3 parts loamy soil: 1 part compost, using the soil from the pit. We have gone as high as 1:1 soil to compost. Allow the bed to settle and absorb rainfall for a week or so. Final soil depth should be at least 10 inches.
4. Sow seed in the fall. It is best to use prestratified seed; scatter seed at a rate of at least 10 per square foot of bed or more (many seed lots have low germination). Press the seeds gently into the soil. Beds can be planted with one- to two-year-old rootlets (seedlings) at 6-inch centers. Mulch bed with 4 inches of leaves, preferably sugar maple.
5. Weed and irrigate occasionally, as needed.
6. Thin to one plant per square foot, two or three years after planting.

7. Wait six to eight years before harvesting, or plant two-year-old rootlets (seedlings) to reduce time till harvest accordingly.

8. Do not replant into the same bed for at least several years after harvesting the first crop.

As far as pests go, white-tailed deer are certainly a source of risk, but deer pressure varies from one location to another. An enclosure or fence may be necessary, though dogs may be effective, too (see chapter 9). Voles or mice, especially in winter, are another potential pest. They can burrow underground and eat ginseng roots. This factor leads some growers to prefer woods-cultivated ginseng, as they can build a raised bed and line it with hardware cloth to deter digging rodents. The only other pest a grower is likely to encounter are slugs, which are as difficult to control for ginseng as they are for any other crop. Multiple strategies should be used, including beer traps, Sluggo, iron phosphate, and other treatments, as well as regular monitoring and removal of slugs.

There is only one other significant pest, and that is the human poacher, who, depending on the location, can be a serious problem. Poaching of ginseng is more prevalent in the South, where harvesting of wild ginseng is more of a tradition. When it comes to poachers, the closer a patch of ginseng is to home the better. One grower in North Carolina has been known to carry a rifle when he suspects that poachers are nearby. The rifle is empty, of course, but they don't know that. Occasionally poachers are arrested (by law enforcement) or even shot at (by growers). Best to keep the location of ginseng patches mum, and consider the potential need to protect them in an overall planning scheme.

ENTERPRISE BUDGET FOR AMERICAN GINSENG

An enterprise budget is a planning tool that a prospective farmer (of any commodity) can use to predict if a single new enterprise is likely to succeed economically. It takes into consideration inputs (expenses and labor) and expected income based on available markets.

As for income from selling ginseng, as previously mentioned, each of the four ginseng methods has its

advantages and disadvantages in terms of time, effort, and potential profitability. Several nationally known herbal products companies obtain raw product by buying forest medicinal herbs from wildcrafters and forest growers. One company publishes a seasonal pricelist for close to one hundred different species and lists the price per pound they will pay to a grower or wildcrafter and will usually offer the same price for both.[17]

Ginseng is treated a little differently from the others. For ginseng the pricelist includes two different categories: (1) "Cultivated" (artificial shade) and (2) "Wild Simulated" or "Woods Grown" (both are forest farmed) (table 6.1). But unlike all the other herbs added for comparison, which have a single specified price, ginseng's price for either source is listed as "market"—a testament to how many different factors influence its value. And just like the price of lobster in a restaurant, the supply is too volatile for the buyer to commit to a guaranteed price point. Pricing of ginseng is a function of supply and demand, which is to some extent influenced by state regulations and mandated by the CITES treaty (as described above). The price is also influenced by state or region and the buyer's assessment of quality. The price for wild or wild-simulated ginseng at any given time and any location is not fixed or controlled by any agency, central buyer, or other collective. Instead it is negotiated by one seller and one buyer at a time.

In general the buyer sets the price for any given transaction based on several criteria, including:

- Shape (branching)
- Color
- Concentric rings (that indicate slow growth)
- Age (as estimated by the number of stem scars on the rhizome (neck)

For instance, in February 2013 in upstate New York, Bruce Phetteplace, the ginseng dealer featured in the case study at the end of this chapter, quoted a price range of $15 to $25 per dry pound for field-grown (artificial shade) ginseng, $300 to $600 for wild-simulated ginseng, and around $600-*plus* for wildcrafted ginseng.

Table 6.1: Comparison of Medicinal Crops and Methods

	Ginseng, Wildcrafted	Ginseng, Wild Simulated	Ginseng, Woods Cultivated	Bloodroot, Woods Cultivated	Black Cohosh, Woods Cultivated	Goldenseal, Woods Cultivated
		Persons (2005)	Persons (2005)	Davis (2005)	Davis (2005)	Davis (2005)
Acreage / Duration	Variable +/– annually	1 ac	1 ac	1.0 ac, 3 yr	1.0 ac, 3 yr	1.0 ac, 4 yr
Labor, hr		1,150	3,900	780	780	3,200
Labor @ $12/hr (2013)		$13,800	$46,800	$9,360	$9,360	$38,400
Planting stock		$2,000				
(seed)	$3,150 (seed)	$3,000	$650	$15,000		
Other expenses		$3,100	$6,150	$3,015	$3,250	$14,500
Yield (lbs)		160	600	1,500	2,200	1,600
Price $/lb (2005)		$350	$200/lb	$9.00/lb	$4.75/lb	
Gross income @ $xx/lb (2004)		56,000	$120,000	$13,500	$10,450	$80,000
Total expense		$17,380	$56,100	$15,375	$13,260	$67,900
Net profit		$38,620	$63,900	–$1,875	–$2,810	$12,100

Note: Data was normalized to 1.0 acre (from ½ ac for ginseng and from 0.1 ac for goldenseal), and hourly wage was normalized from $8/hour to $12.00/hour.[19]

The *plus* refers to exceptionally large or old roots, which may sell for $800 to $1,200 or more, to some extent regardless of size.

Thus, the figures shown are likely somewhat different, as they are based on a different part of the country (North Carolina), a different year, and the judgment of a different buyer. What is consistent in general is that the order of increasing value (price) is field grown followed by woods cultivated, followed by wild simulated and wild.

Minor Medicinal Herbs: Cohosh, Bloodroot, Fairy Wand, Goldenseal, and More

There is no doubt that ginseng is the best-known NTFP of importance in forest farming. No other NTFP rivals its value to a grower who is skilled and patient enough to grow it from seed (or seedling) to harvest. Although not as profitable, goldenseal and other forest plants can also be forest farmed successfully. A "minor"

medicinal is defined here as any forest herb, other than ginseng, that is collected or cultivated primarily for health-related rather than nutritional or ornamental purposes. The main ones of interest are listed in table 6.2 and include mayapple, black cohosh, bethroot (trillium), fairy wand, blue cohosh, goldenseal, and bloodroot. We are inclined to classify goldenseal as a minor medicinal herb because its value is less than that of ginseng.

Black cohosh is a good example of a "minor" forest medicinal herb that is used and marketed for multiple purposes. Today it is commonly used for hot flashes and other menopausal symptoms, while historically it was used for similar maladies and other broader concerns as well. For example, the Delaware Indians used it as a tonic for menstrual irregularities and as an aid to childbirth. The Cherokee Indians also used it for women's reproductive health as well as for relief of pain after childbirth and even to treat depression. Some tribes of the Iroquois Confederacy used it for soaking sore feet and for rheumatism. From 1820

Table 6.2: Uses, Conservation Status, Trade Volume, and Price Paid to Collector or Cultivator

Common Name	Scientific Name	Uses (both traditional/folk and modern clinical)[21]	United Plant Savers Conservation Classification[20]	Trade Volume[21] (kg/yr/dry)	Price (USD) per pound paid to collector or cultivator[22] (unless otherwise indicated)
American Ginseng	*Panax quinquefolius*	Adaptogen, tonic	At Risk	132,500– 350,200 kg	Market (price of ginseng depends on how it is cultivated) (Persons and Davis, 2007)
Black Cohosh	*Cimicifuga racemosa* (*Actea racemosa*)	Menopause and postmenopausal systems	At Risk	259,600– 1,675,100 kg	$3.00
Bloodroot	*Sanguinaria canadensis*	Antibiotic, antiplaque, animal feed	At Risk	58,300– 107,100 kg	$12.00
Fairy Wand, (False Unicorn)	*Chamaelirium luteum*	Diuretic, uterine tonic	At Risk	$700,000 (2001)	$35–50 (Burkhart and Davis, 2007)
Goldenseal	*Hydrastis canadensis*	Antibiotic, haemostatic, stomach ache; laxative, mucous membrane tonic	At Risk	94,300– 583,600 kg	$18
Trillium (Bethroot)	*Trillium* spp.	To treat excessive discharge of the bowels	At Risk		$1
Spikenard	*Aralia racemosa*	Diabetes, TB, wounds, burns	To Watch		
Blue Cohosh	*Caulophyllum thalictroides*	Promotion of menstruation; uterine stimulant	At Risk	10,000– 18,000 kg	$1
Mayapple	*Podophyllum peltatum*	Chemotherapeutic	To Watch		$3.50

until 1926 black cohosh was listed in the *United States Pharmacopoeia*.[23]

North American demand for black cohosh has rebounded since the 1930s: More recent annual trade in black cohosh—over the period from 1997 to 2005—was 259,600 to 1,675,100 kilograms dry weight (Foster, 2014). Although these trade statistics imply a significant demand for these minor forest medicinals, they do not suggest what portion of that demand is supplied by cultivation as opposed to wildcrafting. The fact is that most of the market for these herbs, with the exception of goldenseal, is satisfied by wildcrafting, not by cultivation. With the exception of ginseng, the price paid to the seller (forest farmer or wildcrafter) is the same for either cultivated or wild-collected material, while production costs are much higher for the forest cultivator.

Thousands of pounds of black cohosh, harvested from the wild in North America, are sold to herbal products–related buyers for export to Europe. Accurate estimates of the portion of the trade volume figures that are from forest cultivation are not known with any certainty.

This collection of plants should be seen mostly from a hobbyist perspective, as there is little money to be made from them in current markets. Some of the plants are profitable when wildcrafted, but as they become scarcer in the wild and/or demand for them

Figure 6.11. Mayapple (*Podophyllum peltatum*). Photograph courtesy of Virginia State Park Staff

increases, they may become more common choices for cultivation in a forest farm. This section discusses some of the possible directions for forest farming of medicinal plants.

Not surprisingly, the limited attempts to cultivate medicinal herbs have been undertaken mostly for economic reasons and only secondarily to reduce collection pressure on vulnerable populations. So far income generation through forest cultivation of minor medicinal herbs has been only marginally successful at best and for that and other reasons cultivation has had minimal effect on demand for wild-collected material. Two of the most detailed, well-documented attempts to estimate the potential profitability of forest cultivation of minor medicinal herbs are by Jeanine Davis and her colleagues in North Carolina[24] and another economic analysis by Eric Burkhart in Pennsylvania (Burkhart and Jacobson, 2009). Both Davis and

Burkhart independently collected information from growers and other sources on production expenses, income, and profit for cultivation of several different medicinal herbs. Davis developed enterprise budgets (gross income − expenses = profit) for artificial-shade production (not forest farming) of several medicinal herbs but only one enterprise analysis for goldenseal that was based on forest cultivation. Burkhart, on the other hand, constructed sophisticated economic models for forest cultivation of eight different herbs, using sensitivity analysis that took into consideration the discounted value of money (Net Present Value) over the four-year production cycle.

Despite their differences in methodology Davis and Burkhart came to broadly similar conclusions; namely, that forest cultivation of medicinal herbs, with the exception of ginseng, was not sufficiently profitable to justify the effort and expense. Specifically,

Figure 6.12. Black cohosh (*Actaea racemosa*).

Davis found that cultivation of goldenseal was only marginally profitable ($2,490, over four years with 1,520 hours of labor, ¹⁄₁₀ acre). Very few farmers would be willing to put in so much time and effort for a mere $2,490. The sensitivity analysis approach of Burkhart's, on the other hand, found that forest cultivation of goldenseal was nowhere near the break-even point (that is, what the herb buyer would have to pay the farmer to cover the farmer's expenses was much higher than what the buyer would actually pay). In the case of goldenseal, Burkhart found that only by making the most generous (unrealistic) assumptions (early harvest, no cost for planting stock, no annual costs) was goldenseal even close to being profitable. All the other medicinal herbs, with the exception of ginseng, were even more unprofitable than goldenseal. It should be noted that the Davis study was initially published in 2005 and $8 an hour was used

to calculate the cost of labor, in contrast to Burkhart's study, published in 2009, using a labor rate of $13 per hour. Although many would consider $13 per hour a fair wage for farm labor, closer to $8 an hour is still not uncommon in some areas.

What is most surprising from these two contrasting studies is that despite the high demand from the herbal products industry for most minor medicinals, forest cultivation of these is largely unprofitable. Labor is a significantly greater expense in forest cultivation compared to wildcrafting. This is not to dismiss the amount of time and effort (labor) involved in foraging around in the woods for wild medicinal herbs, but it is usually far less than the labor involved in cultivation of the same species. Other significant production-related expenses for cultivation are discussed below. Burkhart and Jacobson indicated that the fraction cultivated was insignificant for all of the herbs listed, except goldenseal

Figure 6.13. Bethroot or red trillium (*Trillium erectum*). Dmott9, Flickr

Figure 6.14. Fairy wand (*Chamaelirium luteum*). Photograph courtesy of Keri Leaman

and ginseng. According to Greenfield and Davis,[24] the fraction that is cultivated ranges from nearly 0 percent for bloodroot to 20 percent for false unicorn, 15 percent for skullcap, 25 percent for goldenseal, 5 percent for wild indigo, and nearly 0 percent for wild yam.

Before considering several approaches to increase the profitability of forest cultivation of minor medicinals, it is worth asking the question, *Is there any compelling reason for forest cultivation at all?* Asking the question does not necessarily imply that the answer is no (or yes). Let's return to this question after considering several options for enhancing profitability.

ENHANCING PROFITABILITY OF MINOR MEDICINALS

In the case of many medicinal herbs mentioned in this chapter or not, wild crafting or deliberate cultivation is marginally profitable at best. We believe that measures can be taken to improve the prospects for income generation by responsible wildcrafting or forest cultivation.

Improved Cultural Practices

Traditional agricultural practices such as fertilization, irrigation, weed and pest control are less appropriate for farming in the woods because these practices tend to be too resource intensive. The most effective way to optimize production of medicinal herbs in the forest is by matching the requirements of the plant species to the characteristics of the site. Such actors as shade, soil pH, drainage, and organic matter, as well as exposure and rainfall, should all be assessed for a particular site first, before selecting which forest medicinal and other nontimber crops will be grown on the site. More about site assessment will be discussed in chapter 10.

Figure 6.15. Blue cohosh (*Caulophyllum thalictroides*).

Market Medicinal Plants as Garden Perennials

American gardeners spend a considerable amount of money on ornamental perennial species of plants for shade gardening. This includes such garden favorites as hostas, astilbe, marsh marigold, ferns, and orchids. Many of the plants that have been included in this chapter on medicinals are better known to most people as the wildflowers they delightfully encountered in the woods on a spring or summer outing, and they are appreciated more for their appearance or fragrance than for their medicinal value. Some of these perennial wildflowers that are also medicinals could be marketed as landscape perennials for shade gardening. Several forest medicinal herbs are quite striking in appearance, whether one is aware of their medicinal properties or not. For example, black cohosh (figure 6.12), false unicorn or fairy wand (figure 6.14), and blue cohosh (figure 6.15) are attractive ornamentals.

At one Tennessee nursery, the price for online sales of black cohosh, bloodroot, blue cohosh, and goldenseal is $39.99 for quantities of ten. These are sold as roots or rhizomes, rather than as full-grown plants, so costs for preshipment storage and transport to the buyer are minimal. According to Davis, nondormant potted plants of these species sold in nursery containers ranged in price from $3.95 to $10.00 per plant.

Public Education about Forest Cultivation of Medicinal Herbs

The average consumer of medicinal herbs in the form of teas, capsules, tinctures, and so on is probably unaware that most of the products on the market are not cultivated. Instead they are harvested from the wild. Even if the consumer is aware of the plants' wild origin, he or she is probably unaware of the potentially negative ecological consequences of wild harvest of some medicinal herbs; for example, black cohosh and goldenseal. What this book emphasizes is that a permaculture approach to forest farming involves mimicking the natural forest ecosystem, which includes a healthy balance of herbs, shrubs, trees, and vines. Some forest farms strive to accomplish this goal, but even more consider income generation to be at least as important, if not more so. For those forest farmers willing to strike a balance between the two, incorporating eco/agrotourism into their forest farming enterprise may make a lot of sense.

Public education in a forest setting is an opportunity to show visitors the intimate relationship between the natural forest ecosystem and the forest denizens (NTFPs) that have potential economic value. The MacDaniels Nut Grove at Cornell University is dedicated to public education about forest farming and its relationship to nature. Medicinal herbs are illustrative of the understory component of a multistrata healthy natural forest or forest farm. In this setting visitors can come to understand the tension that exists between preservation of the natural ecosystem and the extractive nature of wildcrafting. They can appreciate the triple nexus of forest cultivation, managed wildcrafting, and unregulated wildcrafting. With understanding comes the recognition that preservation has a price, and that's why the price paid for black cohosh rhizome that has been cultivated should be a bit more than that for wildcrafted root.

Unfettered wildcrafting diminishes not only the size of a forest population but also potentially the health of that population and in turn that of the entire forest ecosystem. In a recent encouraging development Davis reported that one major retail herbal products company is selling cultivated organic black cohosh for almost twice what they charge for wildcrafted root, suggesting that the industry is beginning to reward cultivation over wildcrafting.[25]

Direct Marketing of Forest-Cultivated Medicinal Herbs

In both Davis's enterprise budgets and Burkhart's financial models the anticipated wholesale price to a cultivator for one pound of black cohosh ranged from $3.00 to $4.70 and $9.00 to $12.86 per pound for bloodroot. On the other hand, a grower could sell these products at retail more profitably. Unlike retail sales of a perishable commodity such as shiitake mushrooms, where retail sales take place mostly at farmers' markets and other face-to-face venues, retail sales of nonperishable (that is, dried) medicinal herbs can and do take place online at eBay and other commercial websites. There are plenty of Internet-savvy consumers who need no convincing that natural herbal medicinals are healthy alternatives to some mainstream pharmaceuticals. It's the high price of off-the-shelf herbal preparations at a local health food store that puts off many likely buyers. If these potential buyers can connect with a local grower or wildcrafter at a farmers' market or other face-to-face venue, they can purchase unprocessed fresh or dried herbs at more reasonable prices. Internet sales by growers of dried medicinal herbs or value-added extracts are another way to increase profitability. Both bloodroot and black cohosh, for example, are sold on eBay and other Internet retailers. Dried but otherwise unprocessed black cohosh sells online for $30 to $46 per pound and bloodroot for $103, which is far more profitable to the grower than selling to an herbal products wholesaler.

At this rate direct retail sales of either of these herbs by a forest cultivator or wildcrafter to the public over the Internet could be about ten times more profitable than sale to a "middle man" (wholesale) herb buyer. Keep in mind that direct sale to the consumer of unprocessed or value-added herbs generally involves more work on the grower's part than in a bulk sale to a wholesaler, if it's only the time spent at a farmers' market or other face-to-face venue, or the time it takes to package and label small quantities for retail sale. Depending on how processed they are, they may or may not qualify as value-added herbal products. These would include extracts, herb mixtures, and ground product in gel capsules. All must be appropriately labeled as dietary supplements. The Federal Trade Commission publishes an advertising guide for industry,[32] which addresses the kinds of health-related claims that can and cannot be made by marketers.

Value-Added Products

"Value added" in the context of medicinals refers to transforming a plant from its original state as it came out of the ground to a more valuable product. This could include chopping or grinding the dried product, encapsulating it, or preparing an alcohol tincture.

Chopped black cohosh has a retail value of $51.67 a pound, compared to the wholesale value of $1 for 1 pound of dried bulk root sold to a national herbal products company. Goldenseal is another example of how value added can potentially increase profitability. According to Burkhart, a grower could sell 1 kilogram (dry weight) of forest-cultivated root for $59.34 to an established herbal products company, which would grind it, put it into 570 milligram gel capsules, and sell a hundred capsules retail for $300. That is value added, to the tune of a fivefold increase in value.

Planting Stock

Planting stock is grown and sold so that other people can plant it and grow it to maturity. It is often obtained from a nursery, although in some cases planting stock may be obtained from a wildcrafter. It comes in many shapes and sizes, depending on the crop. Planting stock may be either seeds or transplants. Transplants can be seedlings started elsewhere or asexual propagules, including rhizomes, root pieces, bulbs, rooted cuttings, or rooted layers.

For the forest farmer, producing a finished crop for wholesale to the herbal products industry or retail to the end user, purchase of planting stock is a significant expense, but selling it as an NTFP can be profitable. Alternatively, producing transplants for sale to others may be a profitable "end product" for the forest farmer/nursery that specializes in producing and selling planting stock to others who wish to grow it. For those who intend to grow the mature crop for wholesale or retail, planting stock is usually a big part of an overall production budget. Table 6.2, which is based in part on figures provided by Burkhart and Jacobsen (2009), shows that starting a crop from seed is far less expensive than using purchased transplants, but the trade-off is that starting from the less expen-

sive seed requires at least two years longer to produce a harvestable crop.

Seed is by far easier and less expensive to deal with, since sowing seed on a small scale does not usually entail much in the way of specialized equipment, and it takes relatively little time (that is, labor). Prior to planting, seeds of many species can be stored for long periods of time and have a longer shelf life than roots, rhizomes, cuttings, and so forth. In the end the choice of seed or asexual propagule is a trade-off between cost (seed is usually cheaper) and time to harvest (seed is usually slower).

Especially for the beginning forest farmer impatient to generate some income, the reduced time to harvest by planting seedlings or other transplants ("rootlets" in the case of ginseng) rather than seeds may be worth the additional cost (see table 6.2). On the other hand, a beginner may need to keep expenses to a minimum, at least the first time around, so the less expensive seed propagation may be the way to go in such cases.

Normally planting stock is a major expense for most forest growers (see table 6.1 and table 6.2) of herbal medicinals, but in some cases this expense can be avoided or reduced by eliminating the need to purchase planting stock. This can be done by collecting planting stock from the wild (wildcrafting). If the forest farmer has legal access to natural populations of the herb and intends to harvest it sustainably, he or she could collect planting stock for "free" by wildcrafting. For example, in the case of black cohosh a rhizome collected from the wild could be divided into several pieces, each containing a growing point or bud, and grown on to produce two or three new plants from which mature rhizomes could be harvested either for sale or for planting the next crop—or both.

Keeping in mind the emphasis on sustainability, wildcrafting may involve only partial harvest of a given wild population, with sufficient numbers left in the ground to assure rapid recovery of the population.[26] Other aspects of sustainable wildcrafting are discussed below. Of course, this is sometimes easier said than done if the population is visited by more than one collector. Wildcrafting of planting stock for deliberate cultivation may seem exploitive, but it need

Table 6.3. Comparison of Seed vs. Transplant Planting Stock (seedlings, roots, rhizomes, or rooted cuttings)

	Seed		Transplant	
	Cost per ¹⁄₁₀ ac	Years to harvest	Cost per ¹⁄₁₀ ac	Years to harvest
Black cohosh	$170	6	$2500	4
Goldenseal	$2000	6	$10,000	4
Blue cohosh	$1000	8	$2500	6
Fairy wand	$550	8	$10,000	6
Ginseng	167	8	$5000	6

Source: Persons and Davis (2007); Burkhart and Jacobson (2009)

not be repeated annually if the grower is to carefully save either enough seed or enough asexual propagules from the first crop to plant the next or a later crop. The expression "don't eat your seed corn" applies here. Ultimately wildcrafting of planting stock qualifies as "productive conservation" if it is used to establish forest cultivation to meet market demand, thereby ultimately reducing collection pressure on wild populations. One might look at this as a "loan" from the forest, paid back in reduced collection pressure on wild populations.

Hobby Cultivation of Forest Medicinal Herbs

The emphasis in this chapter has been on production of forest medicinal herbs for income. The point is made that growing medicinal herbs other than ginseng and to a lesser extent goldenseal is unlikely to be profitable if they are marketed as medicinals (dried roots, rhizomes, etc.). We have presented several alternative approaches to cultivating these herbs for profit, including nursery production of planting stock rather than finished herbs, or growing them as ornamentals.

Having said that, we understand that there are other nonincome-related reasons for cultivating forest plants. The plants and mushrooms covered in this book can and should be grown joyfully for personal satisfaction, environmental sustainability, personal use, and self-sufficiency. Indeed, we encourage forest farmers regardless of their level of experience and the scale at which they wish to farm or garden in their forest to introduce as much diversity as possible, whether they intend to sell any of it or not. Forest medicinals described here can be grown at any scale and used in various ways other than forest farming as we define it in this book, including shade perennial gardening or forest gardening. If you are starting from scratch, as is often the case with forest gardening, it will be necessary to wait until your initial tree and shrub plantings have grown to the point that they are casting enough shade for shade-loving forest medicinals to thrive.

The section on growing American ginseng in this chapter describes both the wild-simulated method and the woods-cultivated method. The former is well suited to minimal input, long-term sustainable production but is typically implemented at the scale of ¹⁄₁₀ to ½ acre or more. The raised-bed woods-cultivated method is somewhat more intensive, but it can be scaled down to 50 square feet, as described there, or even less.

Table 6.3 is a good place to begin your quest for cultivating medicinal herbs, whether your goal is income, personal satisfaction, or growing the herb for personal medicinal use. Either way, the first consideration is propagation, which is always the place to begin, unless you want to pay someone else to do it for you. For some species it is a matter of choosing to propagate from seed or from vegetative propagules (roots, rhizomes, rooted cuttings) and transplants, which can be of either seed or vegetative origin.

Table 6.3 will help you make this decision. It compares the cost of planting stock for seed versus

Figure 6.16. Goldenseal (*Hydrastis canadensis*).

transplants, and in every case transplants are more expensive, but they have the advantage of shortening the time from planting to harvest. Table 6.4 provides some information about growing conditions necessary for cultivating the herbs. These tables are a good place to start but are by no means a complete grower's guide. For that we strongly recommend the book *Growing & Marketing Ginseng, Goldenseal and Other Woodland Medicinals* by Scott Persons and Jeanine Davis, who have been mentioned in various places in this chapter. The book is full of field-tested recommendations by seasoned professionals. Aside from that, the book is an easy and enjoyable read, well suited for bedtime reading. In the propagation section of chapter 7 the various propagation methods are discussed, and table 7.4 presents pros and cons of each propagation method, including the plants best suited for each method.

Of course face-to-face advice from people in your area will give you location-specific information that can be invaluable, but be aware that no two experts will give you the exactly the same advice. If both people have succeeded in growing the plant you are interested in, neither is necessarily wrong. In the end you will find your own way. It may be difficult to find an experienced forest grower of medicinal herbs in your area, so you might consider asking around local nurseries that may be growing these plants for sale as ornamentals for the landscape trade. The advice they can give you can usually be adapted to smaller-scale personal use.

MANAGED WILDCRAFTING

Wildcrafting is the major source of forest medicinals sold to the herbal products industry. This is and will be true as long as the species of interest are relatively abundant in the wild. Even though the price paid to the wildcrafter for many herbs (excluding ginseng) is quite low, wildcrafting may be more profitable than forest cultivation for the forest farmer because expenses and time are not as much for the wildcrafter. All that is required for the wildcrafter is a sturdy back, a compass and map or sense of direction (don't get lost), and a way to dry the product. On the other hand, the "hidden" long-term cost of wildcrafting, at least in some areas, is overharvesting and increased scarcity in the wild. Some examples of this are goldenseal, as pointed out by Albrecht and McCarthy[28] and Cech,[27] black cohosh by Cech and by Davis (2002), and false unicorn by Persons and Davis (2007).

Many wildcrafters understand and others also should understand the need for sustainable wildcrafting practices, which are being promoted through education and in some cases by individual states (most have departments that deal with natural resource conservation, by various names), as well as by national (US Fish and Wildlife Service) or international regulations (CITES). Of course, wild populations of certain medicinal herbs are at risk not only due to overharvesting (wildcrafting) but also deer pressure and habitat loss through human encroachment.

As introduced above, United Plant Savers includes on its "At Risk" list black cohosh, bloodroot, false unicorn, goldenseal, and trillium (table 6.2). Despite concerns about wildcrafting from individual local populations or regional vulnerability of the species, on an industry-wide basis the supply of wildcrafted herbs satisfies most market demand, but as wild populations diminish, this is beginning to change. As long as the market demand for herbal medicinals is strong and the amount contributed by forest cultivators is small, wildcrafting will persist despite environmental consequences. For example, only 4 percent of black cohosh is cultivated.

One factor that contributes to the predominance of wildcrafting to meet commercial demand is that, with the exception of American ginseng, there is no difference in price for cultivated and wildcrafted medicinal herbs. Wildcrafting involves less investment of time and capital than cultivation, so the preference for wild-

crafting should come as no surprise. One approach to this "problem" is to promote public education to convince people that wild populations of medicinal forest herbs are worth preserving, and to convince them that for this reason they should be willing to pay more for sustainably cultivated herbs compared to wildcrafting. Don't hold your breath.

As long as cultivation of medicinal herbs remains, at best, marginally profitable (see enterprise budgets summarized in table 6.1) and wild populations are not too badly depleted, wildcrafting will persist. Realistically, the question becomes not, "should wildcrafting be replaced by forest farming?" but rather, "are wildcrafting and forest farming compatible, and mutually sustainable?"

Some authors and practitioners have drawn a rather clear line between the two, because the management associated with forest farming is fundamentally different from the functioning of a natural forest ecosystem. On the other hand, we think forest farming and wildcrafting can be better understood as a continuum where management is not necessarily confined to forest farming but can to some extent be a component of sustainable wildcrafting practices. Rather than hope for a dramatic reduction in the cost of forest cultivation or an increase in the price to the grower of cultivated material, it is perhaps more realistic to promote "managed wildcrafting."

Some of the management activities that may have a positive impact on the sustainability of wild populations include timber stand management, canopy thinning, weeding, management of soil health, and partial harvest of a given population. These will be discussed below. Undoubtedly these and other managed wildcrafting strategies are being practiced to a limited extent as we speak, but demonstration research to quantify the impact of canopy thinning, partial harvest, and so on could help to convince traditional forest farmers on the one hand and traditional wildcrafters on the other to adopt this intermediate approach.

Replanting of Wildcrafted Plants
Earlier in this chapter, the legally mandated, conservation-minded requirement of directly replanting ginseng seed in the immediate vicinity of a wild-harvested

ginseng plant was described. The planting of seeds with the intention of harvesting then replanting some of the seeds is an agricultural management practice that goes back to the beginning of agriculture. In the case of ginseng it is the practice that is mandated in all US states that export ginseng. Some 'sang hunters (ginseng wildcrafters) go so far as to carry "exotic" seeds into the collection site, where they are sown in hopes of "replacing" the wild ginseng that was just harvested. This has been criticized by some as a risk of genetic pollution.[29] Some wildcrafters and other experts are highly skeptical about the success of this approach, mainly because there is no way to enforce it.

Management of Light

Overstory light management by pruning overhanging branches can be used to increase the light penetration to an underlying NTFP population. Despite their tolerance of low-light forest conditions, most forest medicinal herbs will respond positively to increased light intensity. Increased light intensity, in this case, usually means *less* shade but not full sun. When Naud et al. thinned the forest canopy they found that the increased light levels stimulated the growth of black cohosh, wild ginger, and bloodroot but not blue cohosh.[30] When thinning is applied on a larger scale to conventional forest management, it is often called timber stand improvement (TSI). Trees deemed less valuable for timber are removed to favor the growth of the remaining trees, due in large part to increased lighting. The increased lighting resulting from TSI, or more localized thinning of branches or removal of individual trees, would also increase the productivity of wild populations of some NTFP, such as black cohosh, bloodroot, or wild ginger. Of course, TSI may have other benefits for forest farming, such as providing pole wood for shiitake mushroom cultivation or mulch for soil improvement (see chapter 10).

Selective Removal of Invasive Weeds

Weeding is another management practice that may improve the growth of small populations of most medicinal plants. Weeding is a common agricultural practice intended to reduce competition for water, nutrients, and light between a crop plant and adjacent noncrop vegetation. This applies equally well to NTFPs in the forest. Oftentimes a plant of interest may be overcrowded to the point that individual plants aren't getting enough light. Thinning of an overly dense stand of target plants (medicinals) to reduce competition is a practice to consider. Weeds that are in competition with forest medicinals include not only herbaceous plants such as grasses and the invasive garlic mustard but also invasive woody shrubs such as multiflora rose, privet, buckthorn, and honeysuckle.

Replanting

When wildcrafting merely involves harvest and removal of an entire plant (leaf, root, shoot), it is extractive, regardless of its abundance. Of course when the population is superabundant, or appears to be, extraction may have or appear to have little impact on the apparent health of a wild population. The reason for equivocating multiple times in the previous sentence is that abundance isn't always what it seems.

For example, in 1787 the intrepid frontiersman Daniel Boone collected and shipped wild-dug ginseng from Kentucky down the Monongahela River to Philadelphia. He wasn't the only one who took advantage of the considerable demand for this herb in China. In a single year 750,000 pounds was shipped to China. By the late 1800s, the plant had been overharvested in North America by greedy wildcrafters and was no longer abundant. In response to this shortage, ginseng cultivation in North America did not begin until about 1880. A similar situation exists today with ramps, which appear to be superabundant in some places, but research has shown that less than 10 percent of a given population can be harvested over a several-year period if wildcrafting of this NTFP is to remain sustainable (see chapter 4). Wildcrafting of ramps, or forest medicinal herbs, can be done sustainably or at least contribute to sustainability, by dividing a plant that is harvested for its root or rhizome (cohosh, etc.). The rhizome of most forest plants is a rootlike structure that is really an enlarged underground stem, and as such, it has at least one bud, like ginseng, or multiple buds, like goldenseal.

A multibud rhizome of many species can be cut into several pieces, each containing a bud, and in the

Figure 6.17. A rhizome is a horizontal underground stem with many buds (although the buds are not readily apparent on these goldenseal rhizomes). Roots are emerging from the rhizome. Photograph courtesy of Catherine Bukowski

interests of sustainability one or more of these can be planted back to the site it was collected from. According to Richo Cech (1998), who is a widely known herbalist, cultivator, and all-around expert on medicinal herbs, "in sustainable wildcrafting, every act of taking is coordinated with an act of planting." Another strategy that he recommends for plants that are harvested for their leaves, such as ramps (some collectors), is to harvest only the tops, so that new leafy shoots can regrow from the roots or rhizome. For some species harvesting only larger rhizomes is preferable, so that smaller ones remain to regenerate (Albrecht, 2006). The rate of regeneration of plants that have been wildcrafted from a given area may be affected by when they are harvested. Goldenseal populations will regenerate faster when harvested in the fall, compared to a midsummer harvest. For this species when the rhizome is harvested pieces of the rhizome and fibrous roots break off and remain in the soil where they can regenerate new plants vegetatively. This process is more likely to succeed when plants are fall harvested, probably because the rhizome and roots have accumulated more storage reserves (starch) by the end of the growing season than earlier.

Another consideration in the timing of harvest is that it should not occur until after seeds are ripened. The legal harvest season for most states is timed so seeds will have ripened by the time the season begins. Immature seeds of any plant are less viable (lower percentage germination) than mature seeds. In addition to when to harvest, another consideration is where to harvest. Ideally a wildcrafter should be harvesting from more than one population (patch), but only one during any given year, then rotate among all the populations over as many years.

This is more difficult to implement successfully if more than one collector is harvesting from any one patch. Of course, not all approaches to managed wildcrafting involve light, timing, and location. Managing soil health is an overlooked and often impractical approach to optimizing the performance and sustainability of wild medicinal herbs. Fertilizing agricultural crops with chemical fertilizers or organically is a common practice, but fertilizing the forest is not often practiced. Nonetheless, soil health can be promoted by conserving rather than exporting organic material produced on-site, such as leaves and topsoil. Vegetation cover tends to stabilize soil and reduce erosion, especially on slopes, so it follows that removal of too much plant cover, which could include the wildcrafter's "target" crop, as well as weeds and other competing vegetation, could lead to erosion that ultimately degrades the site.

All of these considerations, from pruning to increased light to timing and location, are aspects of managed wildcrafting that can contribute to sustainable harvesting of wild populations. The increased scarcity that can result from unsustainable wildcrafting, which may boost the price paid to either the wildcrafter or the cultivator (forest farmer), is often cited as motivation for forest farming, but from an environmental or ethical standpoint, overharvesting from the wild is never justified. Ultimately the stabilization of wild populations through managed wildcrafting is a more reasonable alternative and a positive contribution in relationship to forest farming.

Future Prospects for Medicinal NTFPs

The medicinal herbs classified as NTFPs that are appropriate for forest farming are generally shade-requiring

Table 6.4. Propagation and Growing Conditions of Forest Medicinals

Crop	Seed	Vegetative	Preferred Growing Conditions
Ginseng (Panax quinquefolius)	Fall sowing* Double dormancy Purchase of prestratified seed preferred Purchase 1–2-yr-old seedlings	Cannot be done	Cool, moist soil, well drained N–NE-facing slope Do not replant in the same location as previous crop 70% shade
Goldenseal (Hydrastis canadensis)	Double dormancy Sow immediately after collecting	*Rhizome cuttings Root cuttings, fall	Rich, moist but well-drained, loamy soils Do not replant in same location[3] Zones 4–8
Mayapple (Podophyllum peltatum)	Seed: no information	Rhizome cuttings*	Cool, moist organic soils Zones 3–9
Fairy wand (Chamaelirium luteum)	*Sow in fall Seed heads shatter	Rhizome cuttings	Rich soil Zones 5–8
Blue cohosh (Caulophyllum thalictroides)	Double dormancy 2–3 years to germinate	*Rhizome pieces	Moist soil Zone 3, to Georgia
Blood root (Sanguinaria canadensis)	Don't allow seeds to dry after collection Double dormancy Seed heads shatter	*Rhizome cuttings[4]	Moist, loamy soil Zone 3–8
Black Cohosh (Actaea racemosa)	Sow immediately after collection in fall	*Rhizome cuttings, fall	Moist, high-organic soil Light shade Zones 3–8
Bethroot (Trillium erectum)	Double dormancy Sow in late fall	*Rhizome cuttings, fall	Cool moist soil Zones 4–9

Note: These species are regarded as medicinal, but they may be grown as ornamentals or both.
Asterisk (*) indicates preferred (most likely to succeed) method of propagation.

or shade-tolerant, slow-growing perennials. American ginseng is by far the more valuable because of strong Asian demand. It can be grown at several levels of intensity, from artificial shade (the most intensive) to woods cultivated to wild simulated to wildcrafted (the least intensive). Artificial-shade production is not relevant to forest farming. Wild-simulated production is the most suitable for forest production because of its low input and relatively high price.

Minor medicinal herbs are far less valuable than ginseng. These include goldenseal, black cohosh, blue cohosh, bethroot, fairy wand, and Virginia snakeroot. The principal buyers of the minor medicinal herbs are retail herbal products companies that acquire most of their product from wildcrafters. Because they pay so little to the wildcrafter or forest farmer, there is little motivation to cultivate the minor medicinal herbs. Alternatives to forest cultivation for sale to herbal products companies include (1) growing and selling them as planting stock for other growers; (2) growing them as nursery crops for sale as ornamental shade perennials; (3) direct marketing; (4) managed wildcrafting that includes such things as managing the light environment in conjunction with timber stand improvement, clearing invasive weeds, sustainable harvest, and replanting.

CASE STUDY: BRUCE PHETTEPLACE AND THE HOUSE THAT GINSENG BUILT
CENTRAL NEW YORK

Bruce, figure 6.7, is an unusual and highly successful ginseng grower who lives in central New York State. About ten years ago he was a cooperator on one of our ginseng research projects, but I hadn't seen him since then. As I pulled into his driveway, I was surprised to see that an entirely new house was sitting right where the old one used to be. Bruce took me inside, and as he showed me around I was "floored" to say the least.

Let's take a look around the house where he and his wife live, because this is the house that ginseng built. Speaking of being floored, it was made of boards from six different hardwood tree species in an array of colors and textures. The walls complemented the floors, as did the stairway leading to a balcony overlooking the living room two stories below. Bruce explained that all the wood in his almost entirely wooden house was cut and milled by him from his own 150-acre forest. Bruce has worked for many years as a builder of other people's homes. This experience is reflected in the utmost skill and loving care with which he, and he alone, built his home.

There were a number of eye-catching finishing touches all over the house, like the doorknobs and cabinet knobs that were made from highly polished tree burls. In the living room stood a 15-foot-tall sculpture made of a single tree trunk with four stout limbs projecting out from the top of the trunk, each terminating at about 3 feet. The whole 15 feet of trunk and branches was polished to a mirror finish. What made it art rather than just a polished tree is that it was turned upside down so it rested on the four leg-like limbs. The effect was like some long-necked, four-legged alien from H. G. Wells's *War of the Worlds*. It took a double take for me to realize it was in fact an upside-down tree.

Bruce was born in 1952 on the same property where he lives now. Over the years he has been and is still making his living in many different ways. He started out as a dairy and vegetable farmer, and along the way he owned an antique business, and he was a plumber, then a logger who sold timber and firewood. During summers he built houses, and every fall and winter he buys wild fur pelts from trappers and wild-collected ginseng from 'sang hunters (wild ginseng collectors).

As a buyer of wild-collected ginseng he must be registered as a dealer by the New York State Department of Environmental Conservation. As such he is a key part of the mandated supply chain that passes from the collector to the dealer to the buyers who export nearly all of the North American wild-collected ginseng to China. Similar environmental regulations of all 32 ginseng-exporting states are intended to ensure that wild ginseng is harvested sustainably. For most of his adult life he has not only been a dealer of wild ginseng but also a 'sang hunter himself and a farmer of forest-cultivated ginseng.

Bruce started growing ginseng in 1971, and he's been doing it ever since. Back then there was very little information available about cultivating ginseng, so he learned mostly by trial and error. He told me that for every successful grower like himself, there are many who drop out after a few years and that the main reason for their failure is poor site selection. I asked him what the characteristics are of a good site for growing ginseng. Bruce's answer to my question was, "I know a good site when I see it," although he did allow that the two most important contributing factors were a canopy of sugar maple trees and the presence of two particular understory herbs, spikenard (*Nardostachys jatamansi*) and rattlesnake fern (*Botrychium virginianum*).

By now, Bruce has been growing ginseng for forty years, and until recently some of his roots have been in the ground for that long. That is one of the reasons that I referred to him as unique among other forest farmers, who typically harvest roots after eight to ten years. Bruce has planted ginseng year after year since 1971, but he only began harvesting roots in earnest three years ago. It is possible to find wild ginseng populations that are forty years old or older, but this is almost unheard of in the realm of forest-cultivated ginseng. He considers it his retirement nest egg.

His ginseng "patch" is 7 acres of forest that is predominantly sugar maple. The entire area is fenced in to exclude white-tailed deer that are so prevalent in central New York. Although he has been harvesting roots grown by the wild-simulated method for only the last three years, for a lot longer he has been harvesting two other kinds of ginseng crops that he grows using the woods-cultivated method. I refer to a crop of seed and a crop of rootlets. These are two different ways that other folks who want to expand or start their own ginseng patches can proceed. Bruce also sells "value added" (more expensive) seeds that are prestratified (moist/chilled) for 1½ years before sale. Stratified seed germinates the following spring after fall planting, rather than the two years it takes for freshly harvested seed to germinate. He also sells two- to three-year-old seedlings, known as "rootlets," to folks who can afford to get a two-year jump on harvesting their ginseng. That would be six or seven years (minimum) to harvest rather than the eight to nine years from seed.

During his most extensive production year, Bruce was growing ninety thousand plants, of which twenty-five thousand were older than three years, and many of these were literally decades old. Some of the older plants were intended for production of seeds. Two- to three-year-old seedling rootlets were sold for planting stock, and others he kept for longer-term production of large roots. The younger seedlings intended for sale as rootlets and the older plants used for seed production are grown by the more intensive woods-cultivated method, whereas older plants intended for production of larger roots are grown using the wild-simulated method. Both are described in the section on American ginseng in this chapter.

As far as cultural factors are concerned, Bruce is in agreement with most experts, who have found that fertilizing does more harm than good. By trial and error Bruce learned that commercial inorganic fertilizer ended up killing the plants, although not directly. Ironically, the fertilizer accelerated plant growth, which is the whole point of fertilization, but many forest farmers think that fertilizer, nitrogen in particular, makes ginseng more susceptible to disease, especially *Alternaria* leaf blight and *Phytophthora* leaf blight and crown rot, which are the two most common fungal diseases of this crop. During wet years Bruce does apply fungicide to control these diseases. When fungicides are warranted he rotates among four different kinds so the fungi will not develop resistance to any one of the fungicides.

Like most other experienced growers, Bruce doesn't fertilize ginseng with the usual N-P-K complete fertilizer, which tends to predispose the plant to the disease. The one nutrient that is often applied to ginseng is calcium in the form of gypsum (CaSO4 or crushed wallboard). Bruce has tried fertilizing with gypsum but found it unnecessary because his soil is relatively high in calcium—because the trees in his woods are mostly sugar maple, the leaves of which are naturally high in calcium. Sugar maple to a forest farmer like Bruce is a multipurpose tree, providing calcium for ginseng, maple sap for sugaring, and some of the lumber for his new house.

The unique advantage gained by leaving ginseng in the ground to grow for decades, compared to the more common practice of harvesting wild-simulated ginseng after eight to ten years, is of course bigger roots with more "character" (primarily shape and gnarliness) and considerably more value. When you think of it, there are very few other perennial root crops that are sold at such an advanced age. The thousands of these valuable old plants that Bruce has grown are a great retirement plan, but the plan is not without risk and other consequences. For one thing, as with any long-term investment, it may pay off in the long run, but in Bruce's case it generated no income for nearly forty years.

For Bruce, like many other smart investors, that wasn't a hardship because of the annual income he received from the sale of seeds and rootlets. At about $200 per pound for seed and about $2.00 each for rootlets, he netted about $15,000 each year. As mentioned above, as a registered ginseng dealer, he derives another source of annual income from ginseng by buying and selling wild-harvest roots.

In addition, Bruce has a fur-trading business. Animal traps of all shapes and sizes hang on the walls of the shop where he transacts business. He explained to me that as a registered ginseng dealer he must certify that the seller is in compliance with state laws regarding minimum harvest age (five years in New York State), date of harvest, and quantity sold. After purchasing fresh roots he dries and sorts them into as many as ten grades. When it comes to paying the seller, there is no set price. It's based not only on weight but also on Bruce's judgment as to quality. Quality, he tells me, is based on the surface texture, the root's shape, and the amount of fiber roots. He shows me one very special (and very valuable) root. It is truly a "man" root in shape and worth far more than its weight alone would dictate. It's preserved in a glass jar of alcohol (vodka), and I can easily make out its "arms," "legs," and a "pecker." The latter is what makes it an especially fine specimen of man root. Many experts say that there is no difference in appearance between wild and wild-simulated roots, but Bruce shows me the horizontal striations that distinguish a truly wild root (with striations) from a wild-simulated root.

An obvious reason that Bruce and others grow ginseng is income generation, but as for most responsible growers, conservation of the species in the wild has been an important consideration. Nonetheless, his views on wild harvesting are not based on the assumption that wild populations are necessarily threatened by overharvesting and deer predation. He tells me that "the deer are saving wild ginseng." He should know, because he is both an avid observer of ginseng as well as an avid deer hunter. He says there are two reasons the deer are saving the ginseng.

First, when a deer eats a ginseng plant it only consumes the aboveground portion (stem, leaves, and flowers/fruits) but not the underground root and neck (rhizome) or the dormant bud that will emerge next spring and grow into a new stem, leaves, and flowers. Ginseng is not a preferred food source for the deer anyway, so in any given season the deer don't eat all the visible ginseng plants in a particular wild population. A plant that lost its top to a hungry deer one year is likely to grow anew next year from its dormant winter bud. In the meantime, this fall there are no red berries to attract a 'sang hunter, so the plant escapes harvest, at least until it regrows next year.

Bruce has observed another reason deer contribute to the survival of wild ginseng. When a deer eats a ginseng plant, it usually crops it off near ground level and walks off, munching away, with the top of the plant hanging partially out of its mouth, often with the red berries attached. Inevitably some of the ber-

ries fall to the ground, eventually establishing new plants some distance from the source.

Lest you think that deer and poachers are the only predators of wild or cultivated ginseng, it's time to tell you about the inextricable link between another ginseng-loving mammal and Bruce's marvelous new house. About three years ago hungry mice forced Bruce to cash in early on his decades-old ginseng gold mine, giving him a considerable and largely unexpected increase in his income. Although he had intended to leave the roots in the ground for a few more years, the mice changed everything. Before that mice were a manageable problem that he controlled with the only rodenticide labeled for use on ginseng. But about three years ago it was pulled from the market by the Environmental Protection Agency (EPA). At that point the mice became an unstoppable horde. Over the next three years the mice destroyed over half of his crop. So he made the difficult decision to harvest what was left. It could have been worse, a lot worse.

He dug and sold a very respectable $150,000 worth of ginseng root but lost even more than that to the mice. Here's where the house comes in. He used a great deal of the profit to build the new house. What kind of house can $150,000 build these days? Not much. But in Bruce's case materials cost him very little and labor almost nothing. He gathered most of the materials from his own 100-acre forested property, which he milled into lumber, and applied his skills as a carpenter and contractor to build the house of his dreams. He didn't even need to buy doorknobs because he made his own by collecting and polishing burls from his own trees. Bruce is one of the most well-rounded, hardworking, and deeply committed outdoorsmen that I have ever known.

P.S. Bruce recently told me that he planted a substantial amount of new ginseng last fall. I think he wants to start all over again.

— Ken

Visual Site Assessment and Grading Criteria for Potential Woodland Ginseng Growing Operation for a Northern Forest Prepared by Bob Beyfuss, Cornell Cooperative Extension Ginseng Specialist Circle only one choice for each category.	
Category A: Dominant tree species (50% or more of mature trees)	*Points*
1. Sugar maple (add 5 points if average circumference is greater than 60 inches; add 2 points if there is a presence of butternut); in southern NY tulip poplar is equivalent in value to sugar maple as an indicator tree species	10
2. White ash, basswood, or black walnut (add 4 points if average circumference is greater than 60 inches; add 2 points if there is a presence of butternut)	8
3. Mixed hardwoods consisting of beech, black cherry, red maple, white ash, red oak, basswood, and some sugar maple	5
4. Mixed hardwoods as above, plus some yellow birch, hemlock, and/or white or red pine	5
5. Red and/or white oak	3
6. Ironwood, white birch, aspen	1
7. All softwoods: pine, hemlock, spruce, fir	0
Subtotal	
Category B: Exposure (orientation)	
1. North, east, or northeast facing	5
2. South, southeast, northwest	2
3. West, southwest	0
Subtotal	
Category C: Slope	
1. 10% to 25% slope	5
2. Level	3
3. 25% or greater slope	0
Subtotal	
Category D: Soil and site surface characteristics	
1. Site dominated mostly by very large trees more than 20 inches in diameter or few surface rocks; 75% of site plantable	10
2. Site dominated by medium-size trees, 10 to 20 inches in diameter, or some surface rocks; 50% plantable	8
3. Site dominated by small trees, less than 10 inches in diameter, very stony; 25 to 50% plantable	5
4. No large trees, saplings and shrubs dominate or large rock outcropping, many boulders; less than 25% tillable	3
5. Soil too rocky to plant anywhere or poorly drained, standing water present	0
Subtotal	
Category E: Understory plants (select highest scoring one only)	
1. Reproducing population of wild ginseng	15
2. Sparse wild ginseng	10
3. Maidenhair fern or rattlesnake fern or red or white baneberry	8

4. Christmas fern or blue cohosh or red berried elderberry or foamflower or stinging nettles	6
5. Jack-in-the-pulpit, other ferns, trillium, bloodroot (bloodroot is a much higher scoring indicator plant south of NY), jewelweed, mayapple, herb Robert (a type of wild geranium), true or false Solomon's seal	5
6. Wild sarsaparilla, Virginia creeper, groundnut, yellow lady's slipper, hepatica	3
7. Club moss, princess pine, bunchberry, garlic mustard, pink lady's slipper	0
8. Woody shrubs such as honeysuckle, mountain laurel, witch hazel, barberry, maple leaf viburnum, arrowwood, shrubby dogwoods, alder, lowbush or highbush blueberry, spicebush (spicebush is often found with wild ginseng in southern or midwestern sites and is considered a good indicator plant there)	0
Subtotal	
Category F: Security	
1. Very close to full-time residence of potential grower, with planting site within easy viewing of residence (if noisy, outside dogs housed nearby, add 5 points)	10
2. Forested land less than 440 yards (¼ mile) from grower's residence, patrolled regularly	8
3. Regularly patrolled woodlot within 1 mile of residence	3
4. Nonresident grower or remote woodlot	0
Subtotal	
Total score (add points from each category's subtotal):	
Results:	
50 points or above: Excellent site, great potential 40 to 50 points: Good site, do complete soil analysis 30 to 40 points: Fair site, test soil Less than 30 points: Poor site, look elsewhere	

Visual Site Assessment and Grading Criteria for Potential Woodland Ginseng Growing Operation for a Mid-Atlantic Forest Circle only one choice for each category.	
Category A: Dominant tree species (50% or more of mature trees)	Points
1. Sugar maple or black walnut (add 5 points if average circumference is greater than 60 inches)	10
2. Yellow poplar and white ash (add 5 points if average circumference is greater than 60 inches)	10
3. Mixed hardwoods consisting of beech, black cherry, red maple, white and/or red oak, ironwood, basswood, and yellow poplar	7
4. Mixed hardwoods as above, plus some hemlock and/or white pine	5
5. Red and/or white oak	3
6. Ironwood, birch, hickory	1
7. All softwoods: pine, hemlock, spruce, fir, or willow	0
Category B: Exposure (orientation)	
1. North, east, or northeast facing	5
2. Southeast or northwest	2
3. West, southwest, south	0
Category C: Slope	
1. 10% to 35% slope	4
2. Level to 5% slope	2
3. 35% or greater slope	0
Category D: Soil physical characteristics	
1. Few stones; 75% tillable	10
2. Moderate small stones; 50 to 75% tillable	8
3. Very stony; 25 to 50% tillable	5
4. Large rock outcropping, many boulders; less than 25% tillable	3
5. Soil too rocky to till anywhere	0
Category E: Understory plants	
1. Reproducing population of wild ginseng	15
2. Sparse wild ginseng, wild ginger, black cohosh	10
3. Maidenhair fern, rattlesnake fern, wild ginger, goldenseal, lady's slipper	8
4. Christmas fern, blue cohosh, red or white baneberry, wild sarsaparilla	6
5. Jack-in-the-pulpit, other ferns, trillium, bloodroot, foamflower, jewelweed, mayapple, elderberry	5
6. Virginia creeper, groundnut, wild yam	3
7. Club moss, princess pine, bunchberry	0
8. Woody shrubs such as witch hazel, viburnum, shrubby dogwoods, alder	0

9. None of the above (no ground vegetation)	0
10. Woody shrubs, spicebush, pawpaw	4
Category F: Security	
1. Very close to occupied, full-time residence of potential grower, within easy viewing of residence (if noisy, outside dogs housed nearby, add 5 points)	10
2. Forested land less than 50 to 100 yards from grower's residence, patrolled regularly	8
3. Remote woodlot within ¼ mile of residence, patrolled regularly	3
4. Nonresident grower, remote woodlot	0
Total score (add points from each category):	
Results:	
40 to 60 points: Excellent site, great potential 30 to 40 points: Good site, do complete soil analysis 20 to 30 points: Fair site, test soil Less than 20 points: Poor site, look elsewhere	

7 A Nursery in the Forest

A beginner at forest farming is likely to start cultivating nontimber forest product plants by bringing plants or seed in from the outside, often purchased or wildcrafted, such as ginseng seedlings ("rootlets"), nut trees, paw-paws, and others, if only for the reason that propagating "from scratch" may take a year or more before a plant is ready to be put out in its final location on-site. As the forest farm grows in size and diversity, most experienced forest farmers will want to propagate and grow their own plants in an area set aside as the nursery, rather than obtaining older, more expensive plants from outside sources. Saving money is the most obvious reason for producing plants on-farm, but selection for genetic improvement is another important reason for clonal propagation. Of course, propagation of forest plants can also be a lucrative income-generating scheme itself.

A nursery is a protected, intensively managed section of the farm devoted to raising new plants before they are ready to go either out to their final location on the farm or to be sold as a nursery crop to outside buyers. Saving seed, collecting cuttings for rooting, and grafting scion wood should be on every forest farmer's "to do" list. In this chapter the concepts, materials, and practices associated with the establishment of a nursery in the forest are discussed. To summarize, there are three main reasons the authors advocate for forest nursery production:

1. To start new plants for use on the forest farm. Raising new plants on-site can save the cost of buying plants from someone else's nursery or someone else's wildcrafting.
2. To start new plants for sale off-site as planting stock. Rather than growing plants to the point that they are ready to perform their intended function (bearing), planting stock is sold at an early stage to someone who wishes to grow the young plants to maturity and harvest the final product, such as fruits or nuts.
3. To grow herbaceous perennials to be sold as ornamentals. There is considerable demand for landscape ornamentals for shade gardening. Rather than selling these at a modest price from the nursery at a young age as planting stock for someone else to grow on, ornamentals can be grown to maturity on-site and sold for relatively high prices to gardeners who want "finished" plants.

The entry point for nursery production of new plants is propagation. The next stage in the nursery is "growing on" the plants that have been propagated from seed or from clones or cultivars. Growing on includes the nursery production system (inground vs. aboveground containers or pot-in-pot container production). Plants in the nursery often must be protected from the elements to a greater extent than older plants outside the nursery. This may include wind mitigation, shading, and overwintering, along with irrigation and disease and pest control, and other aspects of nursery management. Finally, the nursery plants are either moved out of the nursery for use in the forest farm or marketed and sold outside as a cash crop.

Nursery Production Begins with Propagation

Regardless of the reason for starting a nursery in the woods, it will never succeed unless the farmer is or becomes a reasonably skilled plant propagator. For some plants it is as easy as sowing a seed. Many others are grown from seed but have more or less complex seed dormancy, which requires the propagator to know how to overcome various types of seed dormancy. For some plants the farmer must know how to root cuttings, and some species are more difficult to root than others. Layering is another strategy for propagating a few species of plants. Many fruit and nut species require the specialized skill of grafting. Plant propagation is where it all begins, and understanding natural plant reproductive strategies shows us the way.

LEARNING FROM NATURE: NATURAL PLANT REPRODUCTIVE STRATEGIES

The wayes of propagation are either Natural or Artificial. Industry and Art may bring Materials, and place them fitly for it, but Nature works them. And therefore, as one sayeth, it is the great Art of Man to find out the Arts of Nature.

— Robert Sharrock, 1660

After propagating plants and teaching others to propagate plants, one comes to appreciate the wisdom of Robert Sharrock, who was the author of an early (1660) text on plant propagation.[1] Sharrock believed passionately that all deliberate plant propagation strategies are based on understanding and modification (if necessary) of natural plant reproductive processes. All of the principal means of propagation that are used today have arisen from natural processes.

Plants have been reproducing sexually (from seed) for 350 million years. Just as evolution has brought about adaptations to changing environments through genetic change driven by natural selection, ancient farmers and modern plant breeders have brought about a parade of useful genetic changes (resulting in useful changes in appearance and performance) by guiding the natural process of seed formation and germination, which has changed little since the domestication of plants began about ten thousand years ago.

As an anecdote, maybe five thousand years ago some protoagriculturist came across a branch of a willow tree

Figure 7.1. Natural grafting occurs when branches or roots of the same tree or separate trees come into contact under pressure.

that had blown down in a storm, landed in the mud on the bank of a creek, and become naturally rooted. From this sort of observation early agriculturists learned how to break off a branch deliberately, stick it in the ground, and thereby propagate figs, grapes, and other edible plants from cuttings. This was limited to relatively few easy-to-root species of plants. Much more recently, the invention of polyethylene covers, mist systems, rooting hormones, and even genetic engineering have enabled us to root far more species than ever before from cuttings.

The "ancients" also observed that shoots and roots from the same or different plants that came into contact with each other became fused together. Three thousand to five thousand years ago, ambitious farmers mimicked the process with apples and other species that were too difficult to root from stake cuttings, so grafting was born.[2]

Like all natural reproductive strategies, deliberate propagation methods derived from them are either sexual or vegetative. Sexual reproduction, which involves seed, results in mixing of genes between male and female parents to create seeds—ultimately seedling offspring—that contain a random mixture of genes from both parents, so that they are not genetically identical to either parent. They are genetically unique individuals. Another way of putting this is that they do not come true to type in most cases.

Actually this is not always the case. The genetic variation associated with seed propagation is a characteristic of trees and other plants that cross-pollinate from one plant to another of the same species. This is called outcrossing. The key to outcrossing is the nature of the pollen transfer either by wind, by bees or other insects, or even by hand pollination by people when another plant or the right pollinators of a same species are not available.

Vegetative reproduction, on the other hand, does not (usually) involve seed. Instead, vegetative propagation involves cuttings, grafting, layering, and division. In each case, new plants are regenerated from a part of another plant. There is no genetic recombination involved in vegetative propagation; all the offspring are genetically identical clones of the parent, and as such they are true to type (identical to the plant they were vegetatively propagated from). When clones are formally named, they are referred to as cultivars (cultivated varieties). Vegetatively propagated cultivars in the United States and most other countries can be protected by plant patents, which give the developer (breeder or selector) exclusive rights to the vegetative reproduction and sale of the patented plant. Propagators should understand the implications of plant patents. Anyone who purchases a patented cultivar cannot legally propagate it vegetatively without paying a royalty to the patent owner, particular if the buyer intends to resell the plants.

SELECTION AND BREEDING FOR GENETIC IMPROVEMENT

Often the plants grown in the forest farm are essentially identical genetically to the same species found naturally or "wild" in the woods. In that sense they can be said to be undomesticated. Ginseng is a good example. There have been very few attempts to genetically improve ginseng by breeding or by selection, so the ginseng that is grown commercially either in the forest or under artificial-shade structures have essentially the same genetics as the wild type. Many other species, of course, have been genetically improved (modified) by breeding or selection. All three of these strategies result in genetic improvement ("genetic gain"), compared to the starting point. Several examples follow.

Selection (artificial selection) refers to discovery of a natural variant that has one or more improved characteristics (larger nut size, sweeter berry, early or later ripening) and propagating that variant asexually by cuttings, grafting, or layering to "capture" its genetics. This is analogous to the process of natural selection except that you decide who is fittest.

Breeding refers to the more deliberate process of crossing (cross-pollinating) two individuals to obtain a superior hybrid individual that performs "better" than either one alone. A hybrid contains genes from both parents. From a population of hybrid seedlings from a single cross, one (or more) particularly promising individual is selected and cloned.

Swarm breeding is a somewhat different approach to genetic improvement that involves whole populations of genetically different plants that naturally cross-pollinate, and some of the resulting hybrids that are particularly well adapted to the site or environment survive, while others do not. This is literally "survival of the fittest." (see case study, ch 4, pg 139)

EXAMPLES OF SELECTION AND BREEDING

. . . from so simple a beginning endless forms most beautiful and most wonderful have been, and are being, evolved."
— Charles Darwin, *The Origin of Species*

Nut Trees
Hickory and other nut trees have been genetically improved either by selection alone or by breeding. In the case of selection someone discovered a wild tree that had larger than normal nuts, then asexually propagated that tree by grafting. On the other hand, deliberate breeding can result in even more

Figure 7.2. Hybridization is a different approach to genetic improvement. On the right is a typical wild shagbark hickory nut. The nut on the left comes from a hybrid between shagbark hickory and shellbark hickory. The name of this particular hybrid cultivar is 'Weiker'.

dramatic genetic gain as shown in figure 7.2, where the nut on the right is a shagbark hickory and the nut on the left is a hybrid between two species (shagbark × shellbark hickory, cultivar 'Weiker'), once again cloned by grafting. In either case the genetic gain (increased size) is fortunate as far as domestication of the species is concerned because cracking out wild hickories with a stone, hammer, or mechanical nut cracker is a bit of a fool's errand because the nutmeat yield with each cracked nut is so small it is hardly worth the effort.

Shagbark hickories at the MacDaniels Nut Grove is a good example of genetic improvement: Lawrence MacDaniels devoted a great deal of his career to testing, trialing, evaluating, and promoting temperate nut trees as a valuable food source for humans. In the proceed-

ings of the Northern Nut Growers Association annual meeting in 1952[3] he described how he developed the site specifically as a variety trial for several temperate nut species, including walnut, hickory, filbert, and Chinese chestnut. Some of the trees were named varieties from earlier breeding/selection programs by others, most of whom were members of the Northern Nut Growers Association. Other accessions were his own selections, which he obtained from the winners of nut contests held at county fairs. Much like today's state fairs, farmers would bring samples of various agricultural products they grew, such as cabbages, tomatoes, and many others, including nuts. MacDaniels was one of the judges for the nut competitions at these affairs, so he observed hundreds of samples of walnuts, hickories, and filberts over the years. MacDaniels would follow up with some of the winners of these contests and arrange to collect scion wood from their champion tree for grafting at the MacDaniels Nut Grove.

Minor Fruit Crops

Blackberries are a good example of genetic improvement through breeding and selection. For some wildcrafters, collecting of wild blackberries from the woods or transplanting wild sucker plants into the forest farm is preferable to buying cultivars. For one thing, many modern cultivars are patented by the breeder and cost more because of that. Modern patented cultivars have been bred and selected for large fruit size, disease resistance, and other desirable characteristics, including thornlessness and yield. The wild types, on the other hand, are less expensive, and some would say they taste better than modern cultivars. From the perspective of the farmer, she could produce greater yield and more profit from the genetically "improved" cultivars. Some modern blackberry cultivars are thornless, which is a blessing for the folks who pick the fruit.

For a "minor" fruit crop, pawpaw is one that has received a great deal of attention by amateur and professional plant breeders because of its uniquely appealing flavor and its potential (largely unrealized) as a commercial crop (see chapter 4). One reason pawpaw hasn't made it into the economic mainstream is that it is not easy to propagate vegetatively, which may be a minor challenge to the breeder/selector. Selection requires cloning (cuttings, grafting, etc.). Pawpaws and some other species are difficult to clone by cuttings, although they can be grafted.

On the other hand, willows, elderberry, and a few other species are extremely easy to root from cuttings. A large "stake" cutting of willow, as large as 2 to 3 inches in diameter by 3 feet long, can be sharpened on one end with an axe, then pounded into the ground. After several weeks the stake will strike roots, and new shoot growth will commence from a bud.

CHOOSING THE RIGHT PROPAGATION STRATEGY

Black raspberry is a good example of a species that is very difficult to root from cuttings, but it does naturally layer when the tip of a current season's shoot (primo cane) touches the ground and forms roots. This is a perfectly natural strategy of vegetative reproduction, which requires no help from us. Unfortunately, for many of the NTFPs used in forest farming, propagation is not so easy. Fruit and nut trees, such as pawpaws, hickory, and walnut, can be propagated by seed, but every seedling is genetically unlike the parents and each other. To maintain the genetic identity of a selected cultivar, such as eastern black walnut, it must be propagated vegetatively (cloned), but most nut trees are difficult if not impossible to root from cuttings, so grafting is the only way left to the forest farmer. Shiitake mushroom might seem out of place in this discussion of plant propagation, but it, too, must be propagated vegetatively from vegetative mycelium (not spores) to maintain the genetics of selected spawn types (see chapter 5).

From these examples it should be apparent that there is more than one reason for choosing one method of propagation over another. One of the most important reasons for choosing one of the vegetative methods over seed propagation is to maintain the genetic identity of a clone or cultivar, assuming that seedlings are not sufficiently true to type. The ease or difficulty of propagation is another important reason for choosing one method over another. Certain species, such as ginseng, can be propagated only from seed, so there is no choice. Some species, such as willow, can be easily propagated from cuttings, but pawpaws cannot, so they must be propagated from seed or by grafting.

Another reason for choosing one method over another has to do with the forest farmer's goals. How and why will the plants be used? If the goals of planting fruit or nut trees is to produce and sell the highest quality fruits or nuts at the highest possible price, then grafting is almost always the way to go (maintaining genetic identity), but if the goal is to provide mast for wildlife, seed is the way to go.

Despite all the methods of propagation that exist, it is no coincidence that seed is what comes to mind when people think about making new plants. Sexual reproduction from seed and the genetic variability associated with it are the basis of evolution of individual species and ecosystems. Without the genetic variability inherent in seed-propagated plants, natural selection would have nothing to select from, and we would all be back in the primordial soup.

However, what's good for natural selection is not always good for agriculture, including forest farming. A farmer is looking for reliability and expects the offspring of a plant to resemble the plant from which it was propagated; that is to say, it must be sufficiently true to type to meet the needs of the farmer. For example, if an early ripening pawpaw cultivar such as PA Golden No. 1 (ripens mid-September) that is well suited to maturing fruit in more northern climes gives rise to seedling offspring, some of which ripen fruit several weeks later (mid-October), and thereby risk frost injury, that is not sufficiently true to type and is not acceptable for growing in higher elevations or higher latitudes. So PA Golden No. 1 must be propagated by grafting. In fact, seedlings of PA Golden No. 1 cannot legally be sold by that name since they are not genetically identical clones of PA Golden No. 1.

Propagation from Seed

Why propagate from seed? Despite the advantages of maintaining genetic identity of clones by vegetative propagation, many types of plants are still propagated from seed. In many cases seed is the only realistic way to propagate the plant. There are several reasons:

1. It's easier to plant seeds than it is to use cuttings, grafting, layering, or division.

Figure 7.3. Bloodroot seed harvested from stock plants at Dave Carman's forest nursery. Photograph courtesy of Catherine Bukowski

Figure 7.4. Seed propagation begins with pollination and seed set, which is usually accomplished with the help of pollinating insects (bees, beetles, hummingbirds, etc.) or wind. After seed collection and storage, dormancy must be overcome (if present) by scarification or stratification. Only then is the seed ready to germinate.

2. It is usually cheaper, especially if the propagator collects seeds from the forest or elsewhere.

3. For most species seed propagation can more quickly generate larger numbers of offspring than vegetative propagation methods.

4. The seeds of most species of plants can be transported easily and be stored for longer periods of time than vegetative propagules (cuttings, rooted layers, scions for grafting, divisions).

5. Even when grafting is the method of choice, propagation of seedlings to be used as rootstocks is usually a prerequisite. This will be discussed later in the section on grafting.

6. There may be no alternative to seed propagation. Ginseng cannot be propagated from cuttings, or by grafting, layering, or division anyway, so acceptably true to type is as good as it gets.

In most of these cases the seeds are not genetically identical to the maternal parent from which the seeds were produced, and yet the offspring were "acceptably true to type"; that is, the variation in performance among the offspring were sufficiently uniform to satisfy the grower or end user of the plants.

Pollination and Seed Set

For most species pollination and fertilization either take care of themselves by wind-blown cross-pollination or by insect pollination, which usually requires no intervention by the propagator. In the case of large fruit and nut orchards, which are beyond the scope of forest farming, insect pollination is not taken for granted, and beehives are brought in at considerable expense to do the job. For 1 acre of apple trees, for example, two to three beehives are recommended. As honeybee populations continue to decline because of sudden collapse disorder, this will become an increasingly serious problem.

On the other hand, a number of important forest-farming species, such as pawpaw, apple, and shrubs (e.g., honeyberry), have a more complex system of pollination and fruit and seed production. These and similar species are self-incompatible, which means that successful fertilization and subsequent fruit/nut set require pollination by another genetically different seedling or different cultivar. "Self" in this context refers to all members of a clone that are by definition genetically identical to each other. Self-incompatibility is sometimes an inconvenience for the farmer, who must plant two or more varieties or seedlings in close proximity to each other. An alternative to planting two different genotypes (independent seedlings or different clones) near each other is to graft a scion from one variety as a source of genetically different pollen onto a tree of another variety. Pawpaw and apple are examples of fruit trees that are self-incompatible.

Seed Collection, Cleaning, and Storage

Seed collection may sound pretty straightforward, but there are a number of considerations that can affect the outcome of propagating seedlings in the nursery. First of all, the seed must be ripe. Unripe seed either will not germinate at all or will germinate poorly (low percentage of germination and/or low vigor). Seed of some species may appear ready for harvesting, but the embryo may not have reached sufficient maturity to germinate. An excellent source of information about the description and timing of seed ripeness, and many other aspects of propagating woody plants from seed, is the US Forest Service online publication *Woody Plant Seed Manual*.[4]

Seed cleaning prior to storage or germination is a matter of separating the seed from its fruit—either pulpy, as in the case of ginseng, walnut, and pawpaw, or dry, such as honey locust. Some seeds, especially woody legumes with hard seed coats, such as black locust and honey locust, can be stored at room temperature for many years, whereas others, such as pawpaw, must be stratified to break dormancy and planted soon after harvest. Most species are intermediate and can be stored for several years under refrigeration or by freezing. This includes most forest tree species, such as oak, maple, and conifers.

Dormancy

The seeds of most species that have evolved in the cool temperate climate will not germinate immediately after ripening and harvest. This prevents them from germinating late in the fall, when the tender seedlings might be killed by low winter temperatures. Seed dormancy can be overcome either naturally or artificially. There are two main types of seed dormancy, internal and external, which must be overcome in different ways, before they are capable of germination even under ideal nursery conditions (table 7.1).

Internal seed dormancy requires a period of *cold* stratification lasting from several weeks to several months. Cold stratification requires that seeds spend a period of time under moist conditions at low temperatures, at a range of temperature between 13°F and 41°F. Refrigerator temperature is just right.

Freezer temperature (−10°F) is too cold. Stratification occurs naturally during winter for some seeds, which fall to the ground and get buried beneath a thin layer of moist soil or leaves. Alternatively, and often more effectively, seeds can be placed in a refrigerator in a loosely closed plastic bag with moist peat or other absorbent material, or sand. The substrates should be moist but not soaking wet for a period of 60 to 120 days, depending on species (see *Woody Plant Seed Manual* for specific recommendations). Moisture is required for stratification. Refrigeration of dry seeds will not break dormancy but is a good way to prolong the storage life of many species. Natural stratification by sowing seeds in the fall may be easier than the refrigerator route, but outdoors rodents may make off with them, or when fluctuating winter temperatures are above 41°F or below 13°F, the stratification clock stops ticking toward the "goal" of 90 days or whatever is the recommended duration.

There is one variation on this "simple" kind of internal seed dormancy that applies to species with seeds that have especially tiny, immature embryos at the time of ripening. These are called rudimentary embryos. Some plants that produce NTFPs, such as ginseng and ramps, fall into this category. Rudimentary embryos need time to mature and increase in size before they can respond to cold stratification. Once the seeds of ginseng, for example, mature in the fall and fall to the ground or are harvested by the forest farmer, they overwinter. Then they need the warm weather (and moisture) of the next growing season just for the embryos to mature (increase in size), before finally undergoing cold stratification the following winter

Table 7.1. Types of Seed Dormancy

	To Overcome	Species
External seed dormancy caused by a hard seed coat	Scarification—abrade seed with sand paper, pruning shears, acid	Hard-seeded legumes such as honey locust, black locust, Kentucky coffee tree
Internal seed dormancy caused by a physiologically dormant embryo	Stratification—moist chilling for several weeks to several months (41°F)	Pawpaw, hazelnuts, walnuts, hickories
Rudimentary embryo dormancy	Warm season followed by cold-season stratification	Ginseng

either naturally in the ground or in a refrigerator, before they can germinate the following spring.

Hence, whereas a seed with a full-size mature embryo that requires only cold stratification will germinate the spring after ripening on the tree (~ six months), a seed with a rudimentary embryo requires at least eighteen months before germination.

External seed dormancy

External seed dormancy applies to seeds with hard, impermeable seed coats that prevent water and oxygen uptake. This type of dormancy is found in woody legumes such as honey locust and black locust. Before germination, the seed coat must be abraded naturally by passing through the acidic gut of an animal or must undergo multiple freeze-thaw cycles. These and other processes that breech the hard seed coat, allowing uptake of water and oxygen, are known as scarification (to scar). Natural scarification can take years, but several deliberate actions can hasten it. Several of these include mechanical means: by carefully clipping off the top end of the seed coat (away from the embryo) with a pair of sharp pruning shears (nicking), or scoring the seed coat with a file or sandpaper, or even soaking it in concentrated acid (this is hazardous). Figure 7.5

shows the effect of several scarification treatments on germination of honey locust seed. The graph in figure 7.6 shows the actual germination percentage two weeks after sowing the seed. The most effective treatment was

Figure 7.5. External dormancy of honey locust seed is caused by a hard seed coat that cannot take up water or oxygen. The seed coat must be degraded (scarified) before germination can proceed. In this student experiment, seeds in each of six rows, left to right have been treated for (1) 30 minutes, (2) 60 minutes, and (3) 90 minutes in concentrated sulfuric acid; (4) boiling water poured over the seed; (5) the seed coat nicked with pruning shears; and (6) the untreated control. Figure 7.6 shows the results graphically.

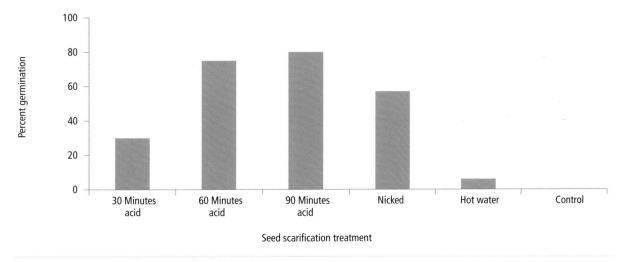

Figure 7.6. Effect of six different treatments to overcome the external seed coat dormancy of honey locust seed. Figure 7.5 shows the experiment underway in the greenhouse. Many student groups enrolled in my Plant Propagation course have done this experiment over the years, and quite often the nicking treatment is as good as or better than the acid treatment. However, nicking one seed at a time with pruning shears is slow and tedious if you are planning to scarify more than just a few seeds.

90 minutes in concentrated sulfuric acid. This and other approaches such as scratching the seed coat with a file or using sandpaper to rub a hole in the seed coat are collectively known as mechanical scarification. Surprisingly, soaking in acid for up to ninety minutes did not destroy the seed but rather etched away the hard outer seed coat, allowing entry of water and oxygen. Pouring boiling water over the seed (but not cooking the seed in boiling water) and allowing the water to cool was only a little effective. This treatment is often more effective in other hard-seeded legume species. Finally, the untreated control seeds remained entirely dormant.

Vegetative Propagation

The terms vegetative, asexual, and clonal propagation are synonyms. The basic unit of propagation, regardless of what method, is the *propagule,* and in the case of sexual propagation it is, of course, the seed. For cutting propagation the propagule is the cutting (duh!): for layering it is an intact branch or stem that will be induced to root while still attached to its original root system; for grafting, a scion is the propagule. There is no standard term for the unit of division, but "clump" will suffice in the case of ramps, and "sucker" in the case of pawpaw and blackberries. Tissue culture (micropropagation) is another type of vegetative propagation that is laboratory-based and will not be covered here. There are tissue culture labs that will do custom work for novices and professionals alike. No other propagation method besides tissue culture has nearly as high a potential multiplication rate (increase in numbers over time).

Cuttings

Species that may be suitable for propagation from cuttings range from those that are very easy to root (willow, elderberry) to moderately difficult to root (sea buckthorn, currant, gooseberry, etc.). Most species of use in the forest farm are difficult if not impossible to root from cuttings and must be left to grafting (pawpaw, nut trees), layering (black raspberries), division (ramps, hostas, golden seal), and even tissue culture (orchids).

For those species that are reasonably easy to root from cuttings, the issues are when to take the cutting, from what part of the cutting donor plant to excise the cutting, and how to induce it to root.

Most easy-to-root species such as willow and elderberry can be propagated as *softwood cuttings* during the period of active growth (spring, summer), greenwood (later summer) or *hardwood cuttings* (after leaf fall, dormant). Softwood cuttings must receive special attention, such as the polyethylene rooting chamber (tent) described in the sidebar on how to construct a poly tent (figure 7.8), since they are extremely "soft,"

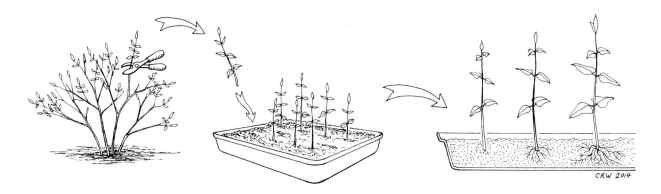

Figure 7.7. A leafy cutting (propagule) is removed from the stock plant with a sharp pair of pruning shears or knife. It may or may not be treated with a rooting hormone, then stuck in a well-drained rooting medium for several weeks. The rooting medium must not dry out, and the humidity must be kept high using a polyethylene tent (figure 7.8) or by other means. Avoiding desiccation is the most important consideration. Illustration by Carl Whittaker

CONSTRUCTING AN INEXPENSIVE ROOTING CHAMBER (POLY TENT)

1. Fill a flat 18″wide × 24″long × 4″deep with a well-drained, but high-water-holding-capacity rooting medium (e.g. peat : sand, 1:1).
2. With a piece of narrow wire fencing or similar material, 32″wide × 24″long, create an arch with a 12″radius from the bottom of the flat to the highest point of the arch, and set this half tunnel directly into the flat of rooting medium.
3. Stick cuttings into the flat/rooting medium.
4. Slide the flat with its wire tunnel into a large translucent garbage bag (not clear or black plastic), and loosely tie together the open end of the bag.
5. Place this poly tent rooting chamber in a location that receives no direct sunlight.
6. Open the tent to check cuttings weekly. Remove any dead leaves, and add water if necessary.

Figure 7.8. Ideal conditions for rooting leafy cuttings can be created by constructing a "poly tent" to increase humidity of the air around the cuttings, which reduces water loss from the leaves (transpiration) and promotes rooting. An arched wire frame is placed over a flat of cuttings stuck in rooting medium, and the flat of cuttings is slid into a large translucent (not black, not clear) plastic bag. Seal the bag loosely. The poly tent should be kept in complete shade (NO sunlight), and water added occasionally if necessary to prevent drying.

that is, succulent, and desiccate rapidly. If a cutting is allowed to wilt for too long it will die. Figure 7.7 shows a progression of the process of rooting cuttings.

Dormant hardwood cuttings taken in late fall, winter, or even early spring are leafless so they don't transpire (lose water from the leaves), and are much less prone to drying out. They are usually collected after leaf drop in the fall or during the winter for rooting indoors, or in early spring before leaf out. Easy-to-root species such as willow can be deliberately stripped of their leaves anytime during the growing season to reduce transpiration, so they behave like hardwood cuttings and can be stuck in a pot or directly into the ground. As long as the ground or the potting soil doesn't dry out excessively, they will root in a few weeks.

From What Part of the Plant Should Cuttings Be Taken?

With respect to the part of the plant from which a cutting is taken, there are root cuttings, leaf cuttings, and shoot cuttings, with the latter being the most common. A shoot cutting is a section of stem typically four to six inches long with or without leaves but definitely with one or more buds or potential growing point (figure 7.7). A cutting without a bud (stem internode or a leaf) may root, but it will never get anywhere as far as new shoot growth is concerned. Leafy cuttings are necessary for successful rooting of some species, but the challenge is to prevent them from drying out before they root. Although the leaves may stimulate rooting by providing sugars from photosynthesis, they draw moisture out of the stem as they transpire. Since the cutting has no root system to replace that water efficiently, it will dry out and die quickly unless moisture loss is minimized.

Drying or desiccation or excess moisture loss—all three terms indicate the same problem—is the leading cause of failure when it comes to propagation from cuttings. "Moisture management" refers to various strategies to minimize moisture loss, particularly from leafy cuttings. This begins with the stock plant, followed by taking the cuttings and transferring them to the nursery. Environmental strategies to minimize water loss from leafy cuttings and hardening off rooted

cuttings so they will survive their time in the nursery or later at their final location will be discussed below. Of course the shortcut is to root cuttings directly at their final location.

Ideally, cuttings should be taken early in the morning on a cloudy day. As they are transported to the nursery, they should be transported in a cooler or wrapped in wet burlap and kept out of the sun.

Here is a typical strategy for rooting leafy cuttings from start to finish (figure 7.7):

1. Use a sharp pair of pruning shears to make a 4- to 8-inch-long shoot cutting from the current season's (i.e., the most recent) growth of the stock plant, preferably in the early morning on a cool day.
2. Transport cuttings to the nursery well protected so they won't dry out.
3. In the nursery remove leaves from the bottom 2 inches of the stem.
4. Treat base of cutting with rooting hormone if the species you are trying to root is considered moderately difficult or difficult to root.
5. Using a well-drained rooting medium (not soil) in a 4-inch-deep flat, stick the cutting at least 2 inches deep.
6. Place the flat in a "poly tent" (high humidity enclosure; figure 7.8 on page 245), in a location that receives NO sunlight.
7. Manage by keeping the cuttings well shaded, and prevent drying. Add water if necessary, but don't overwater.
8. Expect rooting in four to six weeks, less if the species is easy to root.
9. Rooted cuttings are ready to be "harvested" when primary roots (emerging directly from the cutting) begin to branch. After rooted cuttings are potted up they should be kept in the shade for a week, then gradually moved into brighter light at their final location in the nursery or beyond.

Once cuttings are stuck in the rooting medium (usually in flats or beds), and well shaded, there are three moisture-management systems that create the appropriate environment to minimize water loss from cuttings. All of them are based on three environmental factors that minimize moisture loss: shade, high humidity, cool temperature. The three moisture management systems are mist, fog, and a poly tent. The first two require plumbed water and electricity, so they may not be appropriate for a forest nursery. The most practical, most economical, and best suited system for the forest nursery is the polyethylene tent (figure 7.8). The essence is maintenance of high relative humidity, whether that is created by a polyethylene enclosure or otherwise.

LAYERING

Layering is an asexual propagation method that refers to any of a number of methods, which involve rooting of a stem while it is still attached to its original root system. One common type of layering is air layering. It involves girdling a stem, wrapping it in a moist medium, such as sphagnum moss, and wrapping it all in plastic. Within several weeks roots will form above the girdle, and the branch can be cut off just below the new roots, and the new rooted branch can be planted (figure 7.9).

GRAFTING

Grafting is the joining of plant parts from different individuals so they will fuse and grow as a single plant unit. Liberty Hyde Bailey, the founder of the College of Agriculture at Cornell University, around the turn of the nineteenth century spoke eloquently about the essential elements of a successful grafting.

The ways or fashions of grafting are legion. There are as many ways as there are ways of whittling. The operator may fashion the union of the stock and the scion to suit himself, if only he apply cambium to cambium, make a close joint, and properly protect the work.

— Liberty Hyde Bailey, *Standard Cyclopedia of Horticulture*, 1925[5]

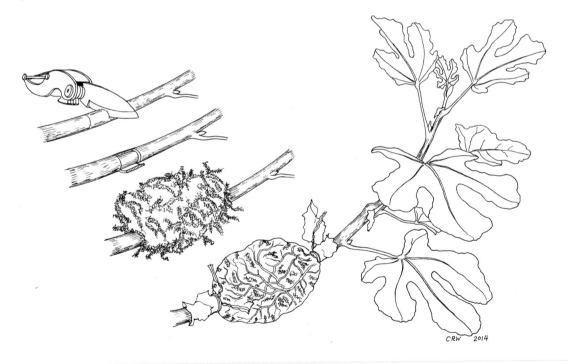

Figure 7.9. Air layering is one of many layering methods that involve inducing rooting on an intact branch. The bark is removed from a short section of the branch, wrapped in damp sphagnum moss and covered tightly with polyethylene, sealed at each end with a twist tie. Within a month or more rooting occurs above the wound, and the branch is cut off just below the roots and planted in a new location. Illustration by Carl Whittaker

Bailey's essential criteria are:

1. Cambial alignment (line up the inner bark of stock and scion)
2. Pressure (tie the stock and scion together tightly)
3. Avoidance of dessication (don't let it dry out)

These three criteria for successful grafting are as true today as they were in Bailey's time (the early twentieth century) and millennia before that. In Ken's course in plant propagation at Cornell, the importance of these critera (and one more) is stressed as critical, using slightly more modern but less poetic terminology. Technically, there is a fourth essential criterion, which may seem obvious at first:

4. Compatibility (stock and scion must be closely enough related genetically)

These are universal criteria for success regardless of the method.

The fourth criterion for successful grafting listed above—compatibility—was not addressed by Bailey, but it needs to be addressed here, because any grafting that is undertaken is doomed to failure if the issue of graft compatibility is not understood. Compatibility, and its converse, incompatibility, refers to the likelihood of success or failure based on how closely related genetically the stock and scion are. A common question from folks who are inexperienced but interested in grafting is some version of, "Can I graft an apple to a pawpaw tree?" or (b) "apple to hawthorn?" (c) "pecan to hickory?" or (d) "oak to maple?" and so on. The correct answer is *no* to apple/pawpaw (families: Rosaceae/Annonaceae) and oak/maple (families: Fagaceae/Aceraceae) but *yes* to apple/hawthorn (both in the family Rosaceae) and pecan/hickory (both in the family Juglandaceae). Notice that the "winning" pairs (successful union of stock and scion) are in the same botanical family and the losers (known as incompatibilities) are not. In some cases combinations that are in the same family are still incompatible. In fact, the degree of relatedness for any particular graft combination for graft compatibly depends on the particular

Figure 7.10. A black walnut cultivar grafted onto a black walnut seedling. Slight change in taper and diameter indicate the location of the graft union.

pairs you are grafting (table 7.2). It's tricky business, and other sources or experienced propagators should be consulted for a particular species before a lot of effort is put into grafting it.

Experience at the MacDaniels Nut Grove speaks to the issue of compatibility. Hickories and walnuts are both in the family Juglandaceae. Hickories (shagbark, shellbark, red pignut hickories) are in the genus *Carya* and black walnut, English walnut, and heartnut are all in the genus *Juglans*. Despite the fact that they are all in the same family, the hickories are not graft compatible with the walnuts. But hickories grafted onto hickories are compatible in some cases, and black walnut grafted onto black walnut are also compatible. The issue of incompatibilities gets a little more complicated when you consider delayed incompatibility, which refers to a graft union that forms more or less uniformly, but then after some time (years or many years in some cases) the

Table 7.2. Effects of Genetic "Distance" (Degree of Relatedness) on Graft Compatibility

Relationship	Likelihood of Success	Compatible Example	Incompatible Example
Intraspecific: Scion/stock are the same species but different clones (cultivars)	High	'Macintosh' apple (*Malus domestica*)/ EM 9 (clonal rootstock cultivar of *M. domestica*)	Red maple clonal cultivars (*Acer rubrum*)/ red maple seedling
Intrageneric: Scion/stock are the same genus but different species	Moderate	Sweet orange (*Citrus sinensis*)/ rough lemon (*C. jambhiri*) Sweet cherry (*Prunus avium*)/ tart cherry (*P. cerasus*)	Almond (*Prunus amygdalus*)/ peach (*P. persica*)
Intergeneric: Scion/stock are the same family but different genera	Unlikely	'Old Home' pear (*Pyrus communis*)/quince (*Cydonia oblonga*) Blue spruce (*Picea pungens*)/ Norway spruce (*Picea abies*)	'Bartlett' pear (*P. communis*, Rosaceae family) / Quince (*C. oblonga*, Rosaceae) Lilac (*Syringa vulgaris*, Oleaceae family)/ California privet (*Ligustrum ovalifolium*, Oleaceae)
Interfamiliar: Scion/stock are different families	0% (considered impossible, despite a few claims to the contrary)		Oak (Fagaceae family)/ Maple (Aceraceae family)

graft union fails gradually and may or may not eventually kill the tree. Table 7.2 summarizes the range of taxonomic (genetic) relationships between stock and scion and their likelihood of forming a compatibile graft union.

These "limits of compatibility" are not the same for different kinds of plants. For example, two species in the same genus are compatible in the case of sweet cherry (*Prunus avium*)/tart cherry (*P. cerasus*), but almond (*Prunus amygdalus*)/peach (*P. persica*) (two species, same genus) are not.

It would only be a small exaggeration to say that there are almost as many reasons for grafting as there are methods, but the most important reason for forest farming is to clone (make genetically identically copies) of plants that are otherwise difficult to propagate vegetatively. This applies to cultivars of many horticultural species, especially woody plants such as apples and pears, hickories, walnut, chestnut, pawpaw, stone fruits, and so on. When species such as these are grafted, the grafting usually takes place in the nursery, but grafting onto an established plant at its final location in the forest farm is appropriate for some applications.

Most grafting is done in the nursery to vegetatively propagate clonal plants that can be moved to their

Figure 7.11. Grafting in the nursery often involves attaching a scion from a stock plant to a seedling understock that has been grown for one or more years in a container or inground. This picture shows a top wedge graft of beach plum in which the understock plant is cut off and the top several inches discarded. Then the stock is split with a knife vertically about 1 inch deep. A V-shaped wedge is cut on the bottom end of the scion, and this wedge is inserted into the split in the stock. The junction is tightly bound, and within several weeks the graft union has healed. Illustration by Carl Whittaker

final location elsewhere in or off the forest farm. Usually grafting involves cutting off a portion (short branch or bud) of one tree and attaching it to another. The upper portion from the donor tree is called the scion. It is usually a short section of stem with several buds on it or a single bud in the case of bud grafting (budding). The scion is attached to the lower part of the grafted pair, called a stock, an understock, or a rootstock. Scions from selected trees (high yield, high quality, disease free), either on the farm or from off-site, are grafted onto seedling rootstocks, which of course must be propagated from seed, starting in the nursery a year or more before grafting. After grafting in spring or bud grafting, usually in the late summer, the plants are typically grown on in the nursery for one or two more years before going out into the forest farm. Grafting is not only a way to clone a desirable cultivar, but grafting a scion from a mature (flowering) tree will come into bearing (will flower and produce fruit) years earlier than an ungrafted seedling.

While all these reasons to graft are appropriate, there are additional reasons listed below for grafting directly onto mature established trees already growing out in the forest farm.

Grafting a Scion That Is Genetically Different

Grafting a scion of the same species that is genetically different from the stock provides a pollen source for self-incompatible stock species such as pawpaw, apple, and some other fruit trees. If more than a few trees are involved, a better solution to overcome self-incompatibility would be to plant several trees in the vicinity of at least two or more genetically different cultivars.

Grafting to Change the Variety on an Established Tree

In the cases of pawpaw, for example, breeders from the University of Kentucky and elsewhere release new cultivars from time to time. A forest farmer could add one or more of the new ones to an established cultivar or seedling tree, which would not only enhance pollination but also give the forest farmer a diverse array of pawpaw varieties even if multiple trees were not practical on that piece of land.

Grafting (Topworking or Highworking) of a Species That Is Difficult to Transplant

Hickory, for example, is extremely difficult to transplant from the nursery to its final location because of its long and difficult-to-excavate taproot. Rather than graft a scion onto a young seedling rootstock growing in a container or field nursery, which must be transplanted—with a low degree of success—it is better to graft a cultivar (scion) onto an older established tree that will serve as a rootstock in its final location; that is, it does not require transplanting.

Grafting for Genetic Improvement

An individual plant with some heritable (genetic) outstanding characteristic such as fruit size, disease resistance, fruit flavor, and so on will be lost if propagated by seed, but if it is propagated asexually, all the clonal offspring will be genetically identical to the source tree. Hickory tree improvement at the MacDaniels Nut Grove is an example.

To graft a scion of a selected variety onto seedling rootstock to multiply the scion variety, a number of different grafting methods can be used, but top wedge grafting is one of the simplest (figure 7.11), and learning this technique can be a stepping stone to other methods and for other purposes.[6]

Top Wedge Grafting

Top wedge grafting is one of the easiest grafts to perform, and one of the methods with the highest success rate. This positive learning outcome makes it an excellent way to learn the general principles of grafting, which can be applied to any other grafting method.

1. At least one year in advance of the actual grafting, grow a rootstock in a nursery container from seed or from a cutting (or purchase a rootstock).
2. Early spring of the following year, before buds have begun to swell, cut a 4-inch section from the top of the scion donor plant (the selected variety you wish to propagate). This should be stored in a refrigerator (in a plastic bag with damp peat) until it is time to graft to rootstock.

Figure 7.12. Topworking a one-year-old scion of a selected cultivar onto ten-plus-year-old understock established in its final location, using an inlay bark graft. Photograph courtesy of Horticulture Department, Cornell University

3. When buds are beginning to swell on the rootstock seedling, which should be about 3 feet tall and ⅜ inch in diameter, cut off the top several inches of the stock with a sharp pair of pruning shears.

4. Using a sharp grafting knife or utility knife, make a vertical cut down the center of the cut surface of the stock to a depth of about 1¼ inches.

5. Using the 4-inch-long scion that you stored in the refrigerator several weeks ago, make two flat angled cuts opposite each other, beginning about 1¼ inches from the bottom of the scion, so they meet at a V-shaped point at the bottom of the scion.

6. Slide the V-shaped wedge down into the cleft at the top of the understock so the top of the cut surfaces of the scion are even with the flat surface of the decapitated stock.

7. Wrap the stock and scion together tightly with a budding rubber (latex rubber band) or polyethylene strip.

8. Keep the grafted plant well shaded until the new growth of the scion is several inches long.

Ideally the stock and scion should be the same diameter (about pencil thick). If the scion diameter is less than the stock, the scion should be on one side of the stock so that the bark of the stock and scion match up along one side. A narrower scion should not be centered in the middle of a larger diameter stock, because the inner bark of each must be in contact.

Nursery Production Systems

Regardless of the size of the nursery you intend to start with, there are four basic nursery production systems to consider. These are field (inground), aboveground container, pot-in-pot container, and raised beds (table 7.3).

FIELD PRODUCTION

Just as it sounds, after propagation (seed or vegetative) and up to a year in a liner (transplant) bed, plants are dug and transplanted directly into the ground, at a

protected location, where they can be grown on from one to several years to a size large enough for transplanting to their final location at the forest farm or sold for planting off-site. Field-grown plants are easily set back because of damage to the root system when they are dug for transplanting. This is the greatest drawback to field production. Balled and burlapped (B&B) is the method commonly used to minimize transplant shock. B&B involves digging the tree with as large a soil and root ball as practical and wrapping the ball tightly in burlap for transport.

ABOVEGROUND CONTAINER PRODUCTION

Figure 7.13 shows a small forest nursery run by Sean Dembrosky in upstate New York. This container nursery, beneath a canopy of pine trees, is part of a

Table 7.3. Advantages and Disadvantages of Three Common Nursery Production Systems

Nursery Production System	Advantages	Disadvantages
Field production	Cost is low. Irrigation is minimal. Root systems are protected from low winter temperatures.	There is transplant "shock" from root damage when plants are dug for outplanting. More labor is involved in transplanting. The system requires more space than container production.
Container production (aboveground)	Containerized plants are more easily moved within the nursery and to outplanting sites or to market. There is little or no root system damage when transplanting or outplanting. Tighter spacing than field production is possible.	The cost of containers is a factor, although large "tin cans" with drainage holes or other homemade containers can be used. Larger shrubs and trees tend to blow over in the wind. Roots at the edges of black pots that are not insulated by surrounding soil can be damaged by heating from direct sun. Root systems not insulated by surrounding soil are easily damaged or killed by low winter temperatures. Because of limited soil volume, plants dry out more quickly and require more irrigation than either field production or pot-in-pot.
Pot-in-pot container production	Root systems are insulated by surrounding soil so root temperatures in summer are lower, resulting in reduced water usage, and less need for irrigation. Root systems insulated by surrounding soil do not get as cold as aboveground containers, so winter injury is avoided. Blowover by wind is avoided entirely. Pots are easily lifted and moved to planting site or to market.	Preparing holes for installing socket pots is labor intensive.

Figure 7.13. A shady forest container nursery at Edible Acres, a permaculture nursery in central New York.

permaculture forest farm where nursery stock, mushrooms, berry fruits, vegetables, and other nontimber forest products are grown.

Container production has certain advantages over field production, including ease of moving plants without disruption of the root system, allowing for more efficient use of space within the nursery, and ease of transport to outplanting sites or to markets. Another distinct advantage is the use of soilless container mixes instead of soil, which allows greater flexibility for optimizing aeration and drainage for individual crops. The growing medium (potting mix) for container production of most nursery crops is a soilless mix consisting of various formulations of fine organic matter to hold moisture and coarse material such as chunks of pine bark or other material to promote drainage. Problems associated with container production systems include

blowover from wind, overheating from direct sunlight in the summer, and freezing injury to exposed root systems in the winter.

POT-IN-POT CONTAINER PRODUCTION

Pot-in-pot is a relatively new nursery production system developed in the southeastern United States.[7] It is usually done with large trees grown in large containers, which have a habit of blowing over in strong winds. Anchoring the tree by burying the pot in the ground is only one of several advantages of pot-in-pot. Wind throw is not as serious a problem with smaller plants likely to be found in a forest farm nursery (herbaceous perennials, shrubs, and small trees). There are still significant advantages to pot-in-pot that will be covered below, but first consider how pot-in-pot works. First an empty pot, called a "socket pot," is buried in the

Figure 7.14a. Small pot-in-pot container nursery where hosta and ferns are being grown in containers recessed into the ground. The pot in hand has been removed from the empty socket pot (upper center). Photograph courtesy of Catherine Bukowski

Figure 7.14b. The hosta in a 2-gallon production pot (nursery container) is being lowered into the socket pot, which is recessed into the ground. Photograph courtesy of Catherine Bukowski

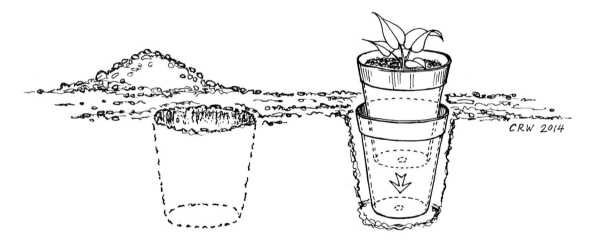

Figure 7.14c. Below ground view of the pot-in-pot container production method showing the empty hole excavated for the socket pot (left) and the plant pot inserted in the socket pot (right).

ground so the rim sits about an inch above the soil line. Figure 7.14a shows an empty socket pot in the ground in about the center of the picture. The "production pot," in this case with a hosta, has been removed by the author (for sale, for propagation by division, or for other reasons). Figure 7.14b shows the plant pot being set back into the socket pot, where it will sit until it is ready for sale. During its time underground, it is insulated from the hot sun, which can overheat and damage the root system of a plant in a pot on the surface. Furthermore, during the wintertime the containerized root system, which is belowground, is protected from freezing injury.

RAISED BEDS

Seedlings or rooted cuttings often spend their first year in a raised bed, before they are transplanted as "liner" for field or container production or sold directly as "liners." They are planted at a fairly tight spacing in the raised beds for a year until they are dug. In its simplest

Figure 7.15. Jack-in-the-pulpit growing in a raised bed containing compost-based growing medium at Spring Haven Nursery, Pennsylvania.

form a raised bed is just soil piled up 8 to 12 inches or more above the ground line, or it may be enclosed within a frame of wood or logs or other framing material. It is usually 4 or 5 feet across so the center can be accessed easily for harvest or cultivating. The soil is usually amended with compost or other organic material. One reason for elevating the soil this way is to increase drainage in an otherwise poorly drained soil.

In any of these production systems, plants must be irrigated as necessary, and woody plants may require pruning. Minimally invasive pest management, disease control, and weeding must be practiced as necessary if you want to end up with useful or salable plants in the end. Protocols for disease and pest control will differ for organically managed vs. conventionally managed nursery. To achieve minimal environmental impact, an important goal of a permaculture approach to forest farming, organic methods are preferred.

Setting Up the Nursery

It's a good idea to start small and grow a nursery with just a few plants to start, to learn the individual needs and habits of the species; many people try to start big and never get to know the particulars of plants

because there are too many to handle. Taking the time to build slowly and deliberately is always in the best interest of long-term success. A forest nursery can be as simple or as complicated as the forest farmer likes. Jonathan Chapman (Johnny Appleseed) is a good example of starting small—very small—and in his case never progressing beyond small. He preferred nurseries that were as simple as one could imagine. He would find a clearing in the forest, or create one, and simply scatter apple seed round about. After sowing the seeds he would move on to another likely prospect, ranging from Western Pennsylvania to the wilderness of mid-Ohio. He would return to each of his nurseries every year or so.

Rarely is nursery production and management so simple, even in the forest farm. Modern apples are almost always grafted onto selected rootstock varieties, for size control, disease resistance, and other desirable qualities. Chapman had religious objectives to grafting (it was "unnatural"), so the apples produced on his trees usually were "spitters" (sour), used mainly for hard cider rather than eating.[8]

STRUCTURES IN THE NURSERY

Aside from the plant systems described below (field, aboveground container, pot-in-pot container, raised bed), there are a number of other components to consider in designing a nursery:

- Structures will be necessary to store materials and provide an all-weather space to work. A simple three-sided lean-to with the opening on the leeward side of the woodlot may be sufficient. You may try to site your buildings near a gap or clearing or at the edge of the forest, especially if the plants you will be propagating will benefit from increased sunlight in their early days, either by planting in the ground, in containers, or by other production options described above.
- A potting shed with a bench and a roof covering may be adequate for some. This space should house some tools and pots and be located near potting medium.
- Composting station.

- Access to water. A rainwater catchment should be considered if a reliable stream or other source of water is not available.
- Potting medium preparation area for mixing of compost and other potting media component for container production.

SOIL AMENDMENTS AND CONTAINER GROWING MEDIA

Containerized plants are not usually growing in soil, because no matter how well natural soil performs "inground" it is usually not sufficiently well drained or aerated for container production. Typical container growing media ("potting mix") is usually soilless, or soil is a minor component. Some common materials used for container mixes include:

- Peat (expensive, difficult to wet, nonrenewable resource)
- Coir (coconut fiber, imported, salt content variable)
- Compost
- Softwood bark
- Sand
- Soil (no more than 10 percent)

Typical mix: 80 percent pine bark, 10 percent peat or compost, 10 percent sand

IRRIGATION

Many forest farms have access to some form of irritation. Shiitake mushroom production, for example, requires water for maintenance soaking and for forced fruiting. Aboveground container crops dry out faster than inground crops because of the restricted container volume and solar heating of the root system. Irrigation for nursery plants can involve toting buckets of water from a nearby stream or well, use of a long (sometimes very long) hose from a plumbed water source, or getting water from a nearby well. Water harvesting from the roofs of buildings or other sources can be stored in relatively inexpensive several-hundred-gallon-capacity high-density polyethylene (HDPE) storage tanks. Pot-in-pot container production reduces to some extent the demand for irrigation of plants in aboveground containers because the cooler soil temperature in the former results in less transpiration.

Specific Production of Forest-Farmed NTFPs

With this background on propagation and other aspects of nursery production and management, it is time to consider in more detail the plants that can be grown in the nursery and why they are grown. The answer to why they are grown has a great deal of influence over what plants are appropriate. The nursery crops grown on a forest farm can be for use on-site, to save money that would otherwise be used for buying nursery stock from outside, and simply to advance the goal of self-reliance. On the other hand, nursery crops are NTFPs that can be grown for sale off-site to a wide range of customers.

BRAMBLES

This is a tale of learning from one's mistakes. Over the last decade attempts have been made to grow several types of brambles (black raspberry, purple raspberry, blackberry) under black walnut trees at the MacDaniels Nut Grove with little success. From the start it's been known that black raspberries do not fruit well under deep shade, but the leafy canopy of black walnut at this site casts only moderate shade compared to most other deciduous trees. Walnut toxicity might also have been a factor. The growth of many species is inhibited by a chemical exudate from walnut roots known as juglone. Fortunately black raspberry is one of those species that is tolerant of the juglone. Hence, it was reasoned that black raspberry would fruit well beneath black walnut (moderate shade, juglone resistance).

Unfortunately that has proven not to be the case. Fruiting has been low, and the few berries that did appear were small and misshapen (cat faced). It took some years of wishful thinking before it was accepted that the black raspberries were never going to produce acceptable fruit beneath our black walnut canopy. However, black raspberry, like many species, produces an abundance of vegetative growth—leafy shoots when

Table 7.4. Species by Method of Propagation for Plants Covered in This Book

Category	Method	Propagule	Advantages	Disadvantages	Examples
Sexual	Seedage	Seed	High multiplication rate Easily transported Easily stored Less expensive	Variability Delayed maturation Dormancy Self-incompatibility	Pawpaw Ginseng Ramps
Vegetative	Cuttage	Cutting, stake, slip	Uniformity Earlier to full size than from seed Easier than grafting Earlier to flower than seedling	Moderately low multiplication rate compared to seed Many species difficult to root	Willow Blueberries Elderberry Seaberry Honeyberry Currant Gooseberry
	Grafting	Scion, rootstock	Uniformity Rapid onset of fruiting Earlier to flower than seedling	Greater skill and more experience to achieve success than other methods Incompatibility Low multiplication rate	Hickory Walnut Chestnut Pawpaw
	Layering	Intact branch	Uniformity Easier rooting of otherwise difficult-to-root species	Very low multiplication rate	Black raspberries Hazelnuts Currant
	Division	Clump, sucker	Uniformity Higher success rate than cuttage or grafting	Very low multiplication rate	Ramps Ferns Blackberries Comfrey Hosta Black and blue cohosh Mayapple, etc.

Figure 7.16. After the tips of the arching prima canes come in contact with the soil, they strike roots and form a dormant bud. This is natural tip layering. Once the tip is rooted and a bud is set, by about early fall it can be cut from the arching prima cane, potted up, and grown on in the pot-in-pot nursery. Photograph courtesy of Kelsey Erickson

grown in the shade. So it finally dawned on Ken that if fruit couldn't be grown, why not raspberry planting stock?

Black raspberry canes are biennial. In the spring new "prima canes" grow from the crown of the plant but bear no fruit until the next year. During this first fruitless year these prima canes elongate and arch up and over until their growing tips touch the ground, where, if the soil is sufficiently moist, they strike roots and form a new bud. This rooted shoot can be cut away, transplanted into nursery containers, and put in the pot-in-pot nursery.

A year later they are ready to be sold as young raspberry plants (planting stock) for growing on at a suitable location in the forest farm or in someone else's forest garden or backyard. The basis of this production system is propagation through natural layering, which like any form of layering involves rooting of a shoot while it is still attached to its parent plant, as described earlier in the propagation section of this chapter.

MEDICINAL PLANTS

With the exception of ginseng and to a lesser extent goldenseal, medicinal plants are mostly wildcrafted, rather than cultivated on the forest farm. The end product of many of these medicinal herbs, such as black cohosh and others listed in table 6.1, are dried roots, rhizomes, and so on. Although there is plenty of demand for these medicinal herbs by the herbal products industry, most of that demand is satisfied by wildcrafters, who are paid very little compared to the ultimate value of the crop. Since there is rarely a premium for cultivated herbs over wildcrafted ones, the time and expense involved in cultivating them on

the forest farm is simply not justified as an income-generating crop. A different approach to generating income from medicinal herbs is for growers to produce these same herbs, not for sales as herbal products (root, rhizome, etc.) but rather as planting stock for others to grow on for local, direct retail sale (farmers' markets, online, and other venues) or personal consumption, or even as garden (ornamental) perennials.

Dave Carman is a grower of forest medicinal herbs, some of which he sells as planting stock and others as an herbal product per se. Read more about Dave in the case study at the end of this chapter.

MULTIPURPOSE ORNAMENTALS

Plants for use as ornamentals are not usually thought of as NTFPs for forest farming. Nevertheless, they are good opportunity for income generation if they can

Figure 7.17. Pot-in-pot nursery at MacDaniels Nut Grove, where hostas and maidenhair fern are grown.

be grown successfully under the relatively low-light regime and other constraints of a forest farm, and if they can be marketed. Ornamentals can be sold to the gardening public for high prices. Even though the range of species that can be grown successfully in a forest farm nursery under shade is smaller than the range of species grown in a typical full-sun nursery, there are many consumers willing to pay high prices for landscape garden perennials well suited to shady conditions, such as hosta, daylilies, astilbe, ferns, and many others.

All the elements of successful large-scale container production of shade perennials are potentially present in the diversified forest farm or garden. Shade is in abundance but sometimes in greater abundance than is optimal for perennials that are shade tolerant but don't absolutely require shade. Hostas and daylilies will grow vigorously under a full or nearly full sunlight but will slow down and perform perfectly adequately in the shade garden. This may require some degree of canopy management, as may be the case for other nontimber forest products in the forest farm, such as berries and pawpaws grown for fruit. Essential components of soil health such as drainage, aeration, and water-holding capacity can be "made to order" in the nursery in part by using soilless mixes tailored to specific crops.

Because the demand for water is reduced for ornamentals growing in a pot-in-pot system (see below), the need for irrigation may be reduced compared to field or aboveground container production, but with high-value ornamentals it is best to have access to supplemental irrigation just in case. For example, at the MacDaniels Nut Grove, hostas and maidenhair fern were grown pot-in-pot for three successive years without irrigation, whereas ginseng in raised beds was irrigated during the hottest part of each summer. Only during the especially hot and dry summer of 2012 did the ornamentals require supplemental irrigation.

Ostrich ferns are notable additions to any ornamental shade garden, but their fiddleheads emerging in the spring are also tasty in a stir-fry. Black cohosh is known for its medicinal value for treatment of symptoms of menopause and menstrual cramps, but it develops a beautiful arching spray of white flowers in midsummer that is a striking addition to any ornamental garden. Wildcrafted black cohosh is sold to end users for impressive prices as an over-the-counter medicinal. At one online site it sells for approximately $7 for a hundred capsules, but the wildcrafter receives a miniscule fraction of that for all his or her efforts gathering it from the woods.

A well-known midwestern herbal products company buys (and sells) over one hundred different kinds of medicinal herbs. They pay $1 per pound for black cohosh root whether wildcrafted or cultivated. On the other hand, the typical retail price paid for nursery-grown 1-gallon containers of black cohosh is about $9.00. Similarly, a well-established, nationally known midwestern herb buyer pays $12 a pound for dried bloodroot rhizome, whereas a forest wildflower nursery in Georgia sells ½ gallon containers of bloodroot for $10.

Other medicinals that also have ornamental character include blue cohosh, fairy wand, trillium, and goldenseal. Ginseng is highly valued as a medicinal crop, but it has largely untapped potential as a garden perennial. Selling ginseng in a container as an ornamental may seem counterintuitive given that ginseng root cultivated and sold as a medicinal is worth several hundred dollars per pound dry weight. But consider that there are usually over one hundred dried ginseng roots in a pound, whereas a single potted ginseng plant could easily sell for $10 or more. Depending on the eye of the beholder, ginseng may not be considered especially ornamental, except for its eye-catching red berries in the fall, but the lore of this species with its exotic association with traditional Chinese medicine and a value far exceeding any other nontimber forest products makes it attractive to shade perennial gardeners, who see more than just what meets the eye.

The low prices paid for dried roots or rhizomes of such forest medicinal herbs is what makes them decidedly unprofitable when cultivated as a forest farming crop, and only marginally profitable when wildcrafted. On the other hand, production of the same species (and others) in a forest nursery can be a lucrative crop indeed.

CASE STUDY: DAVE CARMAN AND HAW POND FARM
PRINCETON, WEST VIRGINIA

Dave Carman's forest nursery is Haw Pond Farm near Princeton, West Virginia. He is a retired telephone industry worker with a love of the woods, where he has spent a lifetime exploring. For the last thirty-five years he has been growing nursery crops, mostly medicinal plants, in the woods behind his house. He told us he thought he grew about fifty different species. My first impression of the nursery was that of an abandoned weed patch (in places), until I started paying attention.

In addition to fairy wand and Virginia snakeroot, another valuable crop that Dave grows in his nursery is American ginseng. As with fairy wand and Virginia snakeroot, he does not grow ginseng for its valuable root, which can only be harvested once. Rather, he grows ginseng as a seed crop, which keeps on giving year after year and is worth on the order of $150 per pound. Although ginseng root grown in a forest garden like Dave's is worth several hundred dollars per pound, it makes sense for Dave to produce and sell seed rather than roots in his relatively small forest nursery because a root must stay in the ground for up to 10 years before it is large enough for sale, but a mature plant produces a seed crop every year.

— Ken

Figure 7.18. Dave Carman's forest nursery in West Virginia specializes in production of planting stock (small rooted seedlings and seed). To the uninitiated visitor (like me) much of this inground nursery appears to be a random, disorganized arrangement of medicinal and other useful plants. This photo shows at least six different species in about 1 square meter of the garden (bloodroot, yellow lady's slipper, Solomon's seal, goldenseal, a lily, and a fern).

Figure 7.19. One of Dave's favorite and most profitable plants is fairy wand (false unicorn). The dried root of this medicinal herb is worth about $60 per pound, but Dave specializes in selling planting stock—both seed and seedlings—for others to grow on. Photograph courtesy of Catherine Bukowski

Figure 7.20. Growing fairy wand begins with germination of hundreds of seeds in a small 2-foot by 2-foot seedbed where seedlings are grown for two years.

Figure 7.21. Two-year-old fairy wand seedlings are transferred to a starter bed (left), where they are thinned as necessary and grown on until they are about 6 years old (right).

Figure 7.22. Six-year-old fairy wand plants are transplanted out into the nursery, where they are grown for several more years, before being either sold as whole plants or grown on in the nursery to produce a valuable seed crop.

Figure 7.23. Fairy wand flower heads are bagged so seed can be collected without scattering to the ground.

Figure 7.24. Virginia snakeroot is another valuable seed crop produced at Dave's nursery. The small vials scattered about are his ingenious way to prevent seeds from falling to the ground and getting lost.

Figure 7.25. To avoid seed loss, an immature seed head is placed into a small aluminum vial, with air holes and a transparent plastic lid. The maturing seed head is placed into the vial still attached to the plant via a narrow stalk (barely visible), which passes through a slot in the side of the vial. When the seed head matures, the seeds drop into the vial, ready for collection by Dave.

CASE STUDY: DAVE CORNMAN AND SPRING HAVEN NURSERY
CENTRAL PENNSYLVANIA

Spring Haven Nursery is a green gem tucked away in the mountains of central Pennsylvania. Although it's not far from where I went to college years ago (Penn State), I got plenty lost. Dave and his dog were both glad to see me and were gracious hosts. While most forest farms I have visited are highly functional but kind of laid back appearancewise, Dave's was immaculate, as if he had been preening for my visit for days, but I don't think so. As we talked near the garage before venturing into the nursery, several charming green and brown cottages (nursery outbuildings, including a potting shed) stood out. And so it began . . .

Unlike Dave Carman's West Virginia nursery (see previous case study), which specializes in producing planting stock (transplants and seed) of forest medicinal herbs, Dave Cornman at Spring Haven Nursery specializes in growing "finished" plants for the ornamental trade. The fact that some of the species he sells as ornamentals happen to be medicinals is a business model that other forest farmers might consider adopting. As indicated above (Haw Pond Farm case study) growing medicinals other than ginseng and a few others for the herbal medicinal trades is unlikely to be as profitable as selling them as ornamentals.

Most of the plants Dave Cornman grows are more or less "strictly" ornamentals, such as maidenhair fern, baneberry, and pitcher plants for water gardening.

The yellow lady's slipper orchid is a thing of great beauty, although difficult to grow in the ornamental garden. Wild lady's slipper orchids are considered imperiled or vulnerable and are listed in CITES Appendix II for that reason. Lady's slipper orchids

Figure 7.26. Spring Haven Nursery lies in the mountains of central Pennsylvania and is a forest nursery that produces a wide range of shade ornamentals. The owners, Dave and Dianne Cornman, built the nursery from the ground up. In addition to the nursery, they built their home, a beautiful display garden, and two or three attractive and functional outbuildings.

were used by herbalists as a sedative, but this is widely discouraged because of their scarcity. Harvest from the wild is legally restricted in some areas and widely discouraged throughout its range. Nurseries typically sell flowering-size yellow lady's slipper orchids for up to $60 per plant. Dave successfully cultivates this valuable NTFP at Spring Haven Nursery, but as a responsible nurseryman, he grows only artificially propagated lady's slipper orchids from seedlings purchased from a tissue culture lab.

In addition to conventional ornamentals some of the plants grown at Spring Haven Nursery can be considered ornamentals or medicinals or both, depending on who's asking. Some examples of these "hybrids" are trillium (red), which is a delightful spring ephemeral wildflower and well thought of as a garden perennial; black cohosh; and blue cohosh.

For example, containerized black cohosh (figure 6.11, page 217) has a beautiful white arching flower stock, making it as an ornamental worth many times the value of the dried medicinal root. Another medicinal, blue cohosh, makes an attractive border plant in the garden, and trillium, known as bethroot in the medicinals trade, is one of Dave's specialties, although in late July, when figure 6.12 was taken, it looks a bit shabby. Although ginseng is not often considered an ornamental, at Spring Hill some shade gardeners use it as a unique specimen plant, not only for its attractive red berries in the fall but also for the lore and tradition associated with it. He sees considerable potential for this species as a garden perennial.

— Ken

Figure 7.28a. The yellow lady's slipper orchid, like all species of lady's slippers, is challenging to grow in the ornamental garden but is stunningly beautiful for all the trouble.

Figure 7.27. The nursery is well organized from the standpoint of efficiently moving plants and other materials, and it presents an attractive forest vista for visitors. Customers can see the same plants that are for sale in the nursery in the relaxing setting of Dave's carefully designed landscape garden.

Figure 7.28b. At Spring Haven Nursery Dave Cornman grows yellow lady's slipper orchids in raised beds from laboratory-grown seedlings.

CASE STUDY: INNOVATIVE PROPAGATION AT FORREST KEELING NURSERY ELSBERRY, MISSOURI

Driving by Forrest Keeling Nursery, in the quiet Midwest town of Elsberry, about one hour north of St. Louis, one wouldn't assume that anything other than the standard nursery operation was going on. The front drive is lined with a planting of oak trees, all looking healthy and vibrant, averaging maybe 6 to 8 inches in diameter. What isn't obvious from the outset is that these trees are growing two to three times faster than they would normally. This is achieved by a process called "root production method" or RPM, developed by the longtime owner, Wayne Lovelace.

Essentially the RPM method is a twelve-step process that varies minutely depending on the species and includes air root pruning, specific soil media, multiple steps in transplanting the trees, and the including of slow-release fertilizers and mycorrhizal fungi. This extra care and intensive management pay off; the trees leave the nursery with an incredibly dense and fibrous root system, which not only ensures the trees' survival but also rapidly accelerates the growth of the tree. Measurements from the University of Missouri on one of the seed orchards I visited confirmed that the oaks were growing an average of ¾ inch *per year*. This is unheard of.

While the company is careful to guard its secret process (which is patented), Wayne was willing to share the basics: good seed stock; a very well-drained

Figure 7.29. This oak seedling has incredible root structure and is just twenty-six weeks old.

Figure 7.30. These Chinese chestnuts were flowering and setting nuts after only eighteen months of growth!

soil (they use about 80 percent sand and the remaining percent in pulverized bark mulch and pine bark and rice hulls); air-root pruning (when roots are exposed to air in the absence of high humidity and are effectively "burned" off, causing the plant to constantly grow new, dense roots), and even a form of minicoppicing that happens during the first stage of growth to encourage vigor.

During my tour I witnessed eighteen-month-old chestnuts that were flowering and almost 4 feet tall, oak trees that were almost a foot in diameter after just ten years of growth, and a tree seed orchard of mostly oak that was a closed canopy, with oaks producing heavy yields in around ten years, with no maintenance, fertilizer, or spraying. The nursery offers a number of tree crops of specific interest to forest farmers, including grafted Peterson pawpaws, selected Chinese chestnut varieties, and Cornell's 'Super Sweet' sugar maple, which produces sweeter sap.

Forrest Keeling has an annual production of over four million plants encompassing more than two hundred species, with a focus on native and nut-producing trees. The nursery works with multiple municipalities and conservation groups and mostly deals in wholesale quantities, unless you visit the nursery in person. I was struck with the potential of other growers to experiment with some of the intensive methods of propagation practiced at Forest Keeling that could result in a higher survivability of plants, which is always a setback in cultivation. Mushroom growers might grow their own oak logs for production in under a decade. Sugarmakers might begin producing from new trees in less than twenty years. If nothing else, the potential for speeding up the growth and development of tree crops might turn some heads and bring more attention to the feasibility of forest farming. One drawback is that the prices are high.

— Steve

Figure 7.31. Wayne Lovelace leans against swamp white oak that is almost a foot in diameter and about thirteen years old. Head of propagation Lupe Rios stands nearby.

Wood Products

The full picture of forest farming is not complete without a look at some nontimber forest products other than medicine, food, and ornamental crops. The following items are not often discussed as part of forest farming, yet they offer the opportunity to expand and develop the possibilities. It's easy to forget that much of the modern world was built on the shoulders of wood resources, not as food or medicine but with a number of functional products that provide warmth and fulfill the daily functions of life. This chapter offers a look at practices and species that should be of great interest to any forest farmer, in an effort to further expand the concept of forest farming and rekindle traditions that have nearly been lost as livelihoods and art forms.

It is not only the functional or practical uses of wood that are important; as society has moved away from depending on trees as a basic material for sustenance, lost with it are a legacy of artisans and craftspeople who developed specialized skills and passed them down from one generation to the next. Largely gone are the charcoal makers, basket weavers, and barrel builders. And gone with these practical and creative artisan relationships to the forest is something even bigger—connection. As Eric Sloane notes in his classic chronicle of people's relationship to forests, *A Reverence for Wood*:

That century of magnificent awareness preceding the Civil War was the age of wood. Wood was not only accepted simply as the material for a building of a new nation—it was an inspiration. Gentle to the touch, exquisite to contemplate, tractable in creative hands, stronger by weight than iron, wood was, as William Penn has said, "a substance with a soul." It spanned rivers for man, it built his home and heated it in the winter; man walked on wood, slept on it, sat on wooden chairs at wooden tables, drank and ate the fruits of trees from wooden cups and dishes. From cradle of wood to coffin of wood, the life of man was encircled by it.[1]

In any ecosystem there are two main flows: energy and materials. In general it can be said that energy enters and leaves the system and materials cycle.[2] For example, sunlight is captured by trees (producers) that store the sunlight in various forms, most notably in a mix with carbon to form the trunk that is the wood of the tree. If the tree is cut and harvested, that relatively stable storage of energy begins to dissipate. Feed the log to shiitake mushroom spawn, and while that energy is being consumed, it ultimately produces mushrooms and cellulose (the decayed log). Humans consume the mushrooms, and their waste then flows to another place (depending on the system), and the log, if left in the forest, feeds the soil biology. Thus, energy is the invisible currency that connects each step of the process, while materials are all the solid, tangible elements—the leaves, the bark, the wood, the mycelium, the mushrooms, the humans, and all the trace minerals and nutrients that are passed from organism to organism.

This chapter examines the potential ways we can integrate energy and materials production from the forest with other forest farming pursuits. Since the largest domestic use of fossil fuels in temperate

MAKING CLOTHESPINS ON THE FLY: A LESSON FROM THE WOODS

On a kayaking trip a few summers ago in the Adirondacks, while setting up camp I became frustrated at the wind that was blowing our damp clothes off the line. "Would've been nice to bring clothespins" was my thought. Then I remembered a wilderness skills school I'd been a part of, where another instructor showed me how to make clothespins from split wood. I looked around and discovered a small stand of striped maple, which is a moist wood and easy to split with a knife when green.

This simple activity was a bit of an epiphany for me, one I've experienced in similar ways over the years from the forest. There is something basic and primal in the experience of needing something, then examining your surroundings and finding it there, staring you in the face. It only takes recognizing that it exists.

There is always the opportunity to see the forest in new ways, and to value the gifts it can offer. It's easy with conventional forestry or agriculture or environmental training to look only for the obvious wealth: mature species of a valuable timber, rich loamy soils, and rare native species. The truth is, every ecosystem has secrets—and gifts to share. It's merely a question of choosing to see them or not. This chapter is about "seeing the forest for more than the trees." While people often think of forest farming as mushroom cultivation or growing ginseng, there is so much more potential for farming the woods.

—Steve

Figure 8.1. Making clothespins from striped maple. The process of meeting needs directly from the woods leads to seeing the forest for much more than just the trees.

climates is for heating buildings, the art and science of burning wood cannot be left out of the nontimber forest products equation. In addition, the managing of forest farms can provide a wide range of materials for art, craft, and functional materials that become useful around the land or can be sold commercially. A large sector of energy consumption comes in the form of materials that we use in everyday life, especially for building. In the current industrial economy, for instance, pressure-treated lumber is much easier to come by than black locust. Plastic and metal, which are energy intensive to harvest, process, and refine, have largely replaced many of the containers, utensils, tools, and day-to-day materials that are necessary for a comfortable life. Animals can play a part in the forest farm, too, as they cycle energy and materials and provide additional yields and benefits in the process.

So in the end, while much of the appeal of forest farming comes in growing food, medicine, and other crops, paying attention to the material and energy flows of the forest is an equally important endeavor, especially in the context of climate change and rising fuel costs. Including wood products in a discussion of forest farming rounds out the possibilities for enjoyment, income, or both.

Calculating Fuel for Home Heating

Wood is the most fundamental source of burnable energy. It preceded any metal ages present in human history by necessity, as fire was in fact crucial to the smelting of metal. Indeed, much of the fossil fuels we burn today are merely the deposits of slowly decaying forests over millions of years. As fossil fuels were discovered, they largely replaced wood in direct and indirect ways. The most obvious replacement, especially in the northeastern United States, was for heat. Oil, coal, and eventually gas proved to be simple and cheap substitutes, and in some cases often cleaner, too (at least in combustion), compared to dirty fireplaces and early woodstoves. And since they were so cheap, who wouldn't want to switch? Today it is estimated that 98 percent of Americans use fossil fuels or electricity for heating, versus just 2 percent using wood. In New England the rate of wood use for space heating, water heating, and cooking is nearly twice the national rate, including close to half of all rural households.[3]

With fluctuating supplies and costs for fossil fuels, coupled with their impact on climate change, people all over the world simply can't afford to continue using fossil fuels for all their energy needs. Further, with the "easy" access to fossil fuels diminishing, new technologies that are more risky and unknown with regard to environmental consequences are emerging, from the tar sands to hydrofracking. While arguments from both sides highlight this possible benefit or that possible risk, the reality is that no one knows the entire extent of possible damages to health and the environment. As one example among piles of them, Dr. Anthony Ingraffea of Cornell University established that Marcellus Shale well casings have failed, meaning fluids, methane, and/or drilling waste have escaped into the ground, at a rate of 6.2 percent in Pennsylvania in 2010 and 2011, based on data from the Pennsylvania Department of Environmental Protection.[4]

The reality with any fossil fuel extraction is that it is extremely resource intensive. Other than the pos-sibility of cheaper energy, it does not give back to the communities it takes from. It is not a technology that is within local means of control and access, so individuals are forced to purchase it from a multinational corporation. It does not improve soil and water quality, nor does it improve wildlife habitat. While good forest management *can* do these things, it is not a given.

If humans do not begin to find "win-win-win" solutions, in which humans, economies, and the environment all benefit from an activity, then society is certainly destined for ultimate failure; that is, collapse. There are plenty of examples of agroforestry systems that benefit communities, individuals, economies, and the environment.[5] And the good news is that by nature these systems are decentralized, localized, with the tools and means of production in the hands of the people; namely, farmers. It is with this context that the importance of fuelwood to forest farming is offered. While this discussion could go much further beyond home heating, exploring the possibilities of biomass, wood gasification, and so forth to provide heat and electricity, for the purposes of this book fuelwood production is primarily considered, as it is the most accessible option to many readers who own or manage woodlots; in the end home heating is a common need for anyone in a cool climate, and while its use is on the decline as homes become more efficient, it still accounts for a significant percentage of individual energy use.[6]

Many people living in colder climates already rely on the woods to keep them warm; indeed most modern-day nontimber management in forests is for firewood. Some even claim a living from harvesting, processing, and delivering wood to their neighbors. And yet while wood use is somewhat common in rural America, much of the subtlety—the "art" of burning wood—is largely lost. This has implications not only for the environment (in the form of emissions) but also in considering the efficiency and economics of burning wood. To design a home-scale wood-burning system, a few questions need to be asked.

HOW MUCH WILL BE BURNED?

The first question to ask revolves around the volume of wood that will be burned. This depends on a combination of three factors: the climate, the size and insulation quality in the structure, and the type of wood being burned. Another, harder to measure variable would be the desired comfort of those who will be in the space, as some like it cooler and some warmer.

Climate = Heating Degree Days

By looking up figures online, a landowner can determine the number of heating degree days (HDD) in her area, which is determined by taking the average difference in temperature of a given day (for example, a day with a high of 40°F and a low of 20°F = 30°F as an average), then subtracting this from 65, the base temperature of a building. Adding up all the days in a given place that need heating results in a total number of heating degree days.

Home Size and Insulation

The sizes of homes are expressed in square feet, which if unknown can be easily estimated by taking the overall square footage of the footprint (length × width) times the number of floors. For the insulation value the home heating index (HHI) is valuable, with ratings from 1 (airtight) to 23 (tent). For our purposes the ratings have been simplified a bit, but more research can help determine the value of a particular dwelling.[7]

Wood Type

The species and how dry the wood is (see table 8.1) has a great effect on how much wood will be needed. The values are based on wood that is reasonably dry, around 20 percent moisture. Note that wet wood will provide significantly less heat value if not allowed to dry out.

Putting It All Together: An Estimate

The combination of factors described above provides a starting point for estimating the amount of wood needed. Follow the chart cross-referencing the heating degree days and square footage for the space that is heated. The different requirements can then be compared based on a relative insulation quality in the house. For example, a 1,000 square foot house in Ithaca, New York would fall into the 7,000 HDD category, meaning that the homeowner would need between 1.5 and 3.3

Table 8.1. Heat Value of Selected Wood (Higher Number = Higher Value)[8]

Species	BTUs when Dry (20%)
Hickory	27.7
Eastern Hornbeam	27.3
Black Birch	26.8
Black Locust	26.8
Apple	26.5
White Oak	25.7
Sugar Maple	24–29
Red Oak	24
Beech	24
White Ash	23.6
Yellow Birch	23.6
Hackberry	20.8
Birch (not black)	20.3
Walnut	20.3
Cherry	20
Black Cherry	19.9
American Elm	19.5
Black Ash	19.1
Red or Soft Maple	18.7
Sycamore	18.5
Box Elder	17.9
Conifer average	16.5
Aspen	14.7
Butternut	14.5
Willow	14.5
Basswood	13.5
Cottonwood	13.5

cords of wood to heat the house, depending on the quality of the insulation and overall "tightness." The average home would need around three cords of wood.

How Many Trees per Cord?

If you are purchasing wood, it's simple from the information above to order the correct amount of cords. But more likely than not, forest farmers will be harvesting at least a portion of their own firewood, if not all of it or even a surplus to sell. The next question

Table 8.2. Estimated Cords of Wood Needed for Various Homes

Rating Based on HHI	Sq Ft of Home	9,000 Annual HDD (Maine)	7,000 Annual HDD (Central NY)	5,000 Annual HDD (West Virginia)
	Ft2	A	A	A
Excellent (4)	1000	1.7	1.3	0.9
	2000	3.4	2.6	1.9
	3000	5.0	3.9	2.8
Average (8)	1000	3.4	2.6	1.9
	2000	6.7	5.2	3.7
	3000	7.6	7.8	5.6
Poor (10)	1000	4.2	3.3	2.3
	2000	8.4	6.5	4.7
	3000	12.6	9.8	7.0

Note: Cords of wood needed based on climate, home heating index, and square footage of the home, using average hardwoods as fuel. A more complete table can be found at the cited source. (Adapted from Fox, 2013[9])

Table 8.3. Estimate of Trees Needed per Cord of Wood

Tree Diameter at 4.5'	Number of Trees per Cord	Cords per Tree
5"	50	.02
6"	20	.05
7"	12	.08
8"	8	.12
9"	6	.17
10"	5	.21
11"	4	.25
12"	3.5	.30
14"	2.5	.40
16"	2	.50
18"	1.5	.65
22"	1	1.00

Source: Based on research by Gevorkiantz and Olsen, 1955[10]

then becomes, how many trees are in a cord of wood? Table 8.3, though of course simplified as an average for hardwood species, offers this information. It's easy in the field to keep a tally of the number of trees felled for a certain diameter, then add up the cordage back at home. To put it simply, for the example home above, if 10-inch trees were being felled, then fifteen would need to be cut in order to supply three cords worth of wood.

Acquiring and Drying Wood

Whatever the volume that is ultimately determined, wood storage should ideally be sized for two to three times this amount, because to really have dry wood, you need to season it for at least two to three years. In a forest farming situation, firewood can simply be the surplus yield of all other activities: harvesting materials for mushroom logs, craft projects, and thinning to improve forest health. Whether you harvest all the wood yourself or purchase some from a dealer, the bottom line is that it needs to be coming from a system that puts ecosystem health first.

Since today much of the forest is either valued for timber or hardly valued at all, the practice of harvesting firewood has become one of little consequence in the mind of many. Unfortunately, this means that choice trees that should be left on the stump may end up in the firebox. Consider the time and energy investment each tree in a forest has put in to grow; trees that are burned should really be the last use of wood, when all other valuable uses are exhausted. Even after applying this principle, there is still, in reality, plenty of wood in the forest for burning.

Besides respect for the subtle nature of each species and the burning qualities it can bring to the table, the other big gap in competent woodburning practices is in taking the time to properly age wood before burning it. Life always seems to get in the way of the best-laid intentions to get to that pile of logs that needs to be moved, split, and stacked, and yet it doesn't save us any time to put off the inevitable task before us; indeed, delaying actually ends up making for more work in the long term.

This is because wood that is not properly dried wastes energy in the burn; the fire must first "dry out" the wood before it can burn, which reduces the heat output. The fact is that it takes a much greater volume of wet wood to provide the same heat as a volume of dry wood, which means more felling, hauling, splitting, and stacking. Better to save time in the long view by getting ahead of the game and making sure that wood is properly dried.

The other issue with wet wood is that it burns less completely. As responsible stewards of forest resources, forest farmers are likely to consume the most wood in a lifetime as firewood; this means that if one goal for management is to reduce negative impact on the environment, the first task should be to get the home system in order.

It is not sufficient to simply fell the trees and leave them lying full length for proper drying to occur. The wood needs to be both split and stacked for it to dry. Wood fit for burning needs to be well below 25 percent moisture content, a level that is difficult to achieve in round form. In addition, piling wood only traps moisture and doesn't facilitate good airflow, which improves drying.

Figure 8.2. Wood stacked at Wellspring Forest Farm. Piles are oriented to face prevailing winds and covered with old metal roofing to shed rain and snow. Wood should be cured for at least twelve to eighteen months after splitting for optimal combustion in the woodstove.

Another common error is covering the pile completely with a tarp or other membrane, to shed rain. This also can trap moisture. Best is to cover just the top of the pile, with a 4- to 6-inch overhang, so that rain sheds but the wood pile can still breathe and benefit from exposure to the elements that aid in its drying. This also equates to doing some thinking as to the siting of a wood stack. Line up your wood so the stack faces the prevailing wind and can also receive good sun exposure (south facing). Each of these details will ensure the fastest drying time possible.

Choosing a Good Stove

In most cases with cordwood, woodstoves are the primary technology to create heat. Yet there are so many different possible stoves to choose from. The criteria for any space and style of stove starts with the following aspects:

AIRTIGHTNESS

The stove compartment should be airtight to avoid leakage, which prevents the user from having complete control over the burn. Stoves that have multiple levers, doors, and pieces are all prone to air leaks and must be checked every season. Better to purchase a stove that is minimal on the gadgetry, well insulated, and airtight by design.

SIZED APPROPRIATELY TO THE SPACE

One major mistake people make is getting a stove that is too big for the space, which ends up putting out too much heat. Many people figure that oversizing a stove is a good idea "just to be safe," but the reality is that a stove that is too big means that the users won't have fires that burn hot and complete, which results in more creosote buildup in chimneys (and potential chimney fires) and greater emissions. If the chimney has to be cleaned each year and there is a heavy amount of buildup, the fire may be burning in a stove that is too big.

SECONDARY BURN TECHNOLOGY

Almost all modern stoves contain some sort of system for recapturing and reburning gases from initial

WHAT ABOUT OUTDOOR BOILERS AND PELLET STOVES?

Other options for home heating (besides those that utilize fossil fuels) are outdoor boilers, which burn wood of all shapes and sizes, and pellet stoves, which burn corn or wood pellets.

An outdoor boiler is essentially a woodstove enclosed in a small, insulated shed outside the home that heats water in a jacket surrounding the stove. The water is then carried to the home via pipes. This system has the advantage of heating water for both space heating and hot water use. It can be especially useful in situations where multiple buildings need heat; for example, a house and a barn.

The issue with outdoor boilers is that they smoke, mostly because, since the stove is enclosed in a water jacket, the wood cannot fully combust. Another factor that contributes to the smoke is the on–off cycling of the system, which creates creosote buildup that is then rapidly burned off when the system starts up again. Simply put, outdoor boilers have a lot of trouble with emissions.[11]

The pellet stove is a system in which a "hopper" is filled with pellets of corn, grain, and/or wood, then fed into a combustion chamber, which is fed by a fan to create airflow. Pellet stoves are appealing because the fuel can be easier to move, store, and use, and the stove can turn on and off via a thermostat, based on the need of the home (no coming home to a cold house).

The biggest drawbacks to a pellet stove are the need for additional electricity (for the hopper and a fan) and, even more important, the reliance on an off-site material for fuel. In climates where heat is an essential function of daily life, relying on the feed store to supply fuel leaves the site very vulnerable. Many locales already experience shortages in supply toward the end of the cold season. Perhaps someday local manufacturing of waste wood into pellets could complement sustainable forestry practices and make this a more viable option.

In the end, good old wood heat in a woodstove offers the most straightforward option, in the authors' opinion, though it does require more labor and planning if it is done in a way that supports a healthy forest and results in a clean burn.

combustion, thereby reducing emissions and increasing stove efficiency. Older stoves may lack this technology. Catalytic converters were the first version of secondary burning, though they tend to make the stove extremely heavy. Newer stoves often have a simple baffle design and reticulate hot air and gases through a series of metal tubes in the stove.

Choosing the right woodstove is an important investment of time and energy. Take time to talk to others that have stoves and to visit as many professionals as possible. Going cheap and not getting the appropriate stove for your needs and space means more problems down the road. As tempting as it may be to get an old stove for a few hundred bucks, in most cases they are simply not that efficient. Modern stoves have come a long way and in the long term are worth any additional investment.

HOW TO START A FIRE

Starting a fire may seem like a basic task, but to avoid problems, such as not getting that initial draft so the smoke travels out of the building and working with wet materials, it is useful to review the basics of starting fires.

Most people think of the necessary pieces to get a fire going as kindling, wood, some paper, and matches or a lighter. However it's useful to have on hand the following if you wish to make a quick, hot-burning fire from scratch.

- Tinder: Dead grasses/goldenrod, grapevine, birch bark, "punkywood," and so on are the materials to light the fire initially.
- Small Sticks: Sometimes called "whispies," these are the fine and resiny sticks from conifers; pines, hemlocks, and so forth that ignite quickly.

Figure 8.3. The key to starting a good fire is to have a variety of sizes of dry material on hand, from large to small. Conifer species tend to work best, as they burn hot and quick.

- Larger Sticks: These form the skeleton of the fire and burn for the first twenty minutes; they are the first "soft" coals of the fire.
- Small Cordwood: Split cordwood that is thin (1 to 2 inches thick) burns without overwhelming the fire; this is the "kindling."

All these items can be collected with ease from any forest without damage to the ecosystem. Be sure to not collect all this material from one location only, as some needs to be left for soil building in the forest.[12] They should all be brittle and dry without appearing or feeling rotten. Collecting dry wood in the forest is a learned skill, so pay attention. Sticks that are in constant contact with the ground are not as good as those hanging from trees, or propped up in some fashion. It is a good idea to stockpile a surplus of these materials in a dry place to make your work easier.

Once the necessary materials are assembled, begin by building a small tepee, using first some larger sticks to build a skeleton, then surrounding the teepee with smaller sticks, leaving a portion open in the tepee as a "door." The whispies are placed in the door, leaving a gap in the middle of the tepee: This is where the tinder can be placed, which is the first thing that is lit.

The Burn

The overwhelming focus in many texts is how much heat value there is in a given species of wood, and while this is a good starting point, it's not the only variable to consider. As in the forest, diversity is key—it's best to have woods that burn hot and fast during the initial burn, transitioning to higher-heat woods that coal well, and finally a third transition to logs that will burn medium-hot and last a long time. These three stages are called the initial burn, the coaling period, and the sustaining period. The respect given to the specific properties of tree species is the biggest hole in proper fire management as a heat source. As a general rule coniferous wood is not appropriate for indoor burning, as the resins in the wood can build up creosote in the stovepipe.

WOOD TYPES WITH EXCELLENT COALING PROPERTIES

According to observation and wood-quality tables, these woods have excellent coaling properties.[13]

- Cherry
- Beech
- Honey locust
- Black locust
- Oak
- Mulberry
- Apple

INITIAL HOT BURN

The first thirty to sixty minutes of a fire should be a hot burn, to bring the stove metal up to temperature and burn residues from the chimney stack. Woods such as basswood, beech, and ash are ideal for this sort of combustion. A stovepipe thermometer can help monitor the activity. Fires should get into the "burn zone" for ideally forty-five to sixty minutes as much as possible, but at least once per week. This helps keep the stovepipe clean and free from the potential of a chimney fire.

COALING

After the initial burn the goal of any good fire should be coaling; that is, creating a bed of solid coals that will sustain the fire. Hands down, black locust is often the preferred species, as a hot fire with black locust will create long-lasting coals that almost "melt" as they burn. Ironwood, hickory, and walnut are also good choices.

SUSTAINING A BURN

A solid bed of coals can last indefinitely with the addition of high BTUs and slow-burning hardwoods, which are sometimes affectionately called "all nighters" for their ability to last out the late hours of the night and into the following morning. When splitting wood, aim to identify oak, hickory, walnut, and hornbeam, and keep some of these logs larger to improve the lasting effect.

In the end wood burning requires experimentation and keen observation on the part of the wood burner to achieve success. A large learning curve for many

STACKING FUNCTIONS WITH A WOOD COOKSTOVE—
WHOLE SYSTEMS RESEARCH FARM, MORETOWN, VERMONT

A long-lost fixture of any American kitchen is the wood cookstove, which provided heat for the home as well as the means to cook and bake for the family. Ben Falk has been rediscovering the role of the cookstove in modern resilient living, which he discussed at length in a recent book titled *The Resilient Farm and Homestead: An Innovative Permaculture and Whole Systems Design Approach*.

Ben acquired a rare Waterford cookstove, which can be hooked up to heat hot water. As Ben writes, "The woodstove has become the logical power center of my own resilient homestead. It's nearly impossible to break a wood-powered heating system, and if it does . . . it's easily repairable by low-tech, often on-site means."

Over the last several years Ben estimates that the stove, which is rated for 35,000 BTUs, heats his 1,500-square-foot home, heats all the hot water needed, bakes and boils their meals, dries clothes, and provides firelight to boot. All of this on about 2 to 3 cords of wood per season. Considering that home heating is such a large part of energy use, this multifunctional approach really gets at the meaning of efficiency in the highest regard.

Figure 8.4. This stove heats the home and hot water while cooking food, simmering tea, and drying herbs and other storage foods. Photograph courtesy of Ben Falk.

involves learning to identify species from a lone chunk of wood among so many in the pile.

An Alternative: The Rocket Mass Heater

There lies in the conventional woodstove design a basic assumption that for wood to combust it must be burned in a system where there is extreme draft, created by the rise in the chimney. While alternative options have some ways to go before being adopted for home use on a common scale, several promising alternatives do exist, if only experimental. As with many topics in this book, ample details exist elsewhere, so here there is just brief mention to bring them to the awareness of readers.

The technology inherent in a rocket mass heater is a wonderful example of how setting good criteria can lead to impressive results. Designers built a stove that uses a fraction of the wood consumed in a regular woodburning stove, all while producing a very low-emission fire and one that is smokeless. Stoves are essentially constructed from bricks and cob, a mixture of clay, sand, and sometimes straw. All this, plus the fact that heat is stored and keeps a space warmer longer, are reasons to pay attention to this technology.[14]

Rocket mass heaters work on the basic tenets of thermodynamics, and the design maximizes a very hot combustion that works in dramatic contrast to the simple woodstove's. The common way to burn wood is simply in a firebox, with a chimney to the outside

Figure 8.5. Rocket stove all cobbed up.

that creates "draft," or the fact that heated air will rise, and when forced through a tight space (the chimney), it gets sucked out at a rapid degree. The problem with this concept is that along with that quick-moving air are gases and particles (smoke) that escape along with the air. The wetter the wood, the more this happens. But even with the driest of woods a simple woodstove may only be 50–75 percent efficient, as heat and energy is a necessary sacrifice based on the design of the stove.

A rocket stove is different, as it operates on the basic idea that the chimney should be contained within the firebox to fully combust as much of the wood as possible. This is done by enclosing the chimney and running gases through a thermal mass, where they are absorbed and able to cool as they leave the building. The exhaust is remarkably cool when it exits, which technically means that combustion is almost complete in this stove.

Another advantage of this construction is that after creating the initial form, the entirety of the stove can be covered in cob, a mixture of sand and clay that will harden and provide thermal mass to store heat. This storage acts like a battery, where a fire "charges" the mass, then releases it over a much longer time period. Types of construction can also be creative, as heated benches and even beds can be incorporated into the design, making for extra cozy spots in the space.

It's good to begin experimenting with these stoves outdoors before committing to utilizing one indoors.

The stoves are incredibly easy and straightforward in their basic construction, but it takes some trial and error to get the details right. They are also rather cheap to build in terms of material costs. Of course, local code may not approve of this stove for a residence, so check with local municipalities before moving ahead with plans. Secondary spaces such as greenhouses might especially benefit from this technology, as one could light a fire and burn it for a few hours while working inside, then leave and allow the thermal mass of the stove to continue heating the space.

Materials from the Woods

A wide range of practical and necessary objects composed of wood and essential to agriculture exists (or used to), from fence posts that keep animals contained to garden stakes for trellising plants to handwoven baskets used to harvest mushrooms. Wood provides considerable domestic and farm products that, with the advent of cheap fossil energy, have been largely replaced by plastic and other materials. Provided here is a basic overview of the possible opportunities of working with wood that are arranged from use of larger trees down to smaller stems. In the framework of a forest farm, the discussion of wood as a material revolves around a scale where wood can be manipulated with simple hand tools or low-cost equipment. Certainly, wood could be worked at a larger scale in shops and mills, but in the forest farm vision the participants are maintaining a mix of growing food, medicine, and wood products. Thus, focus of this section is on woodworking strategies on a small to medium scale, with an aim to be as low tech as possible.

Large-Diameter Logs: Woodworking

Depending on the species, any tree over 8 to 10 inches in diameter and less than 16 inches should be considered a suitable candidate for woodworking. Arguably, trees over 16 inches and those in excellent health and form should be either left permanently on the stump or harvested and sold to a mill. A skilled

TOOLS FOR THE FOREST FARMER

Before getting into specific practices associated with wood products, a small list of some necessary tools will help frame the conversation by giving potential forest farmers a sense of the materials and skills required. Note that the focus here is on making use of small hand tools, which open the playing field for many novices to experiment with these crafts. The projects included here are ones that can be done out in the woods or in a barn with cover, using simple human-powered tools. Of course, woodworking can go much further.

Tools for Harvesting and Splitting Wood

Folding Saw, Pruners, and Loppers—for harvesting small- and mid-size wood.
Chain Saw—for cutting whole trees (safety gear/training a must; see chapter 10).

Ax/Maul/Hatchet—for rough splitting and shaping.
Wedges and Gluts—for splitting out lengths of wood. Wedges made from metal start the split, while gluts, made from wood, hold the split open.
Froe—for precise splitting, traditionally used for woodworking, shingle making, etc.

Tools for Refining and Finishing

Adze—Both hand and larger-size tools "gouge" out wood for bowls and basins.
Drawknife—a sharp, beveled blade and two handles for bark removal and shaping.
Draw Bench—the companion to the drawknife allows for ergonomic working.
Fixed-blade knife—Chisels—a wide range of shapes and sizes for more accurate shaping.
Sandpaper—not really a "tool" but useful to have around for projects.

woodworker should be able to fell a tree and strive to use all but the smallest braches, which can be left in the forest for the benefit of the soil.

In woodworking, a log is taken from its natural state and transformed through a variety of manipulations, each which requires a specific set of tools and the skills to go along with them. Table 8.4 arranges the possible strategies starting with raw materials and rougher

forms of manipulation and working toward finer forms of finishing.[15]

Some products to consider as a starting point for working with wood would prove to be a handy asset to any farm and also offer some direct financial benefit, whether from saving money or from the sales of the materials themselves. The process of taking a whole tree to a finished product is one that takes time and

Table 8.4. Ways to Work Wood

Method	Description	Products	Tools Needed
Sawing/Pruning	Cutting wood to length or removing portions of wood with cutting motion	Every product starts with this!	Folding saw, pruners, loppers, chain saw, carpenter's saw
Splitting	Separating wood fibers along the grain as an end product or to further refine	Split fence posts, stakes, basket and furniture materials, woodworking pieces	Ax, maul, wedges, gluts, froe
Hewing	The old way of "milling" or making square beams and posts	Beams for construction, bridges, etc.	Broadax, adze
Shaping, Chiseling	Removing bark and wood while creating gouges and curves toward a finished product	Tool handles, spoons, bowls	Draw knife and bench, chisels and mallet
Boring	Penetrating wood with a hole	Holes for pins, bolts, pivot points, etc.	Hand or power drill

patience, not to mention practice when new to the work. While more specialized woodworking can be left to neighbors and friends devoted to the skillset, some of the general products listed in table 8.4 are a good "jack-of-all-trades" endeavor.

SPLIT FENCE POSTS

Larger trees can be harvested and split into fence posts; black locust, larch/tamarack, cedar, chestnut, and oak are good candidates.

HEWN BEAMS

This is a way to work wood into a roughly square shape, without the need for machinery. A broadax, a blade with one edge flat and the other beveled, allows for this work to be done effectively. Many hardwoods and conifer species are good for hewing, provided they are straight grained.

TOOL HANDLES

For durability these are best split out from larger logs rather than polewood because the strongest wood is the heartwood of the tree, which has "aged" in the trunk over the years. The advantage to making tool handles on-farm is that they can be customized for more comfortable use.

TOOLS

Many basic useful tools such as mallets, mauls, and wedges can be split out and shaped from the heartwood of larger trees. Mallets, which can be used to pound chisels and froes for shaping and splitting, are best made from the knotty wood at the base of a hardwood tree such as ironwood/hornbeam and hickory. Digging out a root ball from a dead or dying tree makes for a mallet that will last a much longer time.

Figure 8.6. Sean Dembrosky works on splitting a locust log for post material at the Good Life Farm, which sourced all its livestock fence posts from the woods. He also sells the material locally for $8 to $10 per post. Photograph courtesy of Melissa Madden

SPOONS, BOWLS, AND HOUSEHOLD ITEMS

Many of these can be fashioned from the remnants of splitting out logs for other purposes outlined above.

CORDWOOD FUEL

Trees that are dense and hard offer the fire a long burn time but are often tedious to dry out. Species such as oak, hickory, cherry, locust, and walnut should be harvested at a large diameter and split into cordwood, in contrast to some of the softer species that can be harvested at polewood stage for fuel (see Roundwood Fuel on this page).

Medium-Diameter Logs: Polewood

Dimensional lumber certainly has its benefits. In many construction projects, straight edges and uniformity mean that construction can happen quickly and efficiently. Polewood in this book is meant to be wood that is small diameter (4 to 8 inches) and intended to be used in "whole form"—that is, to remain in the round of its natural form, usually with the bark removed.

The practice of harvested polewood is often associated with the practice of coppicing, which is cutting a tree to its base and allowing the trunk to resprout. The shoots are then thinned and managed to a variety of diameters, depending on the end product. One of the beautiful potentials of coppice is that thinning the initial shoots results in material for bentwood work (see below) and at the same time supports better growth of wood into the polewood stage.

The biggest advantage of polewood in comparison to the large-diameter wood above is the labor-saving nature of harvesting many of the same products (fence posts, firewood, and building materials) at a younger age, which saves the need to take time and energy to process them. A forest farmer may be lucky in some cases to have a young forest containing this size wood in significant populations, but often this stage is something that can be managed for, with multiple harvests possible within one's lifetime through coppicing or pollarding (coppicing above the height of browsing animals).[16]

Possible polewood products the forest farmer should take into consideration include the following.

Figure 8.7. Roundwood construction provides beauty and function for this simple arbor constructed from locally sourced materials by Michael Judd of Ecologia Design. Photograph courtesy of Michael Judd

ROUND FENCE POSTS

Fenceposts can be harvested at the right diameter (4 to 6 inches, usually) to avoid needing to split them, which can take a long time. Rot resistance is key, since they maintain contact with the moist soil for the duration of their lifetime and the species mentioned in the previous section are the best for the job.

ROUNDWOOD FUEL

At least a portion of firewood can be from roundpoles, but because not splitting the wood often means it won't dry as fast, extra drying time may be needed. The "softer" hardwoods—essentially *not* oak, hickory, locust, and so on—are better candidates for this type of harvest, since they are not as dense and will tend to dry better in the round.

STAKES

Technically, stakes could be split from large wood, but polewood offers an efficient size to work from when making stakes en masse. The possible uses are endless: staking plants and trellises in gardens, staking logs for terraced beds, and even staking ties for tarps and temporary roofing. These can often be made from the waste materials of other projects. It's nice to have a bunch on hand, ready-made for when the need arises.

MUSHROOM LOGS

A primary yield for the forest farmer from polewood should, of course, be logs for mushroom cultivation. Logs from 4 to 8 inches can be used for bolts and logs from 8 to 10 inches for totems.

Small-Diameter Poles: Craftwood and Bentwood

The threshold for calling a material "bentwood" is simple: If the tree is small enough in diameter that it can be bent without breaking and is either harvested green and worked into a form or worked in place as a living structure, it's bentwood. Wood can also be harvested with the intention to replant as rootstock or scion wood, and even bundled and planted as a method of soil stabilization. In this context the words "bentwood," "greenwood," and "craftwood" are used interchangeably.

Many species of wood can be usefully harvested or grown intentionally for these products. As with polewood, coppicing or pollarding are both exceptional methods to propagate multiple stems, and the savvy forest farmer could keep an eye out for a short-term harvest of bentwood while thinning a cluster of young poles for a future polewood harvest. Polewood, of the three sizes mentioned in this section, has the widest range of possible uses for the forest farm.

BENTWOOD STRUCTURES: TRELLISES, GATES, FENCES

A wide range of simple structures can be constructed using basic bentwood techniques that anyone can do.[17] Wood for such projects should be cut no more that twenty-four hours before a project is to commence. While some wood can be soaked for a few days and still be workable (willow and osier dogwood), it's never as good as fresh. Dry pieces cannot be "rehydrated" and brought back into flexible form. Thus it's best to harvest as fresh as possible.

The idea with simple greenwood structures is to form the structure, then allow it to dry in place. Some

Figure 8.8. An arch and wattle fence made by Erik Phipps, a weaver and sculptor from the Isle of Wight in the UK. Stakes in these fences should be made from rot-resistant woods to preserve the fence. Photograph courtesy of Erik Phipps

species both bend well without breaking and allow for a number of creative designs in the making of gates, fences, and arbors. A wide range of tree species work well for these projects and include aspens and cottonwoods (*Populus* spp.), birch (*Betula* spp.), ash (*Fraxinus* spp.), dogwood (*Cornus* spp.), maples and box elder (*Acer* spp.), oak (*Quercus* spp.), hickory (*Carya* spp.), and willow (*Salix* spp).

From these items, and with a little bit of practice, a few common hand tools already in the forest farm arsenal (hammer, pliers, pruners, loppers) and some small nails and wire, the following creations can easily be made. Though it's easy to make a simple structure, it's in the creative patterns and forms that artistry can emerge. And if they are marketed to local outlets, the forest farmer may find good income from these items for gardens and landscapes.

The easiest place to start is with a simple trellis, which can be created by laying two 10- to 12-foot saplings side by side, about 3 feet apart. Crosspieces tacked to each sapling act as spacers. The tops can then be bent over toward each other and lashed together with wire. From this basic form many variations are possible.

Next, gates and fences require more thorough planning but can be a low-cost alternative to more conventional materials, while adding an amazing aesthetic to the forest farm. Panels for fences can be constructed en masse on-farm, then taken to a location for installing, while arch and wattle fences can be easily constructed inground to create pleasing barriers and borders for plantings (see figure 8.8). Willow, hickory, hazel, ash, and alder are the best species for this project and should be harvested in 6-foot lengths. To increase durability, bentwood constructions can be wired to metal posts or tacked to rot-resistant stakes made from woods such as black locust, tamarack, and cedar, which contact the ground.

Advanced greenwood structures include arbors, which can be used as an entrance feature to a place or even be constructed as a large outdoor seating area. These structures often take longer saplings and more precise layout of the main and support structures. Since considerable time and effort goes into making arbors, it is worth considering the species type relative to the desired use.

All in all, depending on the species and construction methods, these structures can last from five to over ten years. Consider using more rot-resistant species, such as locust, oak, and hickory if a longer life is developed, and especially when building more complicated and time-consuming items such as fences and arbors. A trellis, on the other hand, is quick and easy to build, and replacing it even every five years is a simple chore.

CRAFTWOOD PROJECTS

Another selection of even smaller diameter wood can be harvested for use in a variety of craft projects, including basket making, wreaths, and other decorative crafts that can be sold at local markets. Forest farmers who plan harvests and projects in timing with seasonal holidays and who make the effort to exhibit at local markets could find a niche market that makes these activities worthwhile. Basket making is one of those activities in which a basic basket is an easy task that anyone can do, while making a really high-quality product takes years to master (see Jamin Uticone case study). Harvesting materials from the woods and engaging in a craft project that turns the raw materials into a finished object is a wonderful way to connect kids to the forested landscape, too.

Figure 8.9. Curly willow can be harvested, dried, and sold or used as a decorative item that will last for years in a home. Photograph courtesy of Catherine Bukowski

LIVING STRUCTURES

Another option for working with small-diameter wood comes in utilizing not only its flexible properties but the biological ones as well. Wood can be shaped in place or planted with the intention of molding into shapes, or it can be harvested and replanted as propagation stock for new trees and soil stabilization projects.

Living structures are simply structures made from bentwood that is grown and woven together over time. They can be functional, artistic, or both. Willow offers the best starting point, as it is easy to grow and work with. It can be harvested and "staked," meaning that branches from ½ inch to 1½ inches can be cut into 12- to 18-inch lengths and pounded into the ground. The stakes are best harvested in spring before budbreak or in the fall after leaves have dropped. Pounding the stakes in so the vast majority of the stake is belowground will ensure best success.

This basic concept can be taken in many directions. A living willow trellis offers a pleasing entrance to an

CASE STUDY: JAMIN UTICONE, SWAMP ROAD BASKETS ALPINE, NEW YORK

An old style of basket making is hanging onto existence because of a few devoted artisans and their relationship to a unique tree: the black ash (*Fraxinus nigra*), sometimes called "swamp ash." Logs can be harvested, soaked in water, stripped of their bark, then pounded with a mallet, which crushes the fibers and allows long strips of wood to be peeled off the tree from top to bottom. The strips are cleaned and woven into incredibly durable items such as baskets and backpacks.

Jamin Uticone, along with wife Julia and three kids, came to their land partially because of how rich it was in black ash trees. Over the last decade he has slowly harvested and processed these trees into baskets that will last lifetimes. An average black ash tree will yield several handmade baskets. Jamin learned the craft during a six-year apprenticeship with master basket maker Jonathan Kline, who lives just a few miles away in Trumansburg, New York.

His work earned him the honor of being selected in 2012 for an art showing by the Smithsonian called *40 under 40: Craft Futures*, in which a wide range of emerging artists in America was featured.

Climate change is bringing questions to the table for this basket maker. The largest threat is the spread of the emerald ash borer, which was first seen in the same county where the Uticones live, in Bath, New York, in 2009. Since an infestation can lead to a dramatic population decline in ash trees, the family constructed a small pond that would be able to hold ash trees should they ever need to harvest them wholesale to keep his livelihood going. In addition, Jamin has begun to learn other basket-making techniques, including a form of basket made with poplar, a species predicted to fare better in dramatically changing environmental conditions.

This type of craft, in which a specific method is related to one species, is a rich story of culture and tradition. It's also an indicator of times that are changing, when basketry is not a standard household practice but is seen as fine art, one that may be gone from the state of New York, and even the Northeast, in just a few decades' time. Part of the work of those who glean

Figure 8.10. Finished black oak basket and backpack made by Jamin Uticone of Swamp Road Baskets. Photograph courtesy of Jesse Coker

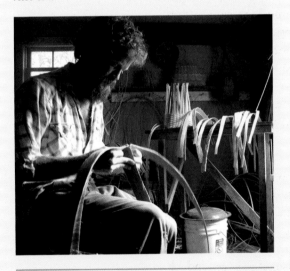

Figure 8.11. Jamin at work weaving in his studio, Alpine, New York. Photograph courtesy of Jesse Coker

a living and livelihood from the forest is to preserve, celebrate, and share these whispers from the past with the voices of the future.

For more information visit: http://swamproad baskets.com.

Figure 8.12. Willow stakes soaking in the seasonal creek at MacDaniels Nut Grove. These were harvested in the spring and kept in cold storage until fall, when they are soaked prior to planting for bank stabilization.

Figure 8.13. One of Pooktre's creations. Photograph courtesy of Pooktre Creations

outdoor space, while a willow fence can provide screening and improve privacy on a site. Willows can be grown in a circle or a spiral and woven together, to form the walls of an outdoor shower. The water from the shower then feeds and supports the willow's growing. In all these instances willows can even be grafted together, simply by tying two pieces together. Sometimes the bark is scraped off where the two pieces meet.

While willow is one of the more versatile species to work with, living structures can be made out of a wide range of other trees, including poplar, alder, and cottonwood. The extent of where this can all head is quite remarkable. Another similar art form is called "tree shaping" where artists use trees as their medium, growing them into circles and forming them into live chairs. A truly remarkable example of this comes from Peter Cook and Becky Northey, who are the founders of Pooktre, where they make tree people, tables, chairs, and more. Some sculptures remain in place while others are grown, harvested and sold as completed pieces.[18] Though the forest farmer may not be the sculptor, he could consider providing high-quality materials for such projects.

FASCINES, FAGGOTS, AND SOIL BIOENGINEERING

Another form of living bentwood comes in the harvesting and bundling of small diameter wood that is bundled for use. There are many variations on the practice that include the bundling of woody materials into faggots and fascines. Faggots are tight bundles of small-diameter brushwood that are bound together, traditionally used as fuelwood for the home. Fascines are similar but don't contain the smallest twigs and are mostly stems from ½ inch to 1½ inches in diameter, with the intent of using them for planting. Some circles generally use the words interchangeably. In the past fascines were often valued for a range of uses, including wood-fired baking because of the high heat output, but today they are mostly used for streambank and erosion stabilization.

CASE STUDY: BONNIE GALE, ENGLISH BASKETRY WILLOWS
NORWICH, NEW YORK

Having an intimate relationship with a tree over many years allows for a deep understanding of the subtle nature between differences many people wouldn't even begin to consider, like the subtle differences in flexibility between wood that is one year old versus wood that is many years old. This intimate knowing of an organism has carried Bonnie Gale, who has worked as a basketmaker and willow sculpture artist for well over twenty years. Her ability to adapt willow to space and place means that the species takes on a new life-form, providing, as Bonnie puts it, "immediate living green three-dimensional structure to an often void and boring space."

Her basic approach to living structures is to take large willow rods and sink them into the ground, weaving horizontal and diagonal pieces to tie them together in a strong "fedge" (fence/hedge). She applies this basic form to make fences, domes, arbors, tunnels, and rooms for a range of locations, including schools, private residences, and public spaces. Bonnie uses the ability of willow to self-root and self-graft in her work. She may include a metal substructure to extend the length and shape of the structure. Her new research focuses on the inclusion of waterproof roofing materials in her living bus and bike shelters.

Applying an appreciation and an eye for beauty to life, Bonnie also creates willow baskets, teaches classes to the public, and engages in landscape design and permaculture. Her eye for form and beauty translates into all of her work, and her dedication is clear from her words: "The process of education is continual and my commitment to learning the techniques of willow work is for this lifetime."

For more visit: http://www.englishbasketry willows.com.

Figure 8.14. A willow tunnel created by Bonnie Gale for a public park in Baltimore, Maryland. Photograph courtesy of Bonnie Gale

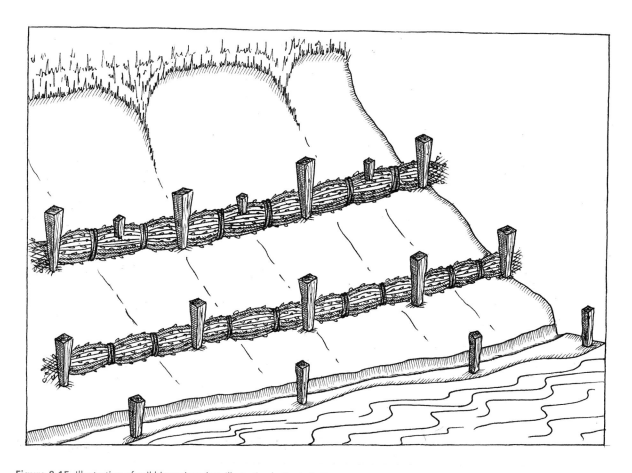

Figure 8.15. Illustration of soil bioengineering. Illustration by Travis Bettencourt

Fascines were recorded as used by the Chinese for dike repair as early as 28 BC. Early Western visitors to China observed dikes and riverbanks that were stabilized with large baskets woven from willow, hemp, or bamboo and filled with rocks. In Europe ancient Celtic and Illyrian villagers had techniques of weaving willow branches together to create fences and walls. Later, the Romans used bundles of willow poles for hydro construction.[19] In more modern times, fascines were carried on the tops of British tanks in World Wars I and II, so that they could be dumped in trenches to enable crossing of a gap in the terrain.[20]

Today a more technical term for a suite of strategies utilizing bundles of wood material is soil bioengineering. The US Forest Service and other government agencies have a number of bulletins on

this practice. One definition summarizes soil bioengineering as:

> the use of live plant materials and flexible engineering techniques to alleviate environmental problems such as destabilized and eroding slopes, streambanks and trail systems. Unlike other technologies in which plants are chiefly an aesthetic component of the project, in soil bioengineering systems, plants are an important structural component.[21]

Depending on the site and intensity, a soil bioengineering project can be as simple as digging a trench and placing material near more complex log and stone structures that are interplanted. Live stakes are often incorporated to hold individual

Figure 8.16. Willow stake that has broken bud only a few weeks after fall planting at the nut grove. As these grow, fascines will be woven in between to further stabilize the bank.

bundles or "mattresses" of bundles in place. This simple practice could have significant implications for projects on a wide range of scales, from stabilizing streams on farms to doing the same on larger creeks and rivers. Success has been reported also using these techniques in road building, to stabilize banks cut during construction.

There is a wide range of materials suitable for projects; it includes willows, ash, hazel, dogwood, cottonwood, sycamore, big leaf maples, spruce, white pine, cedars, aspen, and alders. The best plants are those suited to the locale; these can be determined by exploring hillsides and riparian zones of neighboring landscapes. In addition to bank stabilization, one of the unexplored uses of material would be mitigating erosion in steep forests, where bundles could be staked along contours. Even if they don't sprout and grow

(because of low light), they would provide decent protection of soils.

The biggest gaps preventing this technology from becoming more widely used is in the availability of material and the cultural understanding of how to construct systems properly, depending on specific site context. An entrepreneurial forest farmer could be engaged in this practice by simply growing material for local projects, or by going one step further and doing small installations along driveways and streambanks for farmers and landowners.

The Leftovers of Wood Processing: Biochar and Charcoal

After all the varied uses mentioned in this chapter, one still may find there is some wood left. While it's important to ensure that a good percentage of biomass stays in the woods to assist in soil building, there may still be some residues that can be even further processed for valuable goods, namely into charcoal and biochar.

The reason it is suggested that these two practices are thought of as good strategies to use "leftovers" is that in the authors' opinion, harvesting trees to immediately burn and process them is not a place to begin but a place to end in a sustainable chain of woodland products. The time and energy trees have invested in their growth is simply too valuable to be put directly to this end use. That said, certain circumstances may negate this statement; for instance, if a woodlot is full of very low-value species such as red maple (*Acer rubrus*), which has little value for any of the above practices. Forests that have been "high-graded," that is, picked clean of all the best hardwood species, often become overgrown with red maple, and some circumstances may require wholesale thinning. Another situation would be removing a large amount of invasive or otherwise undesirable brush, which could be easily made into biochar.

The other nice thing about leaving charcoal and biochar to the end is that all of the wood products previously mentioned is this chapter could be potential feedstock for the fire. Spent mushroom logs, broken spoons, and decaying garden trellises are all potentially good options.

Charcoal vs. Biochar

The differences between the two practices are subtle but important. In both cases wood is burned in a low-temperature, low-oxygen environment (called pyrolysis). The main difference comes in the intended outcome, and thus what feedstock (wood) is used in the process. The variables of time, heat, and pressure are managed based on the people making the product, their equipment, and feedstocks available. An easy way to think of the difference is that charcoal uses higher-quality woods (often oak, hickory, apple, and so on) to make an end product that is usable or salable for either backyard barbecuing or, in the case of willow charcoal, high-quality art supplies.

Biochar, on the other hand, uses whatever organic waste materials are around to make stable forms of carbon to bury in the soil as an amendment that creates habitat for microorganisms and absorbs water and nutrients. The good news is that those interested in making both of these products can use simple, homemade technology to do so. Forest farmers are likely to find good interest and markets for locally produced charcoal, an alternative to hardwood charcoals often made from rainforest woods and synthetic charcoal made of concrete and other materials.

Making Charcoal and Biochar

There are two main methods of making charcoal and biochar: direct and indirect. Direct methods involve using the heat from incomplete combustion of the matter; that is, the wood is directly burned, then starved of oxygen before the charcoal itself begins to burn. This method tends to produce more smoke and emit more volatile compounds, not so good for the environment. The indirect method uses an external heat source to more or less cook the wood in a closed, vented chamber. This process has the advantages of being better for the environment and easier to manage, as the wood being turned into charcoal does not have to be monitored as closely.

To illustrate the difference between the methods, imagine using 55-gallon steel barrels to commence a burn. Direct charcoal making would involve loading the wood into the barrel as tightly as possible, then lighting a fire inside the barrel. Indirect would involve loading a smaller barrel (a 16-gallon steel drum or beer keg works well), then nesting it inside the large one, which is packed with wood, then lit. After an initial burn the exterior drum is capped with a lid and flue, which draws air from the bottom of the elevated drum.

The basic system may only yield a few gallons worth of charcoal/biochar at a time, but it could easily be scaled up, depending on the goals. The key elements to consider are the desired end product, the feedstock (as a waste of another process), and how to burn so that minimal gases are released into the air.

Are Charcoal and Biochar Good for the Environment?

Various claims have been made about the production of biochar/charcoal and about the potential impact of sequestering carbon, in particular with biochar, since charcoal is intended for use, and thus will combust, negating any beneficial effects. Since burning and natural decomposition of biomass (including agricultural and forestry waste) adds CO_2 to the atmosphere, with biochar the idea is that the end product can be buried in the soil and be a slowly decaying store of carbon, which in effect "locks" carbon into the ground, potentially for centuries. It is also purported to improve water quality, increase soil health, and raise productivity of crops. Several universities, including Cornell[22] and the University of Edinburgh[23], as well as international non-profit organizations[24] have devoted significant resources to biochar as a major solution to climate change.

Reading the myriad of literature online and in journals could give the impression that biochar is the silver-bullet solution to all the world's problems, including climate change, soil loss, and water quality. While there is certainly promising potential, the devil is, of course, in the details. For example, the Cornell team led by Johannes Lehmann published a paper showing that 12 percent of global emissions could be offset with material from "sustainably obtained" biomass. This is a technical analysis, which demonstrates that the potential is theoretically there. The question is, is there the political will, industry incentive, and so on to actually carry out such a task, which in the current context is an incredibly tall mountain to climb.[25]

The other major issue of concern is the practical application of laboratory research in biochar to field settings. In the lab, conditions are controlled by the researchers to minimize the variables, whereas in the field the differences from place to place are dramatic. The environmental group Biofuelwatch conducted a review of field trials in 2011 to determine how biochar acts in the real world, finding a very small number of actual field studies (five) that tested biochar on a total of eleven different combinations of soil and vegetation. In only three cases did biochar result in additional carbon sequestration.[26]

Another aspect of concern is that a few studies show that the ability for soil to retain carbon is not based on the molecular structure of the carbon, but more on a suite of complex ecosystem properties, meaning that not all soils cycle and store carbon in the same ways[27] (see chapter 3). This means that (not surprising, given the discussion in this book so far) not all ecosystems are the same and that the effects of biochar will be variable depending on the system and the context. Things are always more complicated than they first seem, and if anything, research continues to tell humanity more about what we don't know than what we do.

The takeaway from this discussion is not that biochar and charcoal are not useful practices to engage in at the forest farm level but that practitioners should proceed with caution when proclaiming the benefits to their (and other) landscapes. On a certain (small) scale, making charcoal on-farm from local waste materials does make social, environmental, and political sense. And making biochar *could* improve local soils, especially clay-rich soils, as discussed in chapter 3. In the end the best way to proceed is to produce the stuff, then run comparison experiments on the farm site to compare results. In a quickly changing world, everyone needs to be a skeptical scientist. While general principles and patterns do offer a general direction to head no matter where one is in the world (for example, trees are good), the specifics of time and place are simply too important to ignore. This applies not only to environmental systems but social ones also. Shiitake mushrooms may make a tidy form of income for a forest farming operation in one locale, while in another there may be zero interest in purchasing them. Ultimately, the best forest farmers are those who can observe their surroundings and adapt their practices to the situation presented to them.

Wood Products Expand Forest Farming Potential

One of the underlying goals in the discussion of wood products in this chapter is to encourage forest farmers to view the possible endless value in their existing and cultivated trees. Indeed, much of forestry and forest management today has emphasized in a very extreme way that the only value in the forest comes from timber management. This not only limits the incentives and perception that people can make a livelihood from their woods but also the types of management that do (or do not) occur. For example, a vast amount of acreage in forested land in the cool temperate regions of North America has been logged multiple times, and the forests in existence today are generally young, dense, and emergent. A thinning of small-diameter wood would greatly improve the quality of trees in the long term as they mature, yet few people do these types of thinning because the economic incentive is low; at least that is the story circulating among forest owners. From the above practices it should be rather evident that this perception of low value in forests that don't have timber-quality trees is misguided; it is simply a matter of reframing the possibilities.

Another consideration is that, applying the approach of creating wood products from a variety of stem sizes, the forest farmer can make best use of the whole tree when harvesting. Thus it is best to be on the lookout for opportunities to use as much of the tree as possible, which of course will depend on what practices the forest farmer is willing to adopt. In the end diversity is going to be key for the forest, and for the forest farmer.

9 Animals in the Forest

In the right situations, animals present some interesting possibilities for forest farmers to increase yields from the woods while introducing some of the ecosystem benefits they offer. Ducks, geese, turkeys, and other poultry can engage in soil building and pest control. Pigs and goats can renovate overgrown scrub and help if dramatic shifts in the succession are desired. Cattle and sheep can graze if the canopy is opened sufficiently to support a forage understory. Work animals can log the woods with minimal damage.

Managing animals is in many ways a double-edged sword. Provided the forest farmer matches the correct species and brings the correct number of animals into the woods at the perfect time, the animals can be highly beneficial. It's also easy to do damage—a few hours of loose goats would destroy valuable vegetation, and too many cows for too long could destroy tree roots. In any landscape, leaving animals in one place for too long has a negative effect on the health of the forest (or field).

Figure 9.1. Pigs enjoying a mud bath on a hot summer's day at Cayuta Sun Farm, where they are being employed to help clear dogwood, honeysuckle, and multiflora rose from a dense shrubby woodland. Photograph courtesy of Juliana Nogueira

Designing Animals into the Forest Farm

Design for animals must be well thought out and planned. Taking cues from natural ecosystems helps inform the decisions we ultimately make. In natural systems, animals move through in a "pulse" pattern, in which the land may be browsed or grazed, then left for a time to rest. During their stay, animals offer a boost in fertility as they deposit wastes. If cycles of time and space are properly managed, manure can be a boon to the ecosystem, whereas neglecting the need to keep animals on the move will result in pollution problems.

In any case, the forest can play a beneficial role in an animal's life, most notably in the provision of shade and shelter to buffer effects of wind and temperature.[1] For the forest farmer animals can offer pest control, fruit and nut drop cleanup, and a variety of services that make the farmer's life easier. At the same time, of all the forest farming practices in this book, animals are the topic that is least researched and where poor decisions could lead to irreversible damage. Cautious experimentation is the best way to proceed.

Before getting into specific animals, a few general approaches should be considered by the forest farmer, regardless of the species chosen.

KEEP ANIMALS OUT OF HEALTHY FORESTS

As the old saying goes, "If it ain't broke, don't fix it!" We are in a time when healthy, mature forests are few and far between. Most have been thinned, the best trees "high graded" from the woods. If you happen to have an older, diverse, and healthy stand, it's really best to let it be. Animals should be used in young, recovering forests and especially in situations where there is a large amount of undesirable vegetation, as animals can help accelerate the natural succession of a forest.

The reality with animals is that they can wreak havoc on forests in very short order. In fact, most foresters have been trained that animals should *never* be let loose in the woods, and they are correct. Animals should only be in the woods in carefully designed paddocks and should be moved frequently. Good research into the

Figure 9.2. Older, healthier forests should be left alone when it comes to animals. Make use of more marginal lands that can benefit from their activity.

animals and careful monitoring are critical early on in the development of a forest–animal relationship. When in doubt, it's best to keep the animals in the pasture.

Admittedly, research and development on the relationships between animals and the forest are slim. Many questions remain, and experimentation needs to be carefully planned and executed. For the present, only a few animals have qualities that make their potential role in the forest a worthwhile consideration for the forest farmer. So when in doubt, keep them out!

FORAGING VS. GRAZING

There is an important distinction to draw between foraging and grazing that will have a great impact on the type of system and animals chosen. Grazing is based on cultivating pasture (grass) underneath a forest canopy, where vegetation in a paddock is consumed completely and allowed to grow back, whereas foraging systems utilize a "hunt and peck" strategy on vegetation and also seeds, insects, and other living organisms in the leaf litter and are more sporadic by nature. Both options should be considered by the forest farmer, though if grass is being encouraged this is technically silvopasture (see sidebar, Ruminants and Silvopasture). For this book, only strategies involving animals in foraging situations are emphasized, since silvopasture is a topic all to itself.

TREE CROPS AS FODDER

When plants are deliberately planted or managed to feed animals, it is called fodder. With the mechanization of agriculture and overproduction of grain such as corn and soy, much of the traditional knowledge on fodder species for domestic animals has been lost, and little research has been done on the effectiveness of replacing grain and other inputs with fodder produced on-farm. The desire to reduce these outside inputs is a goal of many farmers, and tree fodder crops may provide a partial means to this end.

An additional advantage of the fodder strategy is that many crops that are difficult and inefficient to harvest by hand for market can be "harvested" by animals, which will gladly spend all day foraging and seeking out food. This makes species such as mulberry, which is hard to harvest and nearly impossible to sell as a marketable end product, a valuable species if fed as a highly nutritious food to chickens or pigs.

Fodder can be loosely designated into two categories: vegetation and fruits/nuts. Vegetation is mostly harvested as leaf material, though some animals will strip bark and gain sustenance from the cambium layer of the wood. Fruits and nuts are obviously harvested as they drop from the tree. Since the cool temperate climate has a season of dormancy, vegetative fodder provides supplements to grass and ground forages and also helps extend the season later into the fall, when grass and herbaceous productivity has declined. Fruit and nut species can offer food sources long after leaf drop and can in some cases be stored for winter use.

The key point to adding fodder species into forest farms is to start with recommended feed rations for the animal, then do minor comparative studies to see how conventional feed can be reduced without compromising animal health. There is certainly useful benefit in simply observing behavior of animals in response to different fodders they are offered. In addition, indicators of health such as weight gain, production (eggs, milk, etc.) and general appearance of the animals should be observed and recorded. In the end, it should never be assumed that an animal can be fed entirely off the landscape. This is a long-term goal, and one that can only be reached with thoughtful observation and record keeping.

In comparing fodder values of tree crops to more traditional forages, alfalfa is considered the gold standard. Research from Michigan State University indicates that several species, including aspen, alder, poplar, black locust, and honey locust, have "a high feeding value," at least when compared to normal feed for cattle.[2]

The honey locust (*Gleditsia triacanthos* L.) offers some of the best evidence of potential multipurpose fodder in temperate forest farms. The leaves are an excellent source of fodder, containing 20 percent protein and a low lignin content, which means better digestibility. When coppiced, the regrowth retains these qualities. Sheep are able to digest the majority of seeds that form in long, spiral pods in late summer. Complete utilization involves machine processing, which make the food accessible for other ruminants.[3]

Rising food costs coupled with the vulnerability of commodity foods for animals make the appeal of fodder grown in place highly appealing. More work needs to be done on the specifics of matching species of plants to animals.

Figure 9.3. Poplar trees are coppiced for fodder at New Forest Farm in Wisconsin. On the left is a tree before animals are given access, versus on the right after foraging has been permitted.

COMBINING FOREST GRAZING WITH PASTURING

As options for utilizing animals in forest farms are considered, it should be noted that in almost all cases it is recommended that animals be given access to pasture in addition to forest. There are simply other benefits to many animals when they are offered open, sunny places in addition to woodlots. Moving animals from pasture to forest ensures they receive a balanced diet and the widest diversity of food options possible, which results in healthier and more productive animals.

FENCING IS CRITICAL

For most situations, fencing is a critical component of animal systems. It allows the forest farmer to control where animals go and for how long, all while protecting them from predators. In reality the only free-range possibility is with birds, because larger animals let loose could cause havoc on neighboring lands. Chickens, turkeys, ducks, and geese in small numbers (less than ten) could be left to forage, though any benefits of pest control, fertilization, and the like will not be as targeted and evenly dispersed compared to the benefits of using a fenced rotational system.

RIGHT PLACE AND RIGHT TIME

There are four key factors in planning an animal system in any conditions: the stocking rate of animals that the overall landscape can handle, the size of paddocks in a rotation, the amount of time animals are left in one paddock, and the amount of time an area is given to rest.

Some animals may only be appropriate to one stage of land development. Bringing domestic animals into a system could be a succession strategy, meaning it might

---

RUMINANTS AND SILVOPASTURE

Within the larger concept of agroforestry, strategies that include maintaining a grass base under the canopy (or planting trees in pasture) is called silvopasture and in a technical sense falls outside the definition of forest farming. Still, just as forest owners should consider multiple strategies and strive to match their land type and goals to the strategies employed, it is worthwhile to look briefly at some of the details of silvopasture in this book.

The simplest way to think of a silvopasture system is to visualize a three-way relationship between a grazing animal, crop trees, and grasses. While other foods such as nuts, seeds, and fodder from trees may be available for the animals, the grasses will form the basis of a stable diet for the ruminant, whether it be sheep, goat, or cow.

A recent Cooperative Extension bulletin notes that the practice is both "deliberate and managed," which is to say that animals are not just released into the woods, but areas of forest are transitioned to a more pasturelike state, or trees are planted and encouraged in pasture.[4]

One of the most tangible ways to "prove" the positive role silvopasture can play in modern farming is to look at the relationship of shade to animal health. According to the University of Missouri's Center for Agroforestry, heat stress reduces the appetite of an animal and can cause reductions in weight gain, decreases in milk production, increased calving morality, and thus increased cost.[5] Of course, some farmers purchase or construct shade shelters, but these are rarely moved around the pasture, leading to areas with more trampled grass and manure buildup. Widely spaced trees in a silvopasture mean that animals, and their manure, are more evenly spread throughout the pasture.

Well-designed silvopasture systems would include a diversity of tree crops. Patches of conifers such as spruces could be grown and provide a winter "living barn" for livestock, while nut and timber trees could provide multiple income streams, and groves of locust and tamarack fence posts and building materials. Trees and pasture also have another beneficial relationship: In cooler temperate climates, pastures consist primarily of cool-season grasses, which decline in production in the hotter summer months. In many cases this decline parallels the time when there is the greatest nutritional need for lactating ruminants.

make sense to bring them in for a year or two, then not again for several years, or ever. Also, since one type of animal is invariably suited for certain situations, the changing of animal breeds over time may also happen.

ADDING ORGANIC MATTER

Bringing animals into the forest can be a boon to the forest farmer in providing additional forages, shade from hot weather, and an increase in fertility. Keep in mind that with an increase in animal activity, nutrients and organic matter are inevitably going to be cycled quicker. This means that, in addition to keeping animals moving to avoid too much of a good thing, in some cases organic matter may need to be added to the system. Think about leaves falling in a hardwood forest; they provide a light layer of mulch that is just enough to cycle through one growing season; that is, by the end of the following summer the leaves have mostly

decomposed. If animals are brought in, this could potentially happen much quicker. Best to have surplus organic matter on hand to supplement as animal systems are optimized in the forest.

The Right Animal for the Job

In this section we cover the basics of raising particular types of animals that are suited to forest farming, but this is of course merely an introduction. There are many fine sources of information on the specific needs of these animals. The intent of this chapter in the context of forest farming is to comment on the possible animals that could fit into a forest farming system and discuss some of the details of this relationship. Table 9.1 offers a basic summary of the animals recommended for forest farmers to consider.

According to a Cornell Cooperative Extension publication titled *Silvopasturing in the Northeast*, the three keys to success with silvopasture are:

1. Adequate sunlight at ground level
2. Establishment of target forage species
3. Adequate rest periods between grazing

For example, it is recommended to aim for about 50 percent shade in a silvopasture to support good growth in cool-season grasses. Warm-season grasses need about 70 percent light penetration.[6] Most grasses should be grazed to 50 percent of their aboveground biomass, then allowed to rest. Rest periods of twenty to forty-five days provide good regeneration of grasses (that range is dependent on local conditions). Usually the goal for days on the pasture is three to five for good forage utilization. A water source should be no more than 600 to 800 feet from the grazing area. These are good guidelines to get started and to tweak as the system progresses.

The challenge of silvopasture is that it is really a combination of good forest management practices (silviculture) and sound rotational grazing practices. The shift to this paradigm requires experimentation, observation, patience, and interaction. Farmers and

foresters who are looking for systems in which they can put animals out to pasture or plant crops with little interaction until harvesting will not be good silvopasture candidates. Those that enjoy watching their animals, observing their plantings, taking good notes, and making small adjustments all season long will reap the benefits of increased economic and ecological health.

Figure 9.4. Goats grazing multiflora rose, a noxious species, in this silvopasture system, with black locust posts growing as the overstory. At Angus Glen Farm, Watkins Glen, New York. Photograph courtesy of Brett Chedzoy

Table 9.1. Considerations for Selected Animals in the Forest Farm

Animal	When and Where Appropriate	Pros	Cons	Cautionary Tales
Goats	Young and Middle stages of marginal forests; clearing of undesirable brush; can tether or fence in hedgerows and for "cleanup"	Will eat almost anything, including thorny invasive species	Fencing needs to be robust; will eat everything, including valuable crops!	Goats cannot survive on brush alone; need supplement
Chickens/ Turkeys	All stages; marginal forests or on quicker rotations so that scratching doesn't do damage	Easy to move and house; can forage much of their own food	Susceptible to diseases; scratching could damage tree roots; turkeys have high rate of mortality	Predator food! Extra caution and care to secure and protect birds is a must.
Ducks/Geese	All stages; better for more mature, choice forests; still need to rotate, but less frequently than scratch birds	Heavy down feathers mean they benefit from shade in hot summer; disease free; won't damage tree roots	Need lots of water for drinking and bathing	While geese can help as guard animals, still vulnerable to predators
Pigs	Best for clearing and reclaiming of marginal lands; may be hard to sustain on forest land every year	Root, dig, aerate, fertilize, and rapidly convert marginal lands	Too much time in one place can be devastating	Pigs are smart! Keep the fence hot, or they will get out.

Goats

Goats are ruminants, which suggests that they prefer a grass-based diet, naturally leading to another agroforestry practice called silvopasture, discussed above, (see sidebar, Ruminants and Silvopasture). Yet they are also often the first animal that forest owners think of when considering the possibilities of animals in the woods. After all, goats will eat nearly anything, which is a good thing if a forest is full of undesirable brush, but a bad thing if the forest has healthy seedling regeneration in the understory. In fact, two research studies in the Northeast shed some light on these dynamics and offer some direction for future research and development.

One study, conducted at Cornell University and informally titled "Goats in the Woods," proposed the idea that "wisely controlled browsing of Northeast woodlands by goat herds could increase revenue and reduce costs to goat owners, decrease woody plant control costs to woodlot owners and reduce the forest area treated with herbicides." Over three years the project grazed goats at Cornell's Arnot Research Forest and in seven other sites where collaborative teams of a woodlot owner, a goat producer, and an agency

person worked on the research together. Specifically, herds of five to twenty goats were grazed in stands and employed to eradicate populations of undesirable species in the understory, most notably striped maple, beech, hemlock, and red maple.

Results of the study indicated that goats could be successfully used to control a significant amount of woody vegetation while maintaining health and weight gains and not damaging stems of commercially valuable species greater than 4 inches in diameter. Critical to this success was the prudent observation of the herd manager to ensure goats were moved at the right time, along with the provision of supplemental feed, at the rate of 2 percent of body weight for adults and 2.5 to 3 percent for juveniles (less than twelve months of age). An unexpected result occurred at two locations where sugar maples grew; the goats appeared to avoid eating the young saplings, making the case for goats having a positive effect on sugarbush management.

Poor performance (defined as less than 30 percent mortality of sapling populations) in goat browse occurred with low stocking rates (five per paddock) or with a poor-quality supplement feed. Damage to

Figure 9.5. Beech trees stripped of their bark by goats during Cornell's research project, which found that well-managed goats could have a positive effect on removal of unwanted vegetation. Photograph courtesy of Peter Smallidge

Figure 9.6. One of the billy goats at Integration Acres standing under a canopy of pawpaw. The goats have no interest in the vegetation or fruit and keep the understory clear for easy harvesting. The farm produces goat's milk and products in addition to selling fresh and frozen pawpaws and a number of other forest products. Albany, Ohio.

desirable vegetation resulted when managers couldn't monitor herds and move them at the right time. As for costs, the study compared the cost of hiring outside help for vegetation control with herbicides to hiring a manager and a herd of goats, finding that a herd manager would need to charge $100 to $150 an acre to make it break even, a rate comparable to the cost of herbicide application. Note that this result does include paying the farmer minimum wage for her time.[7] In this study there were some cases where separate people were farming the land vs. managing the herd.

Many variables make this a venture that should be carefully considered by forest farmers, as it all comes down to the goals and values inherent in this system. If the goats are being raised for the farmer to make a profit off goat's meat, then the forest is not the place to raise them. But if a goat producer could offer landowners the service of clearing unwanted brush in addition to meat sales, some profit potential is there. It is not the case, however, that the meat will pay for the work of the animal. In the end, goats are effective in removal of brush, if that goal is an end that satisfies all involved.

In at least one case, goats can be used to directly support a crop, at least if that crop is entirely toxic to the animals. Since pawpaw (*Asimina triloba*) exhibits a toxin in its vegetation (see chapter 4), goats won't eat it. Farmer

Chris Chmiel and wife Michelle of Integration Acres in Ohio use this fact to their advantage, rotating portions of the milking herd the farm raises to keep the understory clear of other vegetation, making for easy management and harvesting of the pawpaw fruit. In 2001, Chris did some trials in looking at the relationship of grazing animals to both established and new pawpaw plantings, finding that goats (and possibly sheep) were ideal animals for the system because of their size and grazing habits. Though no animals ate the foliage of the trees, large animals would sometime trample younger trees.

Advantages to using goats in pawpaw orchards have been numerous, including brush and grass management, the stacking of crop systems, the addition of fertilizer from the manure, and even possible support of pollination, as the tree is pollinated by flies, and grazing animals could potentially increase fly concentration around the trees. It should be noted that for organic standards there should be at least 90 days between any animal contact with crops that will be harvested. This safety measure means that planning is critical to ensure the timing works out.

Goats offer forest farmers an opportunity to clear land of undesirable brush and maintain plantings in a low-impact way. While the economics of goats for meat or milk can be highly variable and profitably dependent on the abilities of the farmer and local markets, it's easy to compare the cost of a herd of goats with the price of hiring someone to brush hog or remove brush from a property. For one, the goats use less fossil fuel and are less destructive to soil than machinery. The goats further offer fertilizer as they browse on vegetation.

Turkeys and Chickens: "Scratch Birds"

Scratch birds are naturally forest-dwelling species: Chickens come from the jungles of Java in Southeast Asia, while the turkey is descended from the wild bird native to many parts of North America. These are best characterized as "scratch birds" because they use their sharp talons to stir up vegetation and hunt for bugs, seeds, and other tasty items. They can be good contributors to forest ecosystems, especially if there is a desire to aerate soils or integrate organic materials into forest soils. Care needs to be taken to move scratch birds frequently to avoid too much "tilling" of the soil.

Observation of turkeys in the wild shows that they can cover several miles in a day and tend to move through forests, streams, and meadows in search of a wide range of foods. Turkeys wild and domestic prefer acorns, hazelnuts, hickories, and chestnuts, as well as many small fruits. They also feed on insects and occasionally even on small amphibians and reptiles. Chickens enjoy many of the same foods but tend to stick to smaller and easier-to-peck-at items.

Precious vegetation will need to be fenced from scratch birds to avoid trampling or destruction of the root systems. Chickens and turkeys can often be utilized best as a cleanup squad, either before planting in the spring or postharvest in the fall. They can, of course, be allowed to free range among woody shrubs and trees that once they are a few years old will not succumb to any damage. Woodlots that have wild apples and cultivated patches of pawpaws and the like where

DON'T FORGET THE DOGS

While this chapter is mostly focused on livestock, one species cannot be overlooked as a huge benefit to productive forest farms. At Wellspring Forest Farm our dogs Vida (Lab mix) and Sadie (husky mix) are excellent in deterring predators and pests, from the raccoons and fox that are after the ducks to the squirrels and chipmunks curious about mushrooms. Young planted trees and gardens can remain unfenced at the farm because the dogs also keep deer away. Both dogs were not gotten with the intention to offer this "service" to the landscape, but we lucked out (some training helped, too).

— Steve

Figure 9.7. Sadie, one of two dogs roaming the woods at Wellspring Forest Farm. Photograph courtesy of Jen Gabriel

fruit drops may be prevalent provide great food sources for these birds and offer a reduction in pest problems associated with leaving rotting fruit on the ground.

Chicken selection for the forest farm should be limited to egg-laying and "mixed" breeds, as they tend to be better foragers. Turkeys are usually only raised for meat, and again, a heritage breed variety will likely do the best. Poultry raised for meat should arguably be left in the pasture, as there is more food available, and ones bred for heavy meat production tend to loaf a lot more.

In the bigger picture, if the farmer is renovating a forest from scrubland or low-quality species, chickens could be a great assistance in jump-starting the process, following a rotation of pigs, goats, or other ruminants if the system is in silvopasture (see sidebar, Ruminants and Silvopasture). As mentioned above, scratch birds will be happiest if they can make the rounds and get a mixture of pasture and forest in their rotation.

Chickens and turkeys want to roost at night, as high as possible. In theory a savvy forest farmer could employ a system in which birds take advantage of the natural roosting habitat provided by dense shrubs, but this leaves the birds vulnerable to predation. Movable coops are a safer option and should be constructed to

be light and small so they are maneuverable through the forest. Depending on slope, wheeled coops may or may not be appropriate. On steeper sites coops that can be lifted and carried to the next paddock are best. These can be moved once the birds are let out in the morning.

DUCKS AND GEESE

Domestic waterfowl have several advantages over chickens that make them appealing in forest farming situations. For one, they will not scratch and till, as their webbed feet preclude this activity. Instead, these birds root with their beaks, and though they are observed occasionally gobbling on a tree root, they are mostly searching for insects and seeds to devour. Fowl might be best thought of as pest control agents who fertilize the forest at the same time. Keeping ducks and geese in the woods during the hottest summer months

Figure 9.8. Muscovy and Rouen ducks in the woods at Wellspring Forest Farm. The African goose, affectionately named Gary, is with the flock for predator protection, which has worked well over three trial seasons.

Figure 9.9. Ducks and geese love water and will take to bathing and swimming at a very young age if given access. They need it not only for drinking but also to clean their nostrils and feathers multiple times per day. Estimate about 1 gallon per bird per day, with about half that amount in winter if supply is short. The water must be changed frequently, as it becomes dirty quickly with all the activity. Photograph courtesy of Jen Gabriel

Figure 9.10. The beginning frame of the duck housing at Wellspring Forest Farm is a simple 4- by 8-foot structure using old plywood as the base and scrap lumber to frame walls about 2 feet high. This provides 32 square feet of floor space, adequate for ten to twelve ducks. The houses are built on trailers and are narrow enough to be tucked between trees in the woods.

is more enjoyable for them, as they have a thick down coat that offers an advantage in the winter.

Ducks are extremely disease and cold tolerant; farmers don't have to concern themselves as much with mites or foot rot as with the scratch birds (as long as the bedding stays clean). The cold tolerance is key: Duck houses don't need to be heavily insulated come winter, and ducks are happy to get wet and dry out, whereas chickens can easily get hypothermia and need to be kept dry and warm.

The biggest need for ducks is water. While a natural pool or pond is ideal, this is hard to provide in every paddock. Instead, water can be brought to a 10- or 15-gallon tank that is refilled daily. The water gets dirty quickly; ducks and geese root in the dirt with their bills, then clean their beaks and nostrils in the water. They also need water to help wash down food, as they swallow it whole. There is a difference between "somewhat dirty" and "this-needs-changing dirty"; figure about 1 gallon of water per bird per day. If water is limited on the site, this may be the single biggest drawback to raising waterfowl.

Ducks and geese are phenomenal foragers, and many books on the subject mention the possibility of some breeds foraging almost 100 percent of their diet, at

least during the growing season. Because of that down layer, ducks and geese tend to be more efficient in food consumption in the winter, an advantage over chickens. Grain is best given in pellet form; as less is wasted when compared to the granular feed. Combining ducks, geese, and chickens would likely reduce any wastes from feed.

Waterfowl housing is also different when compared to that needed for chickens and turkeys, which like to roost as high as possible at night. Ducks and geese are nesting animals; therefore, housing can be constructed to be shorter—the roof can be just above standing height. Since the birds want to bed down, more floor space is needed, and it is recommended that housing offer 1.5 to 3 square feet of space per duck, if it is raised on pasture.

A fresh layer of bedding (straw works well) should be added every second or third day to keep things sanitary—duck poop is wet and composts best in thicker layers of organic matter. The bedding can be harvested about once a month and composted at another location or added directly to fruit trees and other crops that won't come into contact with it. A trick of the trade is that ducks and geese, unlike chickens and turkeys, can do without water at night. So leave the water outside. It will keep the house much cleaner.

CASE STUDY: CAN DUCKS IN THE WOODS PROVIDE SLUG CONTROL FOR SHIITAKES?
WELLSPRING FOREST FARM, MECKLENBURG, NEW YORK

Before moving to the land where Wellspring Forest Farm is located, Liz and I got into ducks because we wanted to try something different. We'd raised chickens for years and thought, "Why not?" It turned out that this haphazard decision led to a new relationship we never expected. That first season we bought about half a dozen ducks—Indian Runners and Khaki Campbells—and constructed a simple house for them. We set up a pen in the backyard of our rental house that happened to be a wooded spot. And because it was easier to manage two crops with one visit, we also placed about thirty shiitake mushroom logs in the pen, soaking them weekly and harvesting enough mushrooms for personal use.

That season we made an exciting discovery; our pest issues with shiitake (read: slugs) were almost entirely gone! The more we thought about it, it made sense. And as we transferred the mushroom operation to the current one-acre maple grove that was also going to be tapped for syrup, we became intrigued with the fact that a three-way relationship was emerging: a polyculture of a producer (sugar maple trees), consumer (ducks), and decomposer (mushrooms) (see chapter 3).

We wrote a Farmer Research grant for Northeast SARE in December of 2010 to help explore this re- lationship in a more thorough way. In reality it takes funding to make research happen. For one, we would not have invested the roughly $3,000 into new fencing, materials, and ducklings knowing there was a risk we wouldn't make anything back. And further, taking the time to collect and analyze data simply conflicts with the day-to-day needs on the farm and maintaining an off-site job. The funding made it possible to make the research a small "job" where we could be compensated for our time and material costs in exchange for the information we would gather.

After I received word I'd gotten the grant, excitement quickly dissolved into reality. Research turns out to be a decent amount of work. I had to divide my roughly one thousand mushroom logs into three areas: one that would receive no duck activity (the control) and two that would have varying amounts of duck activity.

The main research questions were as follows:

1. Are ducks an effective and reliable slug control in log-grown mushroom cultivation?
2. Is the forest affected in any negative way from the presence of ducks?
3. Are ducks economically viable as an additional income stream?

In addition, duck well-being and happiness was critical. Some people may disagree on the details about raising animals for meat, but we believe it to be a critical element to a sustainable food system and take pride in providing our animals with complete care and access to natural environments. We do not view these wonderful creatures as commodities but as sentient beings that need our respect and admiration. Duck happiness had to always be a must at our farm.

YEAR ONE: 2012

Ducklings were raised in metal stock tanks for two weeks, then given grass forage during the day for two more weeks, during which time they were given free choice of grain. From then on, all the ducks were rationed at 0.4 pounds of feed per bird, per day (the recommended rate for meat ducks). Ducklings were given

Figure 9.11. Ducklings arrived at the farm just a few days old in April/May both seasons and were raised in 250-gallon stock tanks for four weeks before being transitioned onto pasture and forest through several stages. During this time a heat lamp and constant food and water were critical.

free choice of grain during this time, and there were two groups, which would remain throughout the season:

Group #1: 10 Rouen, 15 Muscovy, 1 Chinese
    goose (for protection)
Group #2: 10 Cayuga, 10 Swedish Blue, 1 African
    goose (for protection)

The ducks were purchased as all male, to offer some consistency since we were going to weigh them. The ducks moved into the site on June 10, when we began taking data on mushroom yields, slug damage, duck weights, feed measurements, and any observations made by me or my help, Joshua.

During these months work was limited mainly to feeding (0.2 pounds per duck, two times each day), watering, mushroom harvesting, and observations. Three randomly selected ducks from each breed were captured once per week and weighed. We learned many things about duck behavior and the differences in breeds. The ducks were taken to a local slaughterhouse on October 16. We stretched the kill date this long to see if there was any benefit to weight gain—or if weights would level off.

In the first season the ducks were all sold to a local restaurant, which also hosted a tasting event and brought together sixteen participants, including chefs, farmers, Extension associates, and consumers. Each breed was minimally prepared and served in a blind test in two rounds; round one was breast meat, round two was leg. Participants tasted the varieties and made notes on a worksheet. Everyone agreed that the most surprising element was that there was such a difference in taste between breeds. The Pekin (donated from a local farm) was the consistent favorite, while the Muscovy received poor marks and the three heritage breeds (Rouen, Blue, Cayuga) had positive marks, with many participants noting more interesting flavors, in comparison to the Pekin, which was deemed a "safe eat" for general consumers.

YEAR TWO: 2013

For the second year, our trials were simplified, and several changes occurred. First the size of the paddocks was reduced and restricted to areas right around mushroom fruiting zones. The ducks were

Figure 9.12. The duck tasting revealed that there was a surprising difference in taste among the various breeds. Photograph courtesy of Jen Gabriel

rotated from forest to field to diversify their diet as well as to reduce the impacts from continuous grazing in the woods. Based on the previous year we decided to raise two flocks of twenty-five ducks each; one of Rouen and one of Cayuga. The flocks were ordered as a "straight run," meaning a mixture of male and female. The biggest change overall was that grain inputs were limited and offered at a lower rate while trying to maintain weight gain (0.2 pounds per bird per day, which is *half* of the previous year's input).

We again received ducklings in the mail, raised them in brooders, then transitioned them to pasture and forest in early June. Duck houses were rebuilt to be smaller and more easily movable. The ducks were moved once a week from field to forest. Each of the three mushroom yards got a different treatment; one was a control (no ducks), one had ducks constantly in and around the mushrooms, and one had ducks visiting only twice throughout the season.

A discovery made in both years is that some ducks will make an effort to eat, or at least nibble at, the mushrooms. This was observed in the Muscovy/Rouen flock of year one and in the Cayugas in year two. This means that to maintain a good crop fruiting mushrooms need to be fenced off from the ducks. This is acceptable because the ducks can be rotated around this enclosure to reduce slug pressure, rather than eating the slugs right off the logs. Fencing off the mushrooms also eliminates any concerns about sanitation of manure and associated concerns with food safety.

The second year, ducks were sold to a local meat butcher, who sold out of them almost immediately, directly to consumers as well as to a local restaurant. In our area, at least, there is clear demand and a good market for duck.

RESULTS

1. Are ducks effective and reliable slug control in log-grown mushroom cultivation?

From this study it can be suggested that ducks offer a viable means to reduce but not entirely eliminate slug pressure on a mushroom crop. We were unable to collect conclusive data on whether ducks offer a viable means to reduce slug infestation on shiitake mushrooms because of variables and unpredictability in weather, precipitation, and temperature. However, our observations did lead us to believe that the presence of ducks in the vicinity of fruiting mushroom logs can help but not entirely eliminate slug pressure on a shiitake crop. They should not be seen as the perfect solution, but rather as a supplement to other strategies, including the removal of organic matter from the fruiting area, placement of gravel, use of beer traps, and monitoring. The rise and fall of slug pressure appears to be related to the amount of moisture in the forest.

Figure 9.13. Slugs mating on top of a shiitake mushroom. One positive outcome of the study was that no mating slugs were found in the mushroom area with ducks grazing around versus many dozens found during wetter parts of the growing season in the control area.

2. Is the forest affected in any negative way by the presence of ducks?

The presence of ducks in the forest has one critical impact observed: Leaf litter from the previous fall decomposes much more quickly when animals are in the forest. In some areas of the woods, particularly where water pooled and flowed during heavy rain events, grazing appeared to create bare ground and mild compaction, which led to some minor erosion. This impact was much more dramatic in year one when the stocking rates were high (fifty ducks continuously in a ⅓-acre paddock) versus year two, where smaller flocks of twenty-five were rotated in smaller, ¼-acre paddocks.

3. Are ducks economically viable as an additional income stream?

To be economically viable, duck meat production would need to be a primary goal, not a byproduct of a desire to control slugs and enhance an ecosystem, because you need to raise more ducks than is needed solely for pest control. In addition, the focus would need to be on larger ducks (Muscovy and Pekin) and not on the medium-size breeds (Cayuga, Rouen). The good news for those interested is that the market demand is very high (at least in our region). It should be noted that we sold to restaurants and a retailer who cater to a market of customers willing to pay more decent prices ($5 to 6 a pound) for sustainably raised local meats.

The total costs to raise ducks for meat appear to be approximately:

$20 per bird for feed (for meat breeds that gain sufficient weight)
$5 per bird for ducklings
$4 per bird for slaughter
$3,000 in start-up costs ($300 per year)
$1,200 per year in labor

If an 8-pound bird can be sold for $40 ($5 a pound), then 150 birds would allow for a break-even including labor over ten years, while raising 420 birds would pay off the costs in a single season. The calculations we have provided are based on 50 birds and relatively minimal numbers with regard to labor, material, and so on.

It is challenging to quantify the value of fertilizer the ducks provide to the system, along with the benefits of slug control. We are finding that the ducks are proving valuable not only in the woods but in our gardens and around planted tree crops as well. In 2013 our garden saw little to no slug and bug damage to plants, and the ducks also performed well when given access to cover crops we'd sown as part of efforts to build degraded soils on the farm.

Out of this focused observation and documentation (as opposed to hardline research), the most valuable result was considering how this system affected our goals and values. One of our primary goals in our farm is to try continuously to reduce outside inputs, especially grain feed, as it is energy intensive and also rising each year in cost. One of the initial appeals of ducks was the idea that they could find much of their food needs on the farm, but what we didn't think about was that raising poultry for meat inherently means getting on the "grain train," as there is pressure to get the birds as big as possible in as short a time as possible. We question whether raising poultry for meat markets is inherently unsustainable, especially when compared to ruminants, who can largely be fed from maintained pasture and on-farm feed (hay). Our focus moving forward is to examine the potential to produce eggs and see if we can maximize on-site food production.

The full PDF report from this project can be downloaded at: www.WellspringForestFarm.com.

— Steve Gabriel

Figure 9.14. Overall, the experience with the ducks was positive. Kept in rotation, they can work well in the woods and likely contribute to an overall slug management strategy for mushroom production. Photograph courtesy of Jen Gabriel

Pigs

Pigs are rooters, diggers, and turners, biological "backhoes" that aid in the transformation (or destruction) of space. As omnivores pigs thrive on a varied diet of seeds, roots, acorns, nuts, vegetation, fruit, fungi, insects, and small animals such as snakes. Consequently, the forest would appear to be an ideal place for them. In the forest farming sense they are best used when decent-size portions of a landscape are in need of clearing and where species need to be removed. For farmers and landowners with large tracts of forests, pigs can also be sustained in woodlots

if areas are given sufficient rest periods between rotations. More research needs to be done, however, to work on some of the dynamics of this relationship.

Many people associate forest pigs with acorns; they indeed love the food, and the nuts have been considered some of the finest finishing food for pigs for centuries. The Spanish product Jamón ibérico is considered the most decadent (and expensive) type of ham globally. Iberian black pigs are rotated through open, savannah-like acorn forests during masting years. A cool temperate forest farmer with an oak forest and a good sense of timing could make use of this crop with

Figure 9.15. Pigs at Cayuta Sun Farm in Alpine, New York, are moved around the landscape along with a simple hut that is adequate for their housing (right), offering protection from rain.

pigs, especially when compared to the task of harvesting acorns for human food, by hand.

Another advantage the forest offers pigs is shade, which is important to pigs, which can be easily sunburned in summer or, even worse, die of heatstroke. Many farmers in North America cite through articles and blog posts that they've successfully raised pigs in the woods and that established trees can handle the impacts. Pigs are most appropriate, then, for areas that are low value or for more established forests but perhaps less so for forests that may have species regeneration as a main objective.

Pigs can be trained to a single- or double-wire fence that can be strung up on posts and easily moved. A simple shelter can be constructed and dragged from one location to the next; a popular design is a cattle panel bent and secured to a frame and covered with roofing.

It should be expected that pigs will need to be fed a substantial amount of grain in addition to anything they are able to forage. Many farmers seek to capture local waste streams (whey from cheesemaking, bakery waste, waste from brewing, etc.) to supplement feed. Do *not* expect that the forest will be able to provide an entire diet for pigs, and in fact not providing enough food to the pigs will likely result in more damage, as hungry pigs will dig more roots, chew more bark, and do more long-term damage.

WORK ANIMALS: HORSES AND OXEN

Witnessing a horse- or oxen-logging operation is truly a sight to behold; it feels at once both timeless

CASE STUDY: PIGS CLEAR THE WOODS
D ACRES FARM, NEW HAMPSHIRE

At D Acres farm in Dorchester, New Hampshire, pigs are used as a way to create disturbances in parts of the forest, to open up space and light for more sun-loving species. The 180-acre farm is mostly wooded, and since the beginning animals have played a critical role in increasing the diversity of structures and species found on the farm.

Josh Trought and his crew have refined a succession where forest is cleared in small patches or gaps. The process begins with selective logging by the farm's oxen, Henri and August, and the harvested wood is used for on-site heating needs and building projects. This opened space then became pig habitat, where Dorchester Dalmatian pigs spend weeks happily rooting up soil, turning the earth, and digging up small stumps and rocks. The forage found on-site is supplemented with the waste of fifteen local restaurants and grocery stores, where waste is turned back into food. In some cases the pigs are given range for several seasons before the space enters its next transition.

Once the pigs are removed the land is shaped into rough terraces on contour, then planted with potatoes, which thrive in the loose but still low-quality soil. Compost is mounded around each planted potato; then the beds are mulched. The plants will be hilled, and more compost is added later in the summer. The way the D Acres folks see it, they are both getting a yield and building raised beds at the same time. Cover crops are sown in pathways to reduce erosion. At the end of summer the potatoes are harvested (one estimate of yield in a patch is 7.5 pounds for every 1 pound planted), and the beds are again worked, then planted with garlic as a transition crop to the following season, when they will be planted to annual and perennial crops.

In this way, succession is driven to an earlier stage through a series of disturbances, then built back up again, converting northern hardwood forest to edible-food forest in just a few seasons. The pigs are the engine that drives this system. Once new plantings are established, the pigs don't come back around but instead move to new ground.

Figure 9.16. The same contour beds at D Acres with potatoes coming up. The pigs are just upslope from the beds, preparing the next space for transformation. Photograph courtesy of Josh Trought

Figure 9.17. After the pigs come through, the loose soil at D Acres is formed into contour bedsand in year one planted with potatoes. Photograph courtesy of Josh Trought

and elegant. An experienced teamster (one who manages working animals) can harvest a woodlot with minimal damage; in fact the best ones leave almost no trace they were even there. This is a stark contrast to skidders and other machinery that leave clear and sometimes irreversible evidence of having

been in a wood. Despite this, pressure to log forests faster and the general decline in knowledge of and passion for utilizing animals in working roles has led to an extreme shortage in draft animal power in many parts of cool temperate climates. A forest farmer could easily specialize in this craft and

PINE STRAW: A TREE CROP
FOR MULCH AND BEDDING?

A most curious forest-farmed crop that occasionally shows up at garden centers and farm stores is pine straw, which can be harvested and baled from any long-needled pine tree, though traditionally it is the Southern species (loblolly, slash, and longleaf pine) and regions where this practice has occurred. With an abundance of unmanaged pine plantations in cool temperate forests in the Northeast, this crop could provide incentive for landowners to better utilize their woods.

Pine straw is harvested by raking and baling needles, usually in the fall or early winter. If tree spacing and landform allow, baling equipment for hay and straw production can be used on a larger scale. Bales are often sold as a garden mulch, considered to be more aesthetically pleasing than straw and also slower to break down. The material is also sometimes used as animal bedding.

Concerns with overharvesting have led to some recommendations that pine groves only be harvested up to five times in a twenty-five-year span of time, while others suggest harvesting a portion of plantations every two years.[8] The straw can provide some impressive yields (at least in southern states), where some estimates say one hundred to two hundred bales can be harvested from an acre in a season.[9] Bales that, when used as mulch, cover about 100 square feet 3 inches thick can be sold for as little as $10 per bale up to $60 at some online outlets,[10] so even with a long lag time between harvests, the economic potential is great.

Removing material from the forest, of course, will have effects on nutrient cycling and surface runoff of water.[11] Pine straw plantings could potentially be combined with grazing systems to address some of these issues, but the practice remains marginal, and little research exists on the particulars of management, especially in northern states.

Figure 9.18. A grove that is harvested every two to three years for pine straw at the Horticulture and Agroforestry Research Center of the University of Missouri.

Figure 9.19. A pine straw bale.

provide animal-powered services to neighbors and other landowners.

In the forest the typical use of animals would be for logging, though animals could be utilized for any number of tasks that a plow, tractor, or skidder might do. Horses, oxen, and mules all have their advantages and disadvantages, and the right animal should be selected for the anticipated tasks. A teamster signs up for years, if not a lifetime, of work with such animals; it can take two to three years of training before work can be done efficiently. Yet the benefits of working with live animals couldn't be more numerous and include these:

- A full-grown horse or ox may weigh 1,600 pounds, whereas a skidder weights over 10,000 pounds. This difference has a dramatic effect on soil and root compaction in the woods.
- Start-up costs estimated by the University of Kentucky are $10,000 for horse logging, compared to $100,000 for machine logging operations.
- Trained animals can live and work from ten years (for oxen) to twenty years (for horses and mules).
- Animals can work on steep slopes and in tight spaces, without doing damage to residual trees.

While in reality it will be a small minority that will continue to develop and work with animals at this scale, it is a critical community resource that every locale could benefit from.

Animals in the Woods

The examples provided in this chapter of individuals finding novel ways to work with animals in the woods offer a range of potential options for the forest farm. Of all the systems presented in this book, animals should be approached with the most caution, as there are considerable costs in terms of infrastructure and time to monitor and work on optimizing a rotational system that works for the animals, the land, and the farmer. Farmers of all types who have experience working with animals know that it takes several seasons to figure out the correct timing of management, the scale of an operation, and the associated economics that make or break the system in the longer term.

Figure 9.20. Leo working at the Good Life Farm in Interlaken, New York. As part of the farm mission, horses do all the work a tractor would, including hauling logs, plowing, and moving materials, in the case of this innovative operation a series of movable greenhouses that help extend the farm growing season. Photograph courtesy of Melissa Madden

One overarching consideration that can be gleaned from this chapter is that animals are primarily raised for production or for the services they provide to benefit ecosystems, but not often both, at least successfully. The goals and values of the farmers will ultimately determine the way a system works. Considering the building blocks of ecology that were discussed in chapter 3, it's impossible to ignore the fact that animals play a role in ecosystems of all types. As farmers and land stewards, then, the task is to find a balance point that works for all parts. In doing so, landscapes can be more rapidly restored via animals to greater health and productivity.

10 Designing and Managing a Forest Farm

Equipped with some of the theory and background and the specific methods for growing a wide range of forest farming crops, we may now envision the forest farm as a whole system, putting all the parts together. Engaging in a design process allows forest farmers to match their goals with the forest types they have available. It promotes planning ahead of time to minimize labor, expenses, and energy expenditures. Most importantly, the design process often brings ideas and considerations to the surface the farmer might not have thought to consider.

As much as forest farmers should be interested in individual crops and systems, it is ultimately a whole, integrated system that employs a wide range of species, strategies, and methods that make a successful forest farm. Permaculture (see chapter 2) offers a design process that can aid forest farmers in the development of a site plan that provides the opportunity to make mistakes on paper and work out relationships and decisions before actuating them in the field. The first part of this chapter discusses the specifics of the design process by walking readers through a case study of the design done for the MacDaniels Nut Grove at Cornell University. The second section of this chapter looks at considerations for management, including some starting points in conducting forestry operations safely and effectively.

Permaculture: Design It like an Ecosystem

A core tenet of permaculture philosophy is the use of principles that are extrapolated from natural ecosystems to strengthen the design of an agriculture system.[1] While there are literally dozens of principles, for the purposes of this book a few key principles have been selected that will aid in improving the design of forest farming systems. These principles serve to highlight many of the key concepts presented throughout the book.

MULTIPLE FUNCTIONS

The principle of multiple functions emphasizes that every element of the system perform at least three functions that benefit the system. We see this in trees, as they hold soil, provide shade, and clean air and water as they produce nuts and fruits for consumption by animals and people. The pot-in-pot cultivation system described in chapter 7 is a good example of this principle in action: The system reduces damage to roots from transplanting, protects plant root systems from winter freezing and summer heat, and simplifies propagation work for the farmer.

REDUNDANCY

The principle of redundancy states that for every major function of a system (fertility, pest control, water, income stream) there are at least three elements that support that function. This promotes stability and resilience in a system. In forest farms critical functions include water, fertility, crop type, and income source.

To use income as an example, it is not recommended that forest farmers grow one crop only for an income source (although many do). Rather than planting acres and acres of ginseng, it is better to dedicate a portion of forest to ginseng, a portion to goldenseal, a portion to mushrooms, a portion to fruits, and so on. This guarantees that if one crop experiences a shortfall, the others will soften the blow.

Functional Interconnection

This principle states that the yields or wastes of one element can often serve to meet the inputs or needs of another. By linking these energy flows, problems can often be solved within the system. For example, bringing ducks into a mushroom yard helps reduce slug populations, the major pest of mushrooms. The farmer also accomplishes the task of maintaining multiple "crops" (mushrooms, duck meat and/or eggs) with one visit. These parts, placed together, create a more complex whole that provides many more benefits than if maintained separately.

Stacking in Time and Space

This principle encourages yields to be designed so they maximize a range of times (annual versus perennial versus long-term yields) as well as make efficient use of both vertical and horizontal planting space in the forest. A great example of this principle is illustrated in the "polyculture" concept common in permaculture, where communities of multifunctional plants are designed for functional interconnection. For the nut grove a polyculture was designed to work within "Walnut Island," where the dominant species is logically black walnut, so plants need to be adaptable to some shade, as well as juglone tolerant (see chapter 4). The polyculture also both spreads yields out over time (fifty years to ten years to five years to one year) and stacks vegetation in the understory to maximize production of a given space. Table 10.1 shows one way of achieving the principle of stacking in time and space.

Putting It All Together: The Design Process

There are many good design processes out there. Understanding the basic parts and progression of design provides a road map to discovering the relationships and systems that are best for the site context. Oftentimes an individual gets excited about a particular crop or concept and becomes hooked on

Table 10.1. A Black Walnut Polyculture

Species	Vertical Space Occupied	Years to Mature Yields
Black Walnut	80–100 feet	40–50
Pawpaw	15–30 feet	5–10
Currants	2–5 feet	3–5
Raspberries (propagation)	1–2 feet	1–5
Stropharia	Ground level	1–2

the idea first, leading him to expend extra effort to get the landscape to conform to his vision. Instead, design offers the opportunity to step back, observe, and match practices that fit into the landscape. This idea gets back to the concept of "limiting factors" described in chapter 3, as recognition of these limits provides a guidepost for design and implementation.

For example, the south- and southwest-facing slopes at the MacDaniels Nut Grove create warm and dry conditions, which make production of certain forest-farmed crops unsuccessful. Luckily for the site, MacDaniels planted hickories, chestnuts, and walnuts, all species that thrive on such slopes. If he had tried to establish a sugarbush instead, the grove would not be nearly as successful as it is today.

In 2009, students in the Practicum in Forest Farming class initiated a design for the nut grove that was further refined over the next few seasons. In 2012, the focus shifted from design to implementation, and several projects are currently under way. The basic design procedure used included the following steps:

1. Site assessment, including creating a base map, in which a number of environmental characteristics were examined and documented
2. Goals development, in which the values and intentions of students and staff were articulated and refined to clarify a sense of purpose
3. Schematic design, in which the previous two items were examined and the possibilities for design were "brainstormed" in a range of ways
4. Final design, which brought together all the elements and proposed a road map for implementation

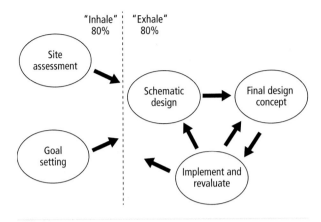

Figure 10.1. There are many variations on the details of the permaculture design process, but the general parts of goals, site assessment, schematic, and final designs are consistent. This is the version taught by the Finger Lakes Permaculture Institute. Design is iterative and ever changing, so as systems are implemented they are evaluated, then redesigned, causing each of the steps to be revisited.

Prior to design, the nut grove had been in existence for roughly five years. It had examples of mushroom, medicinal, fruit, and ornamental production scattered throughout the 3-acre site with a series of random paths connecting elements. Visitors found the site confusing and hard to navigate. Work was tedious and inefficient. Cropping systems were not matched to the appropriate places on the site. These were some of the problems that the design aimed to solve, while also increasing the productive capacity of the site.

STEP 1: MAKING A BASE MAP

The first step for any design is to create a base map for the site. A base map shows the major *existing* features of the site—boundaries, infrastructure, large trees, pathways, water features, and so on—and serves as the "base" of any design work you do, which will not be

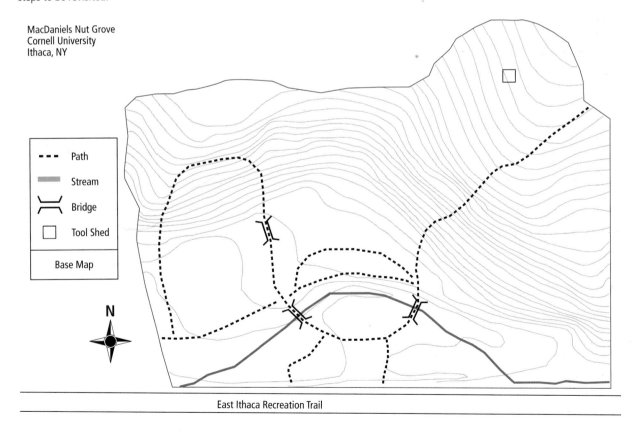

Figure 10.2. A base map of the basic features of the nut grove in 2008. Base maps should be kept simple and only include the features that currently exist and that won't likely change, forming the "base" to make overlays for site assessment and design concepts.

drawn on this map but overlaid with tracing paper (or a "layer" on the computer, if done digitally).

To make a base map, first do some searching for available maps of the site. Aerial photographs of many locations are easily found through the Internet and through local agencies such as soil and water conservation offices, municipal offices, and libraries. (See the sidebar, Obtaining Maps, for specific instructions on locating maps.)

Next, visit the site with an aerial or other map as a base layer, and make some sketches and notes of elements that may not show up on the map. Nothing beats firsthand experience to develop an accurate map of the site. Remember, the goal is to make a map only of *what is*, not what you may want to see happen on the site. This way, a base map can be timeless and useful many years down the line. Design ideas will be layered on top of this map, so it can be preserved.

Some of the elements to consider including on a base map are:

- Property boundaries and easements
- Contour lines
- Ponds, streams, springs, septic
- Pipes and other utilities
- Roads, paths, gates, doors
- Existing buildings
- Downspouts and gutters

STEP 2: SITE ASSESSMENT

In site assessment, designers take a look at important cross sections of the landscape to better understand the site as a whole. The checklist of site elements comes from a modified list originally developed by P. A. Yeomans that was named "the Scale of Permanence,"[2] as elements on the list are arranged from those hardest to change (climate) to those easiest to change (aesthetics).

To conduct a site assessment, take the base map of the site and, overlaying it with a piece of tracing paper for each category, create a sketch that makes both visual and written notations of observations. In some cases related topics can be combined onto a single map (such as water and landform), but it's best to keep

OBTAINING MAPS

In the United States, resources for site maps include:

- The local county soil and water conservation district can provide you with aerial maps, along with information on topography, soils, hydrology, and so on. In New York State find office locations by county: http://www.agmkt.state.ny.us/soilwater /contacts/county_offices.html.
- The local county clerk's office can provide a copy of a tax map. You may need to know the site address and ideally the tax ID number.
- Check with the local library and historical society (county, town, or city), and see what maps may be available.
- Conduct a search online. Google maps: http://maps .google.com and Microsoft maps: http://www .bing.com/maps/ both have different imagery to work with, including aerial and satellite imagery. For those who want a basic mapping software, Google Earth is a free and excellent tool: www .earth.google.com. One interesting feature in Google Earth is the "historical imagery" function, which allows a user to zoom in on a site, then see aerial imagery from historical records, which often shows past land use and vegetative cover.

things as separate as possible, since it will help focus on different aspects of the site. Good site assessment maps are detailed about elements on the site; the point is to get to a summary of each topic area and discover the important considerations for design. The process provides direction and eliminates possible species and techniques. For example, if steep slopes prevent access to the forest with a vehicle, then mushroom cultivation may be limited to the logs that can be harvested from the stand, as bringing in outside material may be cost prohibitive.

In the case of the MacDaniels Nut Grove, students both mapped each category and wrote a short summary description for each site characteristic. The key points were then refined to concise statements that highlighted those key points. Site assessment contains

Table 10.2. Categories of Site Assessment

Category	Checklist
Climate	❑ Latitude and effects
	❑ Plant hardiness zone
	❑ Temperature averages and extremes
	❑ Annual precipitation and seasonal distribution
	❑ Frost-free dates and growing degree days
	❑ Weather patterns
	❑ Effects of wind
	❑ Predicted effects of climate change
Landform	❑ Slope
	❑ Aspect
	❑ Elevation
	❑ Topographic position (midslope, crest, valley floor)
	❑ Depth to water table and bedrock
Water	❑ Watershed boundaries and flow patterns
	❑ Areas prone to flooding
	❑ Existing sources of water supply
	❑ New sources of water
	❑ Pollution sources
	❑ Infrastructure, including culverts, wells, water lines, sewage lines, and septic systems
	❑ Evidence of erosion
Legal/Social	❑ Long zoning laws and building codes (permits needed)
	❑ Property lines, easements, right-of-way
	❑ Interactions with bordering areas
	❑ Well protection and other legal limits
	❑ Site use history and impacts
	❑ Stakeholders
	❑ Events and activities
	❑ Economic factors
Access and Circulation	❑ Points of access by foot, wheelbarrow, and vehicle
	❑ Storage areas for materials
	❑ Circulation of people and materials
	❑ Infrastructure: gates, bridges, stairs, ramps, etc

Category	Checklist
Vegetation and Wildlife	❑ Existing plant species
	❑ Ecosystem structure and character
	❑ Habitat types; food/water/shelter availability
	❑ Animal species: wildlife, domestic, pests
	❑ Old trees
	❑ Invasives/nonnative plants of concern
	❑ Rare and desirable plants
Microclimate	❑ Sun/shade patterns
	❑ Cold air drainage, frost pockets
	❑ Soil moisture patterns
	❑ Precipitation patterns
	❑ Local wind patterns
	❑ Combined effects
Buildings and Infrastructure	❑ Buildings: size, shape, location
	❑ Power lines and electric outlets
	❑ Outdoor faucets, wells, septic
	❑ Underground pipes and infrastructure
	❑ Fences, walls, and gates
	❑ Waste management
Soils	❑ Physical properties: texture, structure, drainage
	❑ Chemical properties, pH (soil test results)
	❑ Biological properties
	❑ Toxins or contaminants
	❑ Management history
Aesthetics/ Experience	❑ Overall setting, mood
	❑ Arrival and entry experience
	❑ View lines and corridors (good and bad)
	❑ Public/private continuum, formal/informal continuum
	❑ Disharmonies; views, noise, spaces, feelings

Adapted from Yeomans,[3] Jacke & Toensmeier,[4] Whitefield[5]

CASE STUDY: EXAMPLE SUMMARY OF SITE ASSESSMENT: MACDANIELS NUT GROVE 2009

CLIMATE

The site is located in hardiness zone 6. Prevailing winds are from the SW (summer) and NW (winter), and storms can come from the N and NE. The site receives around 35″ of rain each year, mostly during the spring and early summer months. Average first frost is mid-October, and average last frost occurs in mid-May. With climate change spring is arriving earlier, and the summers are likely to be hotter and drier, with winters wetter and warmer. The site has ample sun exposure. The bottom half is in a 100-year floodplain, though much is at low risk for more regular flooding, partially due to the large embankment between the site and Cascadilla Creek.

LANDFORM AND WATER

The site has two flatter areas at the NE and NW boundary edges that slope gently and then steeply down S and SW facing slopes to the lower half of the property, which has flat and sunken points. Water enters the site at a rapid pace, notable through the center valley, which runs off from the orchard. There is evidence of extensive sheet erosion on steeper sections of the site. In the lowland areas there are ample opportunities for small catchments, even a pond. An intermittent stream is a nice aesthetic feature but has low reliability for water needs on site. The only infrastructure is a spigot located on Cornell Orchard grounds.

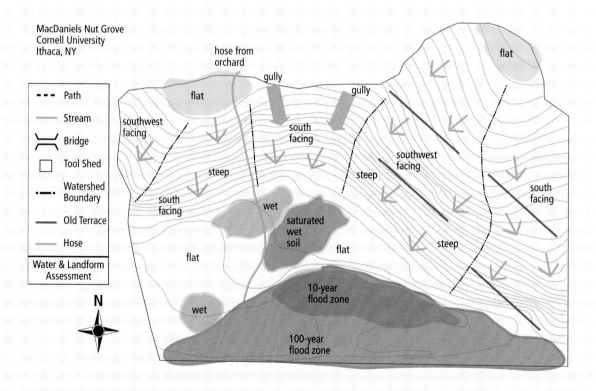

Figure 10.3. Water assessment map for the nut grove. Areas where water was pooling and causing erosion gullies are noted, along with level areas at the top of the slope (good for dropping materials) and the general lack of on-site water for irrigation. The site is overwhelmingly steep sloped with south and southwest aspects. The lower section near the creek is wet and prone to flooding.

LEGAL/SOCIAL

The site is part of CU Plantations Upper Cascadilla Natural Area. Official jurisdiction is delegated to the Cornell Plantations. The grove is managed by Prof. Ken Mudge and a rotating group of students participating in the Practicum in Forest Farming class. A student is also hired part time each summer to work on improvements. Funding comes from an annual endowment originally from Lawrence MacDaniels. Overall funding availability is low. The main function of the site is educational, with a desire to grow marketable products in the near future.

ACCESS AND CIRCULATION

Vehicle access is limited to the perimeter of the site at both the top and bottom. Driving to the site requires either use of CU Orchard lands or the East Hill Recreation Way, both of which require permission before using. Cart access is limited to just a few sections because of landform. Pedestrian access exists for much of the site but is confusing, tedious, and inefficient. Bridges are adequate for pedestrian use but useless for providing cart access. Overall use is sparse, concentrated in a few areas scattered in the grove. Areas include small trail patches for mushrooms, ginseng, berries, and ornamental plants.

VEGETATION AND WILDLIFE

Overall, species and structural diversity is low. Most of the slopes have varieties of hickory with a few maple, oak, and cherry scattered throughout. The lowlands are dominated by black walnut. The shrub layer was previously occupied by honeysuckle, and since removal some herbaceous and groundcover species have arrived. Site overall lacks well-developed understory from the ground up. Small wildlife and deer make use

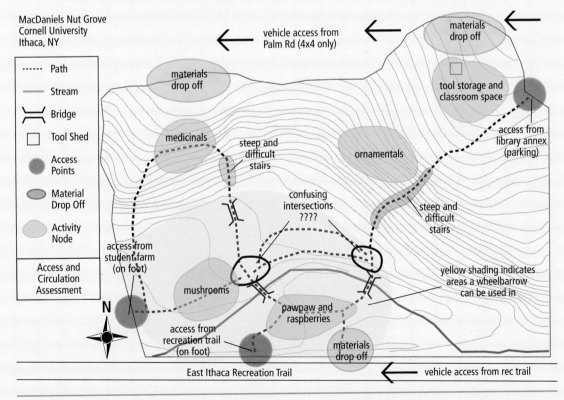

Figure 10.4. Access and circulation is a critical assessment for any forest farm. From the map it is clear that a very small area of the site is used. The steep slopes make it tricky to use a cart or wheelbarrow in much of the site. There is no vehicle access except to drop-off points on the perimeter. Paths are confusing, with awkward intersections and no sense of "flow" to help visitors navigate the site.

of the site, but noticeably few bird and insect species. Little of the cultivated vegetation is thriving.

SOIL
Site has six major soil classifications. Structures include gravelly loam, silty loam, and a little clayey loam. Soils are dehydrated and compacted in most areas from past use and continued sheet runoff down slopes. Best soils are in lowland areas near a creek bed. Middle of site contains a lens of clay that might skew perceptions of depth to water table.

AESTHETICS AND EXPERIENCE
Aesthetics are poor overall. Signage is confusing, sparse, and nondirectional. The site appears to be rarely maintained. No clarity of purpose for any elements is defined. No cohesive story is told. There are several nicer areas to spend time, some more private than others. The site is disconnected from some beautiful natural areas and from Dilmun Hill Student Farm. Entrances are not clear or welcoming.

quite a bit of detective work on the part of those who undertake it; as a rule of thumb this phase of design should be 70 to 80 percent of the total time spent on design. The way to know if the design is on the right track is when assessment (and goals) lead to solutions that start to emerge as an outcome of the process.

STEP 3: GOAL SETTING
The stage of design in which goals are developed is critical: It determines the final elements as they are related to the goals and values of the site participants and stakeholders. Taking time to be clear and specific with goals also helps in defining the priorities of the operation, looking at elements in their parts as well as in relation to the "whole."

Write goals in the present tense—it gives them more weight when others read them. Each goal should express an overall objective for the design. Under each goal describe several criteria that meet the goal—these should become more specific, with the most detailed goals containing numbers that can be quantified. For example:

Goal: Increase production of income-generating crops from the woods
Criteria: Tap 200 maple trees and produce 20 gallons per year
Criteria: Plant 50 pawpaw trees on 8 × 8 spacing along the hedgerow
Criteria: Inoculate 200 mushroom logs each year until 800 are producing

This basic list provides a lot to ponder and consider further, before actually committing to planting, inoculating, or harvesting. For instance, the numbers provided allow for some quick math to determine if the plan is even realistic. Taking the first criteria as an example, the costs for tapping two hundred trees with buckets could be compared with the use of tubing. The price per gallon could be determined by deciding how the syrup would be sold (for example, gallons versus quarts versus pints):

Expenses
    Tapping with buckets: 200 × $20/set = $4,000
    Tapping with tubing: 1000′ of tubing @ $.10 a foot plus 200 taps @ $.25 each = $200
    Wood = free scrap
    Labor = 100 hours each year at $10/hr = $1,000

Income
    Sales of 15 gallons as 60 quarts @ $20 each = $1,200
    Sales of 15 gallons as 120 pints @ $12 each = $1,440

At the end of this rough calculating, the criteria may or may not actually meet the goal (Will syrup even generate income? How long until the break-even point?); then a decision has to be made based on values. In this case, the aesthetic of buckets may be out of the reasonable price range of the forest farmer, so perhaps tubing will be utilized. Sugaring may be abandoned altogether or allocated to a goal that speaks of hobby-type pursuits in the forest farm. Of course, more thorough analysis and calculations than those above should follow if the decision is ultimately to pursue a crop or system for income.

Goals can be organized in a variety of different ways to suit the particulars of a site. For the nut grove the

CASE STUDY: GOALS ARTICULATION FOR MACDANIELS NUT GROVE
2009

1. EDUCATION

Goal #1: Students at Cornell learn about managing the forest for a variety of uses through a series of hands-on activities, focusing on site design and the cultivation, harvesting, processing, and marketing of forest farmed products.

Criteria: 20 students engage in 14 weeks of Practicum in Forest Farming class

Criteria: 4 workdays are held on weekday evenings each semester, from 4:30 to 7:00 p.m.

Goal #2: The general public, including members of the university community, townies, and other visitors, attend work parties, open-house events, and workshops. A self-guided tour and permanent student exhibits about aspects of managing a forest farm encourage recreational and educational use.

Criteria: Workdays mentioned above will be advertised to the general public.

Criteria: An annual fall open house in October provides tours and demonstrations and attracts 100+ visitors.

Criteria: One mushroom inoculation workshop will be held each spring and will be limited to 20 students, who will pay a small fee.

2. RESEARCH

Goal #1: Students and staff grow a wide range of non-timber forest crops (extensive) by taking advantage of the diversity of niches on the site whenever possible, and they grow mushrooms intensively to demonstrate the potential for income generation through forest farming.

Criteria: Five areas of production on a demonstration scale include:

- Nuts (approx. 60 trees throughout the grove)
- Walnut Island (6 mature walnuts and 30 pawpaws)
- The Valley (20 elderberries, 4 ramp beds, and 4 fern beds)
- Medicinal plants (6 beds of ginseng and goldenseal)
- Forest nursery (pot-in-pot production of hostas, ferns, bleeding heart in 4 raised beds)

Criteria: 400 mushroom logs are maintained and soaked weekly in rotation by hired help to produce 10+ lbs per week for sale at the Cornell Farmers Market. Production time, yields, and sales data will be recorded.

3. FACILITIES

Goal #1: Entrances are clearly marked with welcoming features, a site map, and a brochure that outlines the self-guided tour. Directional signage guides visitors to the location of specific exhibits that offer interpretive information with weatherproof signage.

Criteria: The three entrances (from the student farm, rec trail, and orchard) are planted with living willow arbors and include a welcome sign and overview map of the site as well as a brochure for the self-guided tour.

Criteria: The central loop trail will be marked with a color and a symbol and will provide clear directions for navigating the site. Each of the six production areas will house a large sign naming the area and with weatherproof interpretive signage.

Goal #2: Access and circulation flows revolve around the center of the nut grove, where a sheltered classroom is located for event use. Tools and work materials are housed in a more private location in the northeast corner of the site.

Criteria: The classroom is a simple polewood structure with a footprint of 20' × 20' feet and is constructed from local black locust timbers and corrugated tin roof.

Criteria: The northeast corner has a locked 4' × 4' shed for tools and material storage. Wheelbarrows and buckets are stacked behind the shed to maintain cleanliness.

goals were listed under three areas that were determined to be the most critical: education, research, and facilities (see case study on page 317). A well-formed goals document acts like a checklist, in which each of the criteria is an item found on the final design map.

It's important to remember that a goals document is alive. It isn't something that can be written once and forgotten but is a tool that should be revisited and updated on an annual basis, if not more frequently. On projects that have a public presence or multiple stakeholders, the goals articulation becomes the "glue" that holds the project together, as referencing this document literally ensures that all involved are on the same page.

Step 3: Schematic Design

With the design unfolding as site assessment and goals are developed, the process moves from being descriptive to being prescriptive. The schematic phase involves brainstorming ideas to solve problems that the site presents, while keeping in mind the limits of site assessment and the goals of participants. A series of noncommittal activities allow the designer to play with ideas before committing to them. While there are many directions to go with this phase, presented here are a few of the techniques the authors have observed as being particularly useful when teaching students design.

Random Assembly

In this activity participants brainstorm any and all possible elements for the site. At this point ignore potential limitations of time, energy, skills, and finances and just play with the possibilities. Examples of elements in a forest farm could be toolshed, ginseng beds, compost area, mushroom yard, woodworking area, and so forth. Write each desired element on a small scrap of paper.

MacDaniels Nut Grove
Cornell University
Ithaca, NY

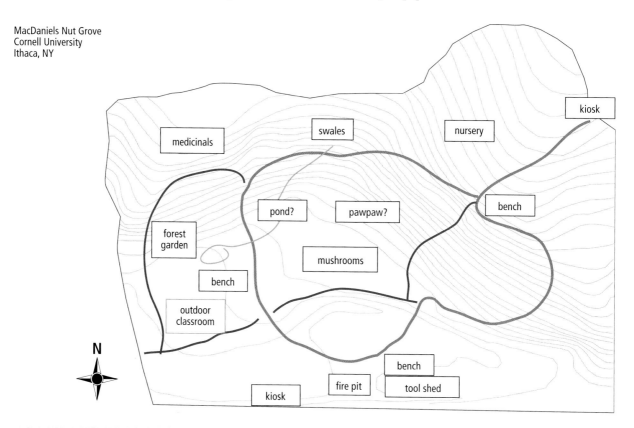

Figure 10.5. Random assembly involves writing desired site elements on slips of paper, then moving them around the site, looking for relationships based on the characteristics discovered in site assessment and for possibilities for functional interconnection.

Overlay a piece of tracing paper on your base map, and spend fifteen to twenty minutes arranging elements on your map to come up with a design concept, considering about how each element relates to the site conditions as well as to one another. Often elements will cluster together and around main centers of activity. Try several combinations and don't immediately allow any of the elements to be locked in place. Finalize this random assembly map by taping down the elements when satisfied with their placement.

Zone Planning

The random assembly can also be combined with the zone planning tool from permaculture (described in chapter 2). To do this, place another clean sheet of tracing paper over your base map and the initial element design. For this activity, it is helpful to have a second identical set of element cards made ahead of time.

Try to locate and draw zones 1 to 5 based on the currently assembly of elements. Remember that zones radiate out from centers of activity and designate different scales of energy expenditure on the part of those who will manage the site. In some cases the zones radiate in a uniform pattern, while in others there is more of a mixture of shapes and blobs that form.

Remove these two layers of tracing paper and start again from scratch, drawing a concept of how zones could look on the site, ignoring any previous placement of elements. It may be worthwhile to produce several of these overlays and try out different combinations of zones with different sizes and shapes. Then cover a completed zone sketch with another fresh piece of tracing paper. Take the second set of elements written on pieces of paper and spend fifteen to twenty minutes placing the elements in relationship to the zones before taping them down. Identify which elements are close to

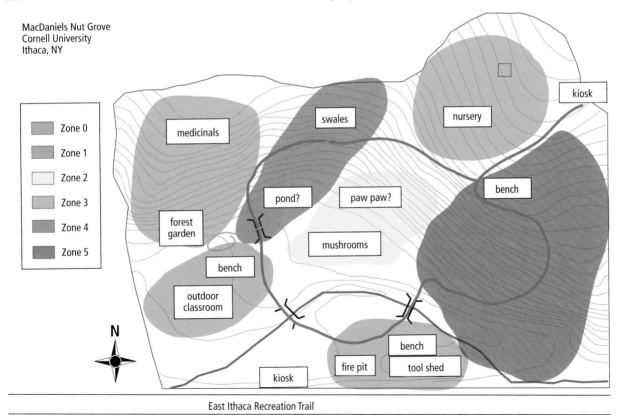

Figure 10.6. Using the zone planning tool to assess the random assembly from figure 10.5. Note here that the way elements are arranged creates split and discontinuous zones, which in this case provides good evidence that elements should be rearranged.

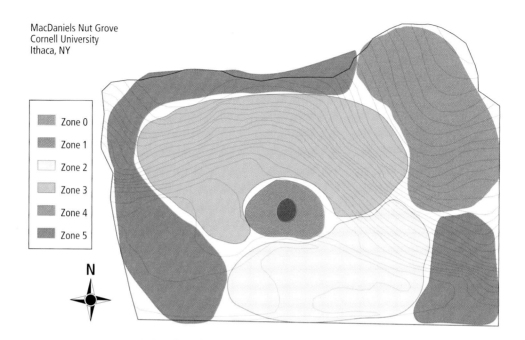

Figure 10.7. Starting with zones before doing random assembly can often help frame a pattern for site layout that results in a better design. Here, it was recognized that the zone 0/1 area in the center was a good place to start because it was the only spot where visitors and students could see the entire site. From there zones are concentrated around the center.

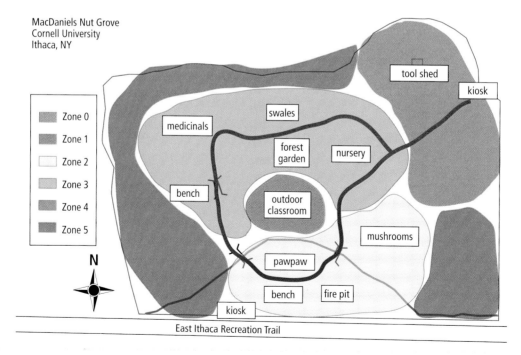

Figure 10.8. Adding in random assembly elements confirms that the site is better organized using zones. The center provides direct connection to the most intensive areas of production: pawpaw/walnut polyculture and mushrooms. The green areas are focused on demonstration of other forest farming techniques. Wild foraging and wild zones are left to the perimeter of the site. The loop trail is refined from the first set of maps (above) to fit within the main zones.

home/zone 0 and which are farthest, and try to cluster elements needing similar amounts of attention within the same zone.

When finished, place both maps side by side. How did the two design tools (random assembly and zone mapping) compare? What differences emerged when one process was done before the other?

Bubble Diagramming

In this exercise the point is to avoid getting hung up on details and instead focus on overall patterns and processes on the site. Start with a clean sheet of tracing paper on the base map. Pick a theme around which to concentrate some thinking. Examples include water, circulation of materials, points of interest, and production versus recreation areas.

Then take time to fill up the paper with bubbles that relate to the theme. For example, if focusing on water, start by concentrating on problem areas. Circle them, and write a brief notation about the area. Then look for ways to connect different bubbles with arrows, notes, and ideas. The point is to fill up the page, then move on to another theme and do the same thing.

After several maps are created they can then be overlaid to look at the relationships. For instance, what is the relationship between water and circulation of materials? As can be seen in the nut grove example, several conflicts emerged, including identifying that several areas where compost and wood chips were stored were also prone to flooding.

Schematic design should use a lot of tracing paper. By the end it's recommended that roughly ten to

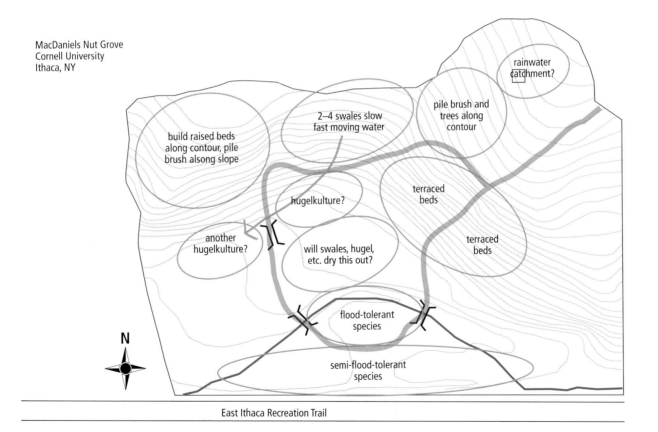

MacDaniels Nut Grove
Cornell University
Ithaca, NY

rainwater catchment?

pile brush and trees along contour

2–4 swales slow fast moving water

build raised beds along contour, pile brush alsong slope

terraced beds

hugelkulture?

another hugelkulture?

will swales, hugel, etc. dry this out?

terraced beds

flood-tolerant species

semi-flood-tolerant species

N

East Ithaca Recreation Trail

Figure 10.9. Bubble diagram related to water. Think of all schematic design as brainstorming on paper. Asking questions, considering possibilities with limitation, and discovering new pathways of understanding are important parts to this part of the process. Make maps until the potential is exhausted.

twenty sketch maps have been made to really flesh out ideas. Try not to get hung up on any one thing, but keep trying different combinations and arrangements.

"Final" Design

After the "kinks" have been worked out in the schematic phase, it's time to get real. Can the proposed design solutions be financed? Are they realistic given the time and skill constraints of the site participants? What design elements are highest priority, versus those that could go?

The goal in this last stage of design is to develop a "final" map that summarizes the way the site will look down the road. Some designers choose to do this in several stages, with an overlay of tracing paper for year one, year five, and year ten site improvements. Or perhaps just one map will suffice, with many elements that might take decades to implement.

Either way, having a map that is clean and visually appealing enables a designer to convey the overall plan to others.

Implementation and Evaluation

While details of this phase will not be covered in depth here, it's important to mention that part of the entire design process comes after the final "pretty picture" and plan are developed. This is when the rubber hits the road—where design is taken into the site and implemented. In most cases the concept immediately begins to shift on-site, both because of considerations that were not apparent before and because attitudes and goals also change for people over time.

So while a final map and description is a really important output of the process, recognize that it is a record of one point in time. Design is iterative, meaning that it will repeat and revisit the various steps of design

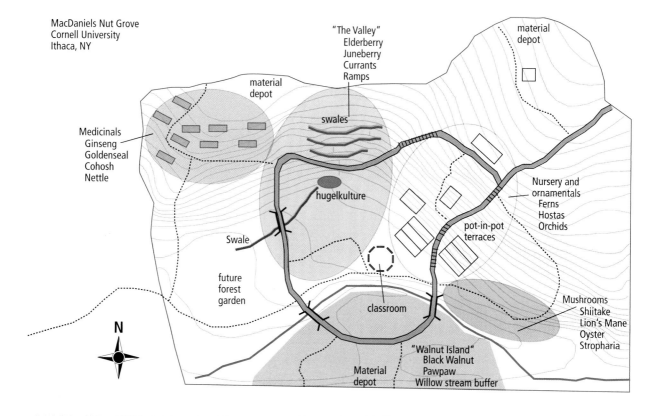

Figure 10.10. Final design for MNG.

CASE STUDY: FINAL DESIGN OF THE MACDANIELS NUT GROVE: PROBLEMS AND SOLUTIONS

The final plan for the nut grove included a map (see figure 10.10) as well as a written summary, which highlighted the main "problems" with the site that were "solved" through the design process:

Problem: Mainly because of the varied and sometimes steep topography, there was not one easy access point where materials (for example, wood chips) could be distributed to the site. Wheelbarrow access was possible in certain zones of use, but one cannot travel easily around the entire site with materials.

Solution: Three main areas to drop and store materials are designed to provide access to materials in different parts of the site. These areas have been sighted to be out of the way of the visiting public for aesthetic considerations.

Problem: A central challenge was the lack of good direction and flow, especially for people unfamiliar with the site. Taking groups on tours results in a lot of back-and-forth, where groups visit an area of the grove, then have to backtrack, which is awkward because of the narrow trails and steep topography. In addition, when working in the grove one often had to take a longer route to get from one area to another.

Solution: A "loop" trail was designed to facilitate movement around the grove that included a stop at each main project area. This trail is clearly marked, and secondary trails for further exploration also facilitate connections for improved work efficiency. The main loop has color-coded signage and is surfaced with wood chips, while the other trails can remain somewhat less defined and maintained.

Problem: The main gathering area was originally located in the northeastern corner of the site, but neither gives a complete view of the grove nor provides opportunities for a variety of events.

Solution: A more central location has been identified as a place to construct a covered pavilion and outdoor classroom. The pavilion will provide sheltered space for classes and other events, seating twenty to twenty-five people for lectures and workshops.

Problem: Different forest crops were grown in random spots and not consolidated into categories, which did not fully and clearly represent the potential of a forest farming operation.

Solution: Having designated areas for each type of production improves efficiency for management and also provides a cleaner picture of the different possibilities forest farming can offer. These exhibits will tie into the loop trail and provide a self-guided tour for visitors.

---

in no particular order. Likely throughout implementation, the decisions made during design will be altered and adjusted. The site assessment, goals, and design will all be revisited. This is good, and it is encouraged that forest farmers revisit and update their designs at least every two to three seasons.

THE LACK OF "DESIGN" IN DESIGNS

Stepping back to take a look at the design process as a whole, one of the most important things to consider is that the steps are more or less ordered in their importance to overall success. In other words, if a person were to choose one thing to do, it's best to engage in site assessment so the constraints of the site can be discovered. Next on the list of importance would be to articulate goals. And so on.

This isn't to say that the design process as a whole isn't important or valuable to complete all the way through. It is simply to acknowledge that in the busy modern world often farmers and landowners simply don't have or don't make the time to do a thorough design. A great irony of the permaculture movement is that while design is often taught in courses and emphasized as very important, few permaculturists have published complete designs for their own sites. It's a bit of the "cobbler's kids don't have any shoes" syndrome perhaps, but it's a chronic problem in the movement as a whole.

So while the process is important, it's equally important to recognize that people all have limits of time and energy. Certainly, anytime something is designed and planned out on paper first, it is almost always the case that it is implemented better on the ground. It also usually saves considerable time, money, and energy on the part of the people working to put the design in place. Even if just a base map is made and some site assessment done or perhaps just a few goals are discussed and jotted down on a napkin, making the effort is always the first step—and never a waste of time.

Management of Forest Farms

There are many aspects to successful management of forest farms. While in previous chapters many of the techniques specific to types of crops were covered, much of forest farming comes down to sound forestry practices. The best ways to learn about the complex nature of forest management include:

1. Reading information about woodlot management from books and Cooperative Extension or state forest agency websites.
2. Asking for a woodlot visit from a trained volunteer, often available through the local Cooperative Extension office. Some states and regions have a "Master Forest Owner" program, which any landowner can enroll in.
3. Asking for a visit from the state forestry agency, which may or may not be free, depending on location.
4. Joining a local forest or woodland owner association and connecting with other woodlot owners to learn how they have approached management.

The goal of this section is to act as a starting point and to highlight some of the key concepts and approaches. The big management error to avoid is mak-

ing rash decisions that can't easily be reversed or revised as you go along. In forestry it's a common saying that once a tree is cut, it can't be put back on the stump. When in doubt, do nothing.

Along these lines, when hiring natural resource consultants, foresters, and loggers it's important to do the homework and find out about the particular professional's values and ethics that frame his or her work. As a general rule, if discussion about the health of the forest is not expressed as the primary objective, a forest farmer may want to find someone else to work with. Don't trust someone just because she has a degree in forestry or talks the talk. As stewards of the land we have responsibility to do the best we can. Through the years we've simply seen too many destroyed forests at the cost of profit, and at the end of the day forest farming is really about preserving our most valuable ecosystems.

PRINCIPLES FOR MANAGEMENT

From the authors' perspective the following principles are fundamental to forest farming management:

1. The health of the forest is always the bottom line. Health is hard to define (see chapter 1). An overall goal to strive for is that a forest can continue to provide an equal variety, quality, and quantity of services and products into the future.
2. Yields should be obtained as a byproduct of forest management activities for forest health.
3. Almost all forests have been damaged at the hands of humans. Forests that are in an untouched or extremely healthy state should remain that way. There is plenty left to manage.
4. For forests that have been mismanaged, human intervention can be a necessary and ethical action to repair the damage.
5. The role of people in the woods is to set a sequence of events in motion, observe how the forest responds, and adjust. People are not the controlling force but are mere players in the whole system.
6. Be alert to seasonal needs and sensitivities of the forest. A good time to fell and move trees is in

the winter; it is safer and does less damage to the residual trees and undergrowth.

7. When in doubt, do nothing. You can't put the tree back on the stump.

FOREST VS. STAND VS. TREE DECISIONS

A useful way to divide up the thinking that goes into forest management relates to scale, which can be classified by the following partitions:

- A *forest* is the tract of woods in the largest sense of the word. It often includes the entire tract under management but could also include adjacent tracts if there are watersheds, wildlife, and other macro considerations.
- Numerous *stands* exist within a forest, which are characterized by a similar species composition, topography, soil type, aspect of the slope, and so on. The combinations of these factors define individual stands. Decisions at this level include characterizing an overall pattern and prescribing a method for management (timber stand improvement, for example).
- Within stands exist *individual trees* that deserve consideration based on their species, overall structure, and health. Tree-level decisions include the final decision of what trees to cut and which ones to leave in the forest.

As an example, the MacDaniels Nut Grove is a 3-acre *stand* sitting among the much larger *forest* that borders Cascadilla Creek. The forest in this case is limited to the 675 acres owned by the Cornell Plantations and designated as a natural area. Within this forest there are multiple stands, including the nut grove, beech–maple, oak–hickory, hemlock–hardwood, maple–basswood, and several aquatic communities.[6] Within the nut grove, management is prescribed as "sanitation thinning," where only trees that are dangerous to visitors, heavily infected with pests or disease, or in clear decline are removed. As much of the overstory is grafted hickory nuts, management leans toward favoring these unique specimens, mostly for their value as historical artifacts.

Keep in mind that these classifications are flexible in scale, depending on the context. Another important item to consider is that, when moving from forest to stand to tree level, management increases in intensity. Most of the work comes down to stand management and individual tree selection, which ultimately impacts the larger forest ecosystem.

Forest Level Decisions: Long-Term Patterning

The overall pattern to emphasize on the forest level is creating a mosaic of diversity in terms of species composition and succession stages that shifts and changes over time. This mix of forest patterns necessitates multigenerational thinking. In addition, considerations on this scale also include considerations for watershed health.

SUCCESSION

Important to this state of development is revisiting the succession concepts presented in chapter 3. Over the course of a forest there should be, over many generations, a shifting mosaic that promotes a mix of early, middle, and climax stages of succession. It's hard to be specific on this point because the character, size, and past history of a given forest largely determine the way that succession dynamics should be played out.

WATERSHEDS

When managing a forest farm, awareness of the watersheds on the site is an important consideration. A watershed is defined as a catchment in which all the water both above- and belowground eventually flows to the same place. The location of a forest at the top, middle, or bottom of a watershed will have implications for both the quality and quantity of water coming through the site. One of the simplest ways to explore watersheds is on foot, following the tributaries, creeks, and other water features to their source. Additionally, county soil and water conservation district offices and online sources can offer watershed maps and more information to help make good decisions.

Figure 10.11. The nut grove is part of the Cascadilla Creek watershed, which flows into the Cayuga Lake Basin. The location of a forest farm in relation to its watershed is an important factor to consider. In this case the nut grove (red arrow) is located below all the tributaries of the watershed; thus, the water coming through may have pollutants or contaminants, depending on the land use upstream.

Watersheds are really a set of nesting catchments that eventually flow to oceans. For instance, the nut grove is more or less in the middle of the Cascadilla Creek watershed (see figure 10.11), which flows to the much larger Cayuga Lake watershed and continues north, eventually through the St. Lawrence watershed to the Atlantic Ocean. Not too far south of Ithaca, the watersheds are completely different, with many flowing south, eventually leaving through the Chesapeake Bay in Maryland. Since these watersheds all nest, then technically the moment a drop falls on the nut grove, it is in the St. Lawrence (as well as Cayuga) watershed.

Depending on location, some forest farms will have great concerns with water, while others won't need to concern themselves with management very much, if at all. For sites that do have issues with flooding and erosion, several strategies can be employed to reduce negative effects. The easiest and most straightforward practice is to keep vegetation in place. Next, pile and lay brush or trees along contours, and try to lay out pathways and cultivation beds across the slope as much as possible. The effect of this pattern can be seen in any forest where a tree happens to fall across a slope. Over time leaves, brush, and other organic matter accumulate and help to slow and absorb water runoff.

Another strategy that may be applicable to some forest sites is having contour swales, essentially ditches that are dug along contours, with the excavated material mounded on the downslope side of the ditch. When water hits the swale, it slows down and spreads along the length of the depression, allowing time for the water to soak into the soil. Care should be taken to design systems to handle the amount of anticipated runoff, and a plan for overflow should always be designed into the system.[7]

In addition, pools that mimic the function of vernal (seasonal) ponds in forests can be excavated to allow seasonal water flows to accumulate and slowly evaporate. Of course, when digging is involved, care must be taken to avoid damaging tree roots. It may not always be appropriate to dig, in which case, mounding materials may have to suffice. In some circumstances, areas that need to be dug may be combined with strategic thinning, which would negate concern for protection of tree roots.

In addition measures can be taken to reduce the erosive effects of seasonal and continuous creeks and

Figure 10.12. At the MacDaniels Nut Grove swales were dug by students to mitigate the effects of seasonal runoff from the uphill Cornell Orchards. Each swale overflows into the next one and eventually flows through a hugelkulture mound and into a vernal pool for eventual evaporation. In this way water is not only managed to avoid causing problems, but the system provides a multitude of wildlife habitat as well.

waterways. Willow stakes (see chapter 8) and riparian buffer resources can provide some further insight into strategies for stabilizing creek beds.

Multigenerational Management

One of the greatest challenges of participating in forest ecosystems is the fact that many of the choices one can make in a lifetime will not show up as consequences until one or two generations later. This is perhaps both why modern agriculture has gone largely toward annual-based monocultures (for the ease of understanding a much simpler system) and why modern human civilization has treated forests so poorly.

If forest farmers are going to devote focus and energy to a system that operates on timescales of fifty-, hundred-, and even thousand-year cycles, then each must ask, "Who will steward this forest after I am gone?" For what is the point of putting in all the effort if the next guy is going to manage in the opposite way? Luckily, there are both social and legal structures that can assist a forest farmer in achieving long-term goals.

Family Ties

Traditionally the way an individual could guarantee continuation of land stewardship was through family: Children and grandchildren were expected to take over the farm eventually and manage it in a way that would honor their family before them. A sense of pride and history governed an intimate multigenerational connection to a particular place.

While there are still remnants of this familial connection here and there in the landscape, the overall cultural pattern has shifted dramatically. Families rarely own land that was passed down through the generations anymore, and often the kids are less interested in inheriting the land from parents. Yet while there may be a deficiency of blood-based relations to ensure multigenerational stewardship of land, some people are beginning to recognize that a similar type of relationship and agreement is still possible.

In the United States the number of young beginning farmers is growing quickly. Aspiring young farmers often have the same hurdle; they are passionate about farming, yet often lack the finances to purchase both land and the infrastructure needed to farm it. This is where those who have land can step in and work toward arrangements that help youth get started, while ensuring that land can be protected in the long term. There is another side to the equation—that knowledge transfer can happen when older foresters and farmers team up with younger ones.

This isn't to say that social relationships revolving around the use of land aren't difficult, whether the people involved are related or not. It takes the right "fit" of personalities, goals, and skill sets to make it work. Many arrangements fail early on, partially because each party wasn't clear or explicit about certain aspects of the arrangement. This is where a clear and well-articulated goals document can help. Clear agreements around finances are also important. While forging lasting relationships around projects may not be easy work, it is arguably very essential if all the effort that goes into forest management is not to be done in vain.

Land Trusts

Another tactic that involves less personality and more legal construct is the idea of putting land into a trust to guarantee its future preservation. Usually this takes the form of an easement that is added to the deed of the property, which legally binds a site to certain uses for the indefinite future. In other words, if the property is sold, the easement is sold along with it.

While some easements take a "hands off" approach and encourage preservation through prohibiting any sort of land use, there are a range of options that can limit use but still allow for production. For example, an easement could prohibit the subdivision of land, the clearing of land of all its trees, or the construction of new buildings. It's often up to the landowner to determine what limits he or she wants to set on a property.

While this process can be time consuming and costly, it's hard to put a price on knowing that a hardworked stewardship effort won't be destroyed by a change in ownership. In reality there are still a lot of nuances with these types of programs, and it's still tricky to ensure the same management patterns persist for decades or longer. It's best to have such conversations with an agency if this concept is of interest. Sometimes protected lands are also eligible for tax breaks, grants, and other financial incentives that may pay off in the long term.[8]

Stand Decisions: Prescribing Management

At the stand level, considerable thought needs to be given to the particular pattern of management that will occur in a stand. A "stand" is defined as a portion of woods that contains a similar species composition, topography, soil, and watershed. As with forest level decision making, the lines one may draw to define a stand are not absolute, and they vary in size, shape, and context depending on unique site features.

Mapping stands is a good design practice prior to management. Sometimes a state natural resource management office will produce such a report for free or for a small fee. The report often will include a map of the property with stands marked. Each stand will be characterized by the species present, and a recommended management strategy will be recommended. Private foresters can also be hired to do this for properties.

Patterns of Management: Even- and Uneven-Aged Management

When deciding on a particular management strategy and the stand level, there are multiple considerations. The starting point is to determine first if the existing forest composition is desired, and if so, which management strategies will support regeneration of the trees

Figure 10.13. Michael Burns, permaculture instructor and budding forest farmer at Cayuta Sun Farm and Homestead, creates a sketch map of his forested property. Identifying the size, shape, composition, and character of stands within a property is one of the first tasks of management.

Table 10.3. Possible Species of Even- vs. Uneven-Aged Management

Uneven-aged management	Trees	Sugar maple, beech, birch, hemlock, basswood, pignut hickory
	Shrubs	Flowing dogwood, mountain laurel, striped maple, witch hazel, ferns
	Wildlife	Pileated woodpecker, flying squirrel, flycatchers, warblers, scarlet tanager
Even-aged management	Trees	Oak, eastern white pine, black cherry, paper birch, white ash, tulip poplar, aspen, eastern red cedar
	Shrubs	Hazelnut, staghorn sumac, raspberries, sweet fern
	Wildlife	Red-tailed hawk, indigo bunting, deer, eastern bluebird, cedar waxwing, cottontail

Source: Adapted from USDA Northeast Regeneration Handbook[9]

in that forest (see table 10.3). Uneven-aged management supports more shade-tolerant species and tends to make smaller interventions that don't dramatically alter light conditions, whereas even-aged management creates more open conditions that support growth of sun-loving tree species. Both forest types can be beneficial to the regeneration of different trees and to other plants and wildlife. It is helpful to think about the difference between the two management strategies as a continuum of disturbance in which with uneven-aged management there are small and frequent disturbances to patches of trees, whereas with even-aged management larger and more dramatic disturbances occur less frequently. A decision as to the type of management comes down to the type of forest present, along with the goals of the landowner.

The bottom line with all forestry practice is promoting regeneration or the growing or regrowth of seeds and stumps of desirable trees. There are three main factors that affect the ability of a tree to regenerate in a forest: creating the ideal conditions for the desired species to seed, protecting new seedlings from competing plants, and protecting new seedlings from deer and other browse animals. Proper management involves

assessing the forest before, during, and after any cuts, to ensure that the intended results are achieved.

TECHNIQUES FOR MANAGEMENT

Following are a variety of techniques for management. While the patterns above focus on general approaches (strategies) based on current and desired species, these techniques get into what to do when it comes time to mark and cut trees. These are not the only approaches to this, of course, but a few of the most common ones used by landowners.

Timber Stand Improvement

The basis of need for a timber stand improvement (TSI) thinning occurs in younger forests that have often either regrown from abandoned agricultural fields or been high graded (all the high-value timber trees removed) in the past. These forests are often very dense, overgrown, and highly susceptible to pest and disease outbreaks. One can think of this as a stand of stressed-out trees, mostly of a consistent height and diameter.

Timber stand improvement is a technique in which the trees that are showing decline are removed.

Figure 10.14. In timber stand improvement diseased and deformed trees (shaded grey and black) are removed to provide more light to the residual trees. Illustration by Travis Bettencourt

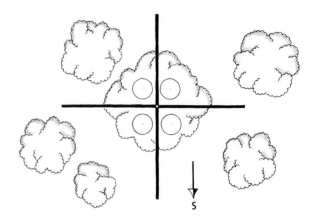

Figure 10.15. Crop tree management identifies the best trees, then thins to support their maximum growth. In this case, trees on the south side would be more beneficial to cut first, as it's not wise to cut everything in one season. Illustration by Travis Bettencourt

Evidence of decline comes in many forms: The tree could be crooked or have a split crotch or could show evidence of dieback in the branches. Diseased trees and those that are dead or dying are the most likely candidates for removal.[10]

In a timber stand improvement individual trees and in some cases groups of trees are removed to change the dynamics in the forest in a more subtle way compared to the even-aged management techniques. This is usually best done over several seasons, so that thinning doesn't open up the forest too quickly. The two-thirds rule is a good one to apply here; start by thinning at most one-third of the stand each season, until the thinning is complete.

Crop Tree Management

This strategy is useful for stands with trees from 6 to 11 inches in diameter, where there are older, healthy trees that with some thinning could really thrive and succeed into old age. Crop tree management (CTM) as practiced is in some ways opposite from timber stand improvement in that the *best* trees are marked, and choices of thinning occur to support the health of the crop tree.[11] This is done by imagining crosshairs running through a particular crop tree (see figure 10.15). Adjacent trees can be assessed first for the level of impedance they are offering to the crop tree—those that are suppressing growth the most should be cut first. In addition, trees that are on the south side of the tree, if cut, will have a much more dramatic impact than those cut on the north side. The overall goal of

CTM is to "leave the best" in the forest and provide them with optimal conditions for growing.

Shelterwood

In some cases, especially where there is a desire to establish fruit and nut trees within a forest, a shelterwood cut may be desired in a forest farming situation. In this type of stand management, a significant and relatively even disturbance is applied to the canopy to increase light exposure to the forest floor. This can spur the regeneration of seedlings from sun-loving species while maintaining some canopy cover and protection from wind and other elements. The concern is that opening the canopy in this way may also open the possibility of undesirable species coming into the forest as well.

Coppice

The practice of coppicing is one that pervades the temperate forests of Europe but has yet to make a big showing in the forests of North America. Coppice management rests on the astounding fact that all deciduous trees when cut to the ground and given adequate light will resprout and regrow new shoots.[12] This fact offers forest managers another option for stand management—and a new form of shorter rotational forestry.

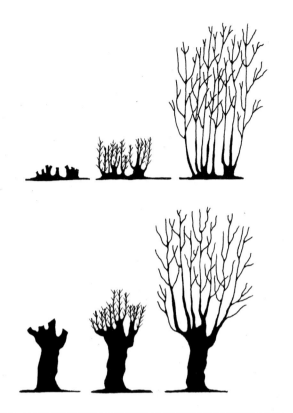

Figure 10.16. The difference between coppicing, above (cut to ground level) and pollarding, below (cut above browse height). Illustration by Travis Bettencourt

WHAT ABOUT CONIFERS?

It should be mentioned that the concept of coppicing can be extended to coniferous species, though it's a practice even rarer than coppicing of hardwood trees. The technical term is "stump culture," and the method is to harvest a conifer but leave the bottom two or three layers of branches (called whorls). This preserves the health of the root system and with adequate light conditions the tree can resprout and regrow. The practice is said to work with many conifers, including firs, pines, and spruce, though evidence is very scarce.[14]

While all deciduous trees coppice, some do it better than others. Examples of species that coppice well include poplar, alder, locusts, hazel, and chestnut. In addition to the importance of selecting the appropriate species for coppice, it's important to plan so that adequate light will hit coppiced stumps and therefore maximize the growth of new shoots. In one sense then, the management of stands through coppicing is technically a form of clear cutting, as the area necessary for good stem regrowth in a dense forest would be about ¼ of an acre. In a pasture or woodland situation a general rule of thumb would be to ensure that the stump receives at least a half-day of sun.

Pollarding

Another variation on coppicing is known as pollarding, or the practice of cutting trees not at ground level but above the level of browse by animals, be they wild, such as deer, or domesticated, such as goats.[13] If a tree's trunk is cut above browse height, the tree can regrow without being suppressed. Pollards were common in European pasture systems, with the harvested woody material often going to feed animals directly. Pollarding is also a choice strategy in woodlands where deer browse is a chronic problem.

A further benefit of pollarding can be that the farmer can use the tree as a place to store hay grasses cut during the growing season for later consumption. In this scenario, the tree becomes a source of shade, food, and material storage all in one.

Tree-Level Decisions: To Cut or Not to Cut?

While to a large degree forest- and stand-level decisions are about implementation of general "patterns" for management, when down to the tree level we get into the nuts and bolts of forest management. This is when the time comes to decide on individual trees and if they should be cut or not.

The best procedure is to follow the adage, "When in doubt, do nothing," Once a tree is off the stump, it cannot go back on. Choose wisely, and be sure before taking action. As a general rule, it is best on one day to walk a woods that is slated for thinning, assessing qualities and marking those that seem to be the best candidates for cutting. Then on another day, come back to the woodlot and review the plan once more before revving up that chain saw. Good forestry never

Figure 10.17. A forester in Pennsylvania marks trees in a stand for thinning. Foresters, who do the overall planning and ensure that the forest health is protected in logging jobs, often tag the trunk of the tree as well as leave a mark at the base, so that they can return after the logging job and ensure that the only trees cut were those marked. Photo courtesy of Penn State

Table 10.04. Life Span of Selected Temperate Species

Species	Average Lifespan
Maple	300
Oak	100–300
Ash	250
Hemlock	450
Hickory	200–250
Black Locust	60

Note: Potential lifespan means the longest a tree may survive under ideal conditions.[15]

happens when one walks around, chain saw idling, sizing up trees, and making decisions on the fly.

While the variables and context make the process of decision making rather confounding, the following questions are good to ask as individual trees are assessed. The bottom line with selecting trees for thinning is that not all trees that establish in a forest can survive to maturity. Another way to say this is to determine the likelihood of a tree's living out its full life potential; that is, will the tree be likely to live until it dies of old age? This assessment should also take into account the products desired by the forest farmer.

How Is the Structure of the Tree?

The first question is all about the shape, form, and growth habits of the tree. Is the tree straight? Does it have a single dominant stem reaching toward the light? Is the crown full of leaves or are there numerous dead or dying branches? Crooked trunks and split-crotch trees are some of the first candidates for removal, for the simple fact that gravity is likely to bring them down before they grow old.

Is There Evidence of Disease or Pest Damage?

Trees that exhibit signs of disease and insect or bird damage are also likely to live a shorter life. While there are numerous signs of defect from infestation that might be outside the range of the average forest farmer, some general observations of the leaves, branches, and trunk, especially in comparison to healthy species, will lend to some decisions in many cases. One of the most common diseases that is easy to note in a tree is called nectria canker, a catchall for a family of fungi that colonize and open up trees to further problems down the line. While the canker is common in forests, some trees show more susceptibility than others. The cankers are specific to a species, so maple will only transfer to maple, walnut to walnut, and so on.

Will Cutting the Tree Benefit the Residual Stand?

After assessing a tree for its overall structural integrity and health, the next question revolves around how cutting the tree will affect the rest of the stand. "Residual" is a name for the remaining trees after a thinning. Often removing a tree will open up light for another, healthier tree. Keep in mind that opening up light for trees on the south side is more effective than in other directions. If there appears to be no benefit to other trees, then care must be taken to think about how the tree will be felled to minimize damage to remaining trees. Sometimes it is worth leaving or girdling a tree if felling it means damaging another, excellent tree close by.

Figure 10.18. A split crotch is where the main stem of a tree splits in two, which can lead to problems as branches get bigger and heavier as the tree matures. The higher up the trunk the crotch is located, combined with the angle of split, determines how vulnerable a tree may be to splitting. Photograph courtesy of Jen Gabriel

Figure 10.19. Nectria canker on a black walnut tree, a species that is highly susceptible to infection. Trees should be observed for resistance to canker, and those with the worst occurrences should be removed to slow its spread. Photograph courtesy of Jen Gabriel

GIRDLING

The process of girdling can be an effective way to thin out trees without having to cut them down. They also make good standing deadwood (snags) for wildlife. The practice involves removing the bark and cambium layer of a tree, in a strip at least 4 inches wide and all the way around the trunk. Some simply cut two rings in the trunk with a chain saw, but certain species of trees can actually grow back together quickly from this treatment.

The recommended method is to make two rings that go 1 to 2 inches into the trunk around the tree, about 4 inches apart. Then, with a drawknife (preferred) or hatchet, scrape away the bark and inner cambium layer.

Are There Any Benefits to Leaving the Tree on the Stump?

The final question to consider is if there would be any benefits to the forest if the tree were left as is. For example, a large-diameter tree such as poplar that could damage other desirable trees if felled might be better left standing as a potential habitat tree. As it dies it will offer habitat for a diversity of life. Standing dead trees (often called snags) are critical to overall forest health.

The Final Test

After considering the previous questions, the final test is to double-check your thinking. If you can't justify at least a few good reasons for cutting a tree, perhaps it isn't a good choice. If you find no evidence of structural problems or the presence of disease, it's best to leave the tree. As always, when in doubt, leave it be.

CUTTING TREES

After going through a detailed and thorough experiment as described above, it's time to cut. Out of all the practices a forest farmer might engage in, the felling of trees is by far the most dangerous. According to the US Bureau of Labor Statistics, logging is the most dangerous job out there. It pays to ensure you are well equipped with the proper safety gear and training to remove trees safely. When in doubt, don't cut, or hire someone else to cut for you.

Saw and Safety Gear

Saws used to cut trees should be a modern style and have the most up-to-date safety features. Grandpa's old saw found in the barn is best to leave as an artifact. A safe chain saw should have the following components:

- Working chain brake (so the chain can be stopped and locked when not in use)
- The *Chain Catcher* (metal or plastic guard designed to prevent a broken or derailed saw chain from striking the operator)
- Safety throttle
- Antivibration feature
- Exhaust: muffler and spark arrester

In addition, an operator should be wearing:

- Helmet specifically for chainsawing, including ear protection and a face guard
- Gloves to reduce vibration
- Chaps that cover the entirety of the legs from hip to ankles

Before cutting any tree, the operator should be fully aware of any hazards and also have a clear plan for how he will go about felling the tree. A good tool to remember for planning is the "HELP" method, borrowed from the Game of Logging trainings, which are a series of courses offered worldwide, which teach general logging safety and the techniques for harvesting trees in a variety of contexts.[16]

Figure 10.20. The safest way to start a saw is on the ground. Be sure that the chain brake is engaged and that the operator is wearing all the necessary safety gear. Photograph courtesy of Jake Delisle

Hazards

Identify and remove any hazards, such as limbs, branches, and small trees that may be in the way of being able to easily maneuver around the tree. One additional hazard to always be aware of is the "kickback" zone on the saw, which when it comes into contact with wood will tend to kick backward, which can result in a loss of control.

Decide on Your Escape Route

Decide which way is best to make an escape once the tree begins to fall, which is usually at a 45-degree angle to the direction the tree will fall. The topography, vegetation, and lean of the tree will all influence this decision. The escape route should be clear for exiting for at least 10 to 20 feet.

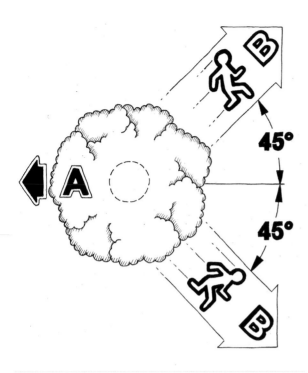

Figure 10.21. When felling a tree, the best escape route is at a 45-degree angle in either direction from the direction the tree will fall (A), usually away from the lean of the tree (B). Illustration by Travis Bettencourt

Assess the Lean

Assess the tree to determine if it has forward, backward, or side lean relative to the desired direction of tree fall.

Don't Start Unless You Have a Plan

Go over the full plan of felling before starting: where to stand, which direction the hinge will take the tree, and what steps will be carried out from cutting to when the tree is on the ground.

CUTTING METHODS

It's remarkable in most cases how much wood needs to be removed from a tree before it will fall down. Novice tree cutters often are surprised that even 70 or 80 percent won't do. It's a result of the fact that tree fibers are some of the strongest bonds around, biologically speaking. When getting into the practice of tree felling, the basic question is, "How do I remove almost all of the wood in a trunk safely, and get the tree to fall the way I want?"

Figure 10.22. Before running the saw, it's important to line up the cut perpendicular with the bar. Most saws have a line across the top of the body that can prove useful. Illustration by Travis Bettencourt

"Chase" Felling

By far the most commonly taught and practiced form of cutting a tree is to cut a wedge from the front of the tree aimed toward the desired direction of fall. This cut usually eats into about 50 percent of the trunk. The saw is then taken to the back of the tree and a horizontal cut toward the wedge results in enough wood being removed that the tree begins to fall. All but about 10 percent of the wood is removed, and the remaining "hinge" is what attempts to drive the tree down in a particular direction toward the forest floor. The reality with this method is that it is the lean of the tree that directs the fall, which can sometimes be disastrous when the feller wants it to go another way.

This might be best described as "chase" felling because as one cuts the tree, the bond holding the tree upright becomes looser and looser, until the point when the tree falls. In other words, the saw is running full bore, and the operator is right there, in the most dangerous zone, until the very last second.

This common method is mediocre to poor at best. While in many cases it can work just fine if the tree exhibits a decent lean in the desired direction of fall and if conditions are not windy, it does leave the feller vulnerable to many hazards. In the end the chain saw operator has little control over what happens. It's the equivalent of crossing his fingers.

Directional Felling

There exists a method of felling in which the operator has complete control of the tree, and the tree is solidly locked in place until the operator decides he or she is ready for it to fall. It also allows for a tree to be felled in almost *any* direction, allowing for pinpoint accuracy while maintaining control. While it might sound too good to be true, it does exist.

Directional felling is the process in which the wood is removed from the tree while maintaining control.[17] This is accomplished with a "bore cut"—the saw is actually plunged into the tree, allowing for the tree to remain attached at the hinge and at the back. It certainly takes time and practice to master, but the outcome is well worth the effort.

For Trees over 10 Inches

Large trees are treated with the same technique, regardless of the species. Keep in mind that the diameter of the tree vs. the length of the saw may make it necessary to make multiple cuts. For instance, a 12″ bar won't make it through a 28″ trunk.

1. As always, first survey the scene and develop a HELP plan.
2. Determine the size of the hinge. As a general rule of thumb, use 10 percent of the diameter of the tree at breast height (DBH) for the thickness of the hinge. It should have equal width across the stump. For example, a 10-inch tree should have a hinge 1 inch thick. The length of the hinge should be 80 percent of the tree's DBH.
3. Begin by sawing a steep face cut, at about 70 degrees. Cut first from the top down, minding your

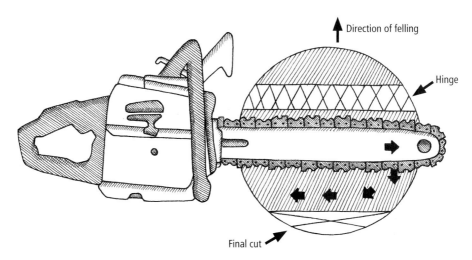

Figure 10.23. Large trees are felled with a steep front cut followed by boring through the tree starting just behind the hinge and moving toward the back of the tree, removing all the wood while still leaving a piece attached in the back, so wedges can be set and the tree can be felled in a controlled manner. Illustration by Travis Bettencourt

angle, then finish the cut with a horizontal slice, keeping the hinge size in mind.

4. Line up your saw with the flat, bottom part of the first cut. Keeping this height, move the saw back 1 to 2 inches from the wedge, and bore into the tree, keeping the saw as level as possible. Drive the saw with the lower corner first, as leading with the top edge will cause kickback.

5. Bore through the entire trunk, ensuring your cut is parallel to the hinge. At all cost avoid boring into the hinge or reducing its consistency of width.

6. Work the saw backward, cleaning out all the wood in the center of the tree, leaving a section of intact wood at the back that is sufficient to ensure the tree doesn't fall until you make the final cut (1 to 2 inches), directly opposite the hinge.

It's amazing because at this point over 80 percent of the wood is gone, and one could stop the chain saw, and walk away. The tree is fixed in place with wood in the front and back.

Depending on the lean, you can set wedges on either edge to help the tree fall properly. Then when you are ready the back connector is cut, and the tree falls, if leaning in the direction the hinge faces. With a tree where

you are working against a back or side lean, wedges are simply pounded until the tree falls.

Trees under 10 Inches

For smaller-diameter trees, the same basic principles apply—but with a slightly different twist.

1. Cut the shallow, steep wedge as mentioned before.
2. Instead of boring from the side, *bore directly through the hinge,* being careful to keep the saw level, straight, and centered.
3. Drive a wedge into the tree from the back side, and snug the wedge.
4. Using the attack (bottom) corner of the saw on the leaning side, make a cut 1 to 2 inches below the wedge so that a hinge is formed.
5. Cut just past the wedge: Care must be taken not to cut the supporting wood under the wedge.
6. Repeat the process on the good side.
7. Drive the wedge through the tree. This should result in the tree's falling in the desired direction.

LEARN FROM OTHERS: TAKE A COURSE

It's almost an epiphany when experiencing directional felling for the first time—especially if a chain saw operator is used to more traditional techniques. The

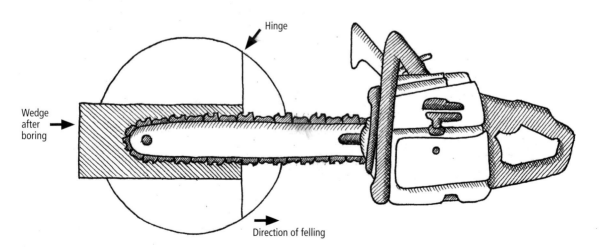

Figure 10.24. Small trees are felled using the same principles as large ones, but instead of boring behind the hinge, the chain saw operator actually bores through the hinge. This method takes some extra practice to master but is well worth it, as small trees can sometimes be the most dangerous and unpredictable. Illustration by Travis Bettencourt

ability to fell a tree with calmness and confidence, versus finger crossing, is a wonderful thing. Of course, keep in mind that the technique takes practice and that extra caution should be taken to learn and practice new methods of felling.

It cannot be overemphasized that chain saw work is by far the most dangerous task that the forest farmer will engage in. There is no substitute for hands-on training with an experienced instructor. Taking a chain saw safety course is literally the best insurance policy if any trees will be cut.

Working the Woods

Design and management of forest farms are where all the pieces come together in the "big picture." Before deciding on crops, where things will go, and so on, it's best to engage in a design process, in which the site is assessed for multiple factors, goals for the participants are clarified, and some thinking is done on paper. While the design process will help forest farmers solidify their ideas into action, it's important to remember that it is an ever-changing process.

Forest farmers should seek more information through reading, having foresters visit their woods, and connecting with other woodlot owners, so they can better learn the best management strategies for their area. Decision making happens on multiple scales, from the long and wide-angle view of the forest, to the patterns that affect stand regeneration, to the task of selecting individual trees for thinning or girdling. The chain saw is one of the most dangerous tools in the forest farmer's arsenal, and care should be taken to learn proper felling techniques and to purchase and wear modern safety gear.

The Last Word

Writing this book has been a revelation of sorts. Both Steve and I have been involved in forest farming as part of a larger commitment to agroforestry and sustainable living, but neither of us has experience with all of the many facets of forest farming, let alone agroforestry as a whole. We have come to appreciate that forest farming does not stand alone with respect to the other temperate agroforestry practices, but rather it is often part of an integrated agroforestry system that involves forest farming and one or more of the other traditional agroforestry practices. Nothing has brought this lesson home to me more than the amazing, multifaceted Wellspring Forest Farm that Steve and Liz are developing near Ithaca, New York.

As for me, my introduction to forest farming was at the formative age of eleven years, when I borrowed a saw, a candle, and a pack of matches from my dad's shop and climbed a big old beech tree in the woods behind our house. Way up in that tree I sawed off a branch and tried to graft for the first time in my life, with guidance only from a picture in the Boy Scout manual. The scion was from an apple tree. Unfortunately the manual didn't mention the critical issues of graft compatibility or cambial alignment, so the graft failed miserably. Ironically, I now teach grafting at the MacDaniels Nut Grove, where grafting played a major role in the structure of the forest farm as it is today. Since my early introduction to forest farming, I've observed forest farming in Africa, Madagascar, Ceylon, and elsewhere, and I've come to realize that forest farming involves both innovation and emulation (of nature and of other forest farmers).

Many of the chapters in this book deal with individual classes of nontimber forest products (food, ornamentals, medicinals, etc.), but we hope that you as the reader have not only embraced the chapter that suits your particular interests and goals but that you have gone on to read some if not all of the other chapters. We hope you have come away with a larger, more integrated picture of forest farming and its role in the wider world. Perhaps you have discovered another class of NTFPs that you have or may pursue in addition to your original NTFP. Beyond that, however, we hope you have come away with the understanding that forest farming is not just a collection of NTFPs. Long-term success requires a commitment to forest health that goes beyond trees and extends to other woody and herbaceous plants, as well as all the other biotic and abiotic factors that interact in a healthy forest ecosystem.

Why did we write this book? If you have read it, and if we have managed to succeed in our task as authors, you already know the answer. We wrote the book not only to encourage better understanding of and wider adoption of forest farming by private forest owners, academics, policy makers, and the general public, but also to encourage appreciation of forest farming in the context of natural forest ecosystems. At present forest farming in the broad, inclusive sense is only practiced by relatively few folks who own or have access to forest land. This is not a matter of lack of forested space. Individual states and provinces, as well as most of eastern North America, have more forested land than at any time since the turn of the twentieth century. Just as it is a common practice for sugarmakers to lease maple trees from landowners with an abundance of them, the practice of leasing land for forest farming should be encouraged.

The key to our approach to adoption of forest farming lies in the concepts and practices of permaculture—earth care, people care, and fair share. Earth care and people care is the core of what we are calling productive conservation, which is no more than striving to enhance or at least not degrade forest health while satisfying the needs of the individual, the family, and the community. There are three components of productive conservation. These are increased profitability, increased well-being for the forest farmer, and increased forest health. It is up to the individual to decide how to address each one. Here are some approaches that we have touched on in this book, and that you can consider as you develop your forest farm:

- Management of the light environment
- Management of the perimeter
- Ecotourism
- Integration of forest farming with other agroforestry practices
- Enhancing the profitability of low-value NTFPs by adding value through processing, direct marketing, and (nursery) production of planting stock for others to grow on, and appealing to the demand for perennial ornamentals for shade gardening
- Genetic improvement of NTFPs through selection and cloning of superior individuals by grafting or other vegetative propagation methods

It has been said that those who strive to classify things into orderly categories, such as plant taxonomists, coin collectors, and even we who write about and practice forest farming, are either lumpers or splitters. We think that those academics and others who have written about agroforestry over the last forty years have tended to be splitters, drawing distinct borders or rigid lines between one agroforestry practice and another. These traditional temperate agroforestry practices include alley cropping, silvopasture, riparian buffers, windbreaks, and forest farming.

Steve and I, on the other hand, tend to be lumpers. The distinction between windbreaks and forest farming may seem pretty clear cut, but planting a hedgerow of fast-growing trees upwind from a shiitake mushroom laying yard certainly blurs the distinction between windbreaks and forest farming. Similarly we regard raising multipurpose ducks (for meat and slug control) in a forested shiitake laying yard as both a silvopasture and a forest farming activity on the same piece of ground. Another nontraditional forest farming practice, suggested originally by Jeanine Davis, is to cultivate and market as perennial ornamentals some of the forest herbs that are traditionally thought of as medicinals. In our experience the most efficient way to cultivate shade perennials, whether they be ornamental, medicinals, or even fruit crops (brambles, pawpaw, etc.), is to use some aspects of modern container nursery production practice, which is not usually considered forest farming.

Perhaps our biggest departure from conventional wisdom about agroforestry has been to suggest that forest gardening, popularized by Jacke and Toensmeier, should be considered as a sixth temperate agroforestry practice. We accept that this will be challenged by some traditional agroforesters, most likely by some of the academics who write about forest farming, rather than the practitioners of agroforestry.

We hope this book has encouraged you to try your hand at forest farming and other agroforestry practices. If you are already growing mushrooms or ginseng or some other NTFP, perhaps you will consider broadening your palette. We hope that the book has encouraged innovation and enriched your enjoyment of the forest. Last, we welcome the give-and-take that will certainly occur with those who have a different point of view.

— Ken Mudge

Notes

INTRODUCTION

1. Wikimedia Foundation. Forest bathing. Wikipedia. 6 July 2014. http://en.wikipedia.org/wiki/Forest _bathing.

CHAPTER ONE

1. Steppler, Howard A. 1987. ICRAF and a decade of agroforestry development. Agroforestry:13.
2. Lundgren, B. O., and J. B. Raintree. 1983. Sustained agroforestry. Nairobi: ICRAF.
3. National Agroforestry Center. 2014. About. 15 May 2014. http://nac.unl.edu/#about.
4. Garrity, D.P. 2005. Forestry in agriculture: The vision of landcare. pp. 47–52. In: A.G. Brown (ed.). Forests, wood, and livelihoods: Finding a future for all. Record of a conference conducted by the ATSE Crawford Fund Parliament House, Canberra. 16 Aug 2005.
5. Gold, Michael A., and Harold E. Garrett. 2009. Agroforestry nomenclature, concepts, and practices. In North American agroforestry: An integrated science and practice 2nd edition, American Society of Agronomy, 45–56.
6. Blackshaw, Judith K., and A. W. Blackshaw. 1994. Heat stress in cattle and the effect of shade on production and behavior: A review. Animal Production Science 34 (2)2:285–295.
7. Chamberlain, J.L., D. Mitchell, T. Brigham, and T. Hobby,L. Zabek, and J. Davis. 2009. Forest farming practices, pp. 219–254. In: North American Agroforestry: An integrated science and practice (2nd ed.). Madison, Wis., American Society of Agronomy.
8. Jacke, Dave, and Eric Toensmeier. 2005. Edible forest gardens: Ecological design and practice for temperate-climate permaculture. Vol. 1. Chelsea Green Publishing, White River Junction, Vt.
9. USDA Agroforestry strategic framework—fiscal year 2011–2016. 2011. Forests and Forestry | USDA. 3 January 2014. http://www.usda.gov/wps/portal/usda/usdahome?navid=FOREST_FORESTRY.
10. Personal communication with co-author Ken Mudge, 1991.
11. Smith, John Russell. 1929. Tree crops: A permanent agriculture. Harcourt, Brace and Company, Inc., Rahway, N.J.
12. Douglas, James Sholto, and R.A. de J. Hart. 1984. Forest farming. Towards a solution to the problems of world hunger and conservation. Intermediate Technology Publications, London, UK.
13. Hill, Deborah B., and Louise E. Buck. 2000. Eight forest farming practices, ch 8. In: H.E. Garrett, W.J. Rietveld and R.F. Fisher (eds.) North American Agroforestry: An integrated science and practice, American society of agronomy, ASA, Madison, Wisc.
14. AFTA—Association for Temperate Agroforestry (AFTA). What is agroforestry? 3 January 2002. http://www.aftaweb.org/.
15. Mudge, Ken. 2010. Forest farming. Arnoldia. 67(3):26–35.
16. Kolb, T.E., M.R. Wagner, and W. Wallace Covington. 1994. Concepts of forest health: Utilitarian and ecosystem perspectives. Journal of Forestry 92:10–15.
17. Leopold, Aldo. 1970. A Sand County almanac, 1949. Ballantine, New York.
18. Balvanera, Patricia, Andrea B. Pfisterer, Nina Buchmann, Jing-Shen He, Tohru Nakashizuka, David Raffaelli, and Bernhard Schmid. 2006. Quantifying the evidence for biodiversity effects on ecosystem functioning and services. Ecology Letters 9.10:1146–1156.
19. Society of American Foresters. 1916. Proceedings of the Society of American Foresters. New York. Vol. 11.
20. Hart, R.A. de J. 2008. Forest gardening. Green Books, UIT, Cambridge, UK.
21. Lindberg Nutrition. Japanese knotweed (Polygonum cuspidatum, Reynoutria japonica, Fallopia japonica). 3 January 2014. http://www.lindbergnutrition.com/ns/DisplayMonograph.asp?StoreID=1C7A08050B8F4419BFFBA945004CA5D1&DocID=basic-interactions-polygonumcuspidatum.
22. Klinghardt, Dietrich K., and Bellevue Wa. 2005. Lyme disease: A look beyond antibiotics. Explor Infect Dis 14.2:6–11.
23. Davis, Mark A., M.K. Chew, R.J. Hobbs, A.E. Lugo, J.J.. Ewel, G.J. Vermeij, J.H. Brown, et al. 2011. Don't judge species on their origins. Nature. 474. 7350: 153–154.
24. Wolfe, Benjamin E., V.L. Rodgers, K.A. Stinson, and Anne Pringle. 2008. The invasive plant *Alliaria*

petiolata (garlic mustard) inhibits ectomycorrhizal fungi in its introduced range. Ecology. 96.4: 777–783.

CHAPTER TWO

1. Vale, T., ed. 2002. Fire, native peoples, and the natural landscape. Island Press, Washington, D.C.
2. Woodlief, Anne. 1782. Negotiating nature/wilderness: Crèvecoeur and American identity in letters from an American farmer. 15 May 2014. http://www.vcu.edu/engweb/crev.htm.
3. Cronon, W. 1983. Changes in the land: Indians, colonists and the ecology of New England. Macmillan, New York.
4. Denevan, W.M. 1992. The pristine myth: The landscape of the Americas in 1492. Annals of the Assoc. of Amer. Geographers. 82.3:369–385.
5. Perlin, J. 1997. Forest journey: The role of wood in the development of civilization. (4th ed) Harvard Univ. Press, Cambridge, Mass.
6. Sargent, C.S. 1884. Report on the forests of North America (exclusive of Mexico). Vol. 3. US Government Printing Office, Washington, D.C.
7. Smith, B.E., P. L. Marks, and S. Gardescu. 1993. Two hundred years of forest cover changes in Tompkins County, New York. Bul. Tor. Bot. Club. 229–247.
8. Burgess, D. 1996. Forests of the Menominee—a commitment to sustainable forestry. The For. Chro. 72.3: 268–275.
9. Gilbert, M. Thomas P., D.L. Jenkins, A. Götherstrom, N. Naveran, J.J. Sanchez, M. Hofreiter, P.F. Thomsen, et al. 2008. DNA from pre-Clovis human coprolites in Oregon, North America. Science. 320.5877:786–789.
10. Butler, E.L. 1948. Algonkian culture and use of maize in southern New England. Bull. of the Arch. Soc. of Conn. 22:2–39.
11. Voisin, A. 1989. Grass productivity. Island Press, Washington, D.C..
12. US Forest Service. 2000. Wildfire—an American legacy. Fire Management Today. Summer.
13. Abrams, M.D., and G.J. Nowacki. 2008. Native Americans as active and passive promoters of mast and fruit trees in the eastern USA. The Holocene. 18.7: 1123–1137.
14. Davies, K. Indian Agroforestry. Indian Agroforestry. 14 April 2013. http://www.daviesand.com/Papers/Tree_Crops/Indian_Agroforestry/.
15. Shawangunk Ridge Biodiversity Partnership (SRBP). Fire in the Gunks. Fire in the Gunks. 14 June 2013. http://www.gunksfireplan.org/.
16. Sexton, Tim. 2000. Constantly Looking For Ways to Improve Program. 66(4):51. http://www.fs.fed.us/fire/fmt/index.html.
17. Cronon, W. 1983. Changes in the land: Indians, colonists and the ecology of New England. Macmillan, New York: p. 63.
18. Wikimedia Foundation. Usufruct. Wikipedia. 3 January 2014. http://en.wikipedia.org/wiki/Usufruct.
19. Smith, J.R. 1987. Tree crops: A permanent agriculture. Island Press, Washington, D.C.
20. Archive.org. 2009. Tree Crops. 4 January 2012. https://archive.org/details/TreeCrops-J.RussellSmith.
21. Elliott, S.B. 1912. The important timber trees of the United States: A manual of practical forestry. Houghton Mifflin, Boston, Mass.
22. Wilson, A. 1996. Silvopastoral agroforestry using honeylocust (*Gleditsia triacanthos L.*). WANATCA Yearbook 20:58–66.
23. Dupraz, C., S. M. Newman, and A. M. Gordon. 1997. Temperate agroforestry systems, p. 181–236. In: Temperate agroforestry: The European way. CAB International, Wallingford, U.K.
24. Bryan, J.A. 1995. Leguminous trees with edible beans, with indications of a rhizobial symbiosis in non-nodulating legumes. Yale Univ., New Haven, Conn., PhD Diss.
25. Ferguson, Rafter Sass, and S.T. Lovell. 2013. Permaculture for agroecology: Design, movement, practice, and worldview. A review. Agronomy for Sustainable Development. 1–24.
26. Jacke, D. and E. Toensmeier. 2005. Edible forest gardens. Chelsea Green Publishing, White River Junction, Vt.: p. 112.
27. MacDaniels, L. H., and A.S. Lieberman. 1979. Tree crops: A neglected source of food and forage from marginal lands. BioScience: 173–175.
28. MacDaniels, L. H. 1971. Nut growing in the northeast. Cornell Univ. Ag. Exp. Sta., Ithaca, N.Y.
29. Mollison, Bill, and D. Holmgren. 1978. Permaculture One. Morebank, NSW Australia: Transworld Publications.
30. Benjamin, Tamara J., P. I. Montañez, J. J. M. Jaménez, and A. R. Gillespie. 2001. Carbon, water and nutrient flux in Maya homegardens in the Yucatan peninsula of Mexico. Agroforestry Systems. 53.2:103–111.
31. Kumar, B. Mohan, and P.K. Ramachandran Nair. 2004. The enigma of tropical homegardens, p. 135–152.In: New vistas in agroforestry. Springer, Netherlands.
32. Fernandes, Erick C.M., A. Oktingati, and J. Maghembe. 1985. The Chagga homegardens: A multistoried agroforestry cropping system on Mt. Kilimanjaro (Northern Tanzania). Agroforestry systems. 2.2:73–86.
33. Pushpakumara, D. K. N. G., A. Wijesekara, and D. G. Hunter. 2010. Kandyan homegardens: A promising land management system in Sri Lanka. Sustainable use

of biological diversity in socio-ecological production landscapes. 52:102.

CHAPTER THREE

1. Wikimedia Foundation. Limiting factor. Wikipedia. 20 December 2013. http://en.wikipedia.org/wiki/Limiting_factor.

2. Cramer, Wolfgang, D. W. Kicklighter, A. Bondeau, B. Moore III, G. Churkina, B. Nemry, A. Ruimy, and A. L. Schloss. 1999. Comparing global models of terrestrial net primary productivity (NPP): Overview and key results. Global Change Biology. 5.S1:1–15.

3. Klass, C., and M. Hoffman. Beneficial insects | lawn, garden & landscape resources. Lawn Garden & Landscape Resources. 12 June 2012. http://blogs.cornell.edu/horticulture/about/basic-gardening-info/garden-beneficial-insects/.

4. 2011. Eric Mader, Matthew Shepherd, Mace Vaughan, and Scott Black in collaboration with Gretchen LeBuhn. Attracting native pollinators: Protecting North America's bees and butterflies: The Xerces Society guide. Storey Publishing, North Adams, Mass.

5. Winfree, R., N.M.. Williams, J. Dushoff, and C. Kremen. 2007. Native bees provide insurance against ongoing honey bee losses. Ecology Letters 10.11:1105–1113.

6. Jeffries, P., S. Gianinazzi, S. Perotto, K. Turnau, and J.M. Barea. 2003. The contribution of arbuscular mycorrhizal fungi in sustainable maintenance of plant health and soil fertility. Biology and Fertility of Soils 37.1:1–16.

7. Van Der Heijden, Marcel GA, R.D. Bardgett, and N.M. Van Straalen. 2008. The unseen majority: Soil microbes as drivers of plant diversity and productivity in terrestrial ecosystems. Ecology Letters. 11. 3:296–310.

8. Perry, D.A., R. Oren, and S.C. Hart. 2008. Forest ecosystems. Johns Hopkins University Press, Baltimore, Md.

9. Maser, C., J.M. Trappe, and R.A. Nussbaum. 1978. Fungal-small mammal interrelationships with emphasis on Oregon coniferous forests. Ecology, 59:779–809.

10. Maser, Chris. 2009. Truffle in the forest. 14 December 2013. http://www.chrismaser.com/truffle.htm.

11. Jacke and Toensmeier. 2005. pp. 239–290.

12. Burby, R.J., T. Beatley, P.R. Berke, R.E. Deyle, S.P. French, D.R. Godschalk, E.J. Kaiser, et al. 1999. Unleashing the power of planning to create disaster-resistant communities. J. of the Amer. Plan. Assoc. 65.3:247–258.

13. Morrow, R. 2006. Earth user's guide to permaculture. Kangaroo Press, Portland, Ore.

14. Yamamoto, Shin-Ichi. 1992. The gap theory in forest dynamics. The botanical magazine= Shokubutsu-gaku-zasshi. 105.2:375–383.

15. Zobel, B. R. U. C. E. 1992. Silvicultural effects on wood properties. IPEF International. 2:31–38.

16. Shepard, Mark. 2013. Restoration agriculture: real-world permaculture for farmers. Acres U.S.A., Austin, Tex.

17. Climate Change 2013. 2013. IPCC—Intergovernmental Panel on Climate Change. 18 December 2013. http://www.ipcc.ch/report/ar5/wg1/.

18. Bonan, Gordon B. 2008. Forests and climate change: Forcings, feedbacks, and the climate benefits of forests. Science. 320.5882:1444–1449.

19. Climate change faster than predicted: Scientific American. 21 November 2013. http://www.scientificamerican.com/article.cfm?id=climate-change-faster-than-predicted.

20. David Suzuki Foundation. David Suzuki Foundation | Solutions are in our nature. 3 December 2013. http://www.davidsuzuki.org/.

21. Gillis, J. Heat-trapping gas passes milestone, raising fears. New York Times. 28 October 2013. http://www.nytimes.com/2013/05/11/science/earth/carbon-dioxide-level-passes-long-feared-milestone.html?_r=0.

22. Morello, L., and C. Wire. 350 Science. 2012. 350.org. 19 December 2013. http://350.org/about/science.

23. Xie, X., and M.J. Economides. 2009. The impact of carbon geological sequestration. Journal of Natural Gas Science and Engineering. 1.3:103–111.

24. Harmon, M.E. 2001. Carbon sequestration in forests: Addressing the scale question. Journal of Forestry. 99.4:24.

25. Pretzsch, H. 2005. Diversity and productivity in forests: evidence from long-term experimental plots, Forest Diversity and Function Ecological Studies Volume 176, 2005, pp. 41–64; Springer: Heidelberg, Germany.

26. Morgan J.A., R.F. Follett, L.H. Allen, S.D. Grosso, J.D. Derner, F. Dijkstra, A. Franzluebbers, R. Fry, K. Paustian, and M.M. Schoeneberger. 2010. Carbon sequestration in agricultural land of the United States. J. Soil Water Conserv. 65:6A–13A.

27. Dixon R.K. 1995. Agroforestry systems: sources or sinks of greenhouse gases? Agroforest Syst. 31:99–116.

28. Quinkenstein, A., C. Böhm, E. da Silva Matos, D. Freese, and R.F. Hüttl. 2011. Assessing the carbon sequestration in short rotation coppices of *Robinia pseudoacacia L.* on marginal sites in northeast Germany, p. 201-216. In: Carbon Sequestration Potential of Agroforestry Systems. Springer, Heidelberg, Germany.

29. Liski, J., D. Perruchoud, and T. Karjalainen. 2002. Increasing carbon stocks in the forest soils of western Europe. Forest Ecology and Management. 169.1:159–175.

30. Jandl, R., M. Lindner, L. Vesterdal, B. Bauwens, R. Baritz, F. Hagedorn, D.W. Johnson, K. Minkkinen, and K.A. Byrne. 2007. How strongly can forest management influence soil carbon sequestration? Geoderma. 137. 3:253–268.

31. Torn, M.S., S.E. Trumbore, O.A. Chadwick, P.M. Vitousek, and D.M. Hendricks. 1997. Mineral control of soil organic carbon storage and turnover. Nature. 389. 6647:170–173.

32. Domm E., A. Drew, R. Greene, E. Ripley, R. Smardon, and J. Tordesillas. 2008. Recommended urban forest mixtures to optimize selected environmental benefits. EnViro News. 14.1.

33. B.K. Gugino, O.J. Idowu, R.R. Schindelbeck, H.M. van Es, D.W. Wolfe, J.E. Thies, and G.S. Abawi. Cornell soil health assessment training manual. 2009. Cornell Soil Health. 24 March 2014. http://soilhealth.cals .cornell.edu/extension/manual.htm.

34. Peter Donovan. Soil Carbon Coalition. 2014. Soil Carbon Coalition. 24 March 2014. http:// soilcarboncoalition.org/.

35. USGCRP. 2009. Global climate change impacts in the United States. T.R. Karl, J.M. Melillo, and T.C. Peterson (eds.). United States Global Change Research Program. Cambridge University Press, New York.

36. Prasad, A.M.L. Iverson, S. Matthews, M. Peters. New climate change tree atlas—a spatial database of trees in the eastern USA. ND. 9 July 2013. http://www.nrs .fs.fed.us/atlas/tree/tree_atlas.html.

37. Sullivan, K. 2009. Climate change impacts on your woods and wildlife." Lecture, ForestConnect Webinars from Cornell University Cooperative Extension, 18 March 2009. http://www2.dnr.cornell.edu/ext /forestconnect/web/schedule.htm.

38. Prasad, A.M.L. Iverson, S. Matthews, M. Peters. New climate change tree atlas—A spatial database of trees in the eastern USA. 24 March 2014. http://www.nrs .fs.fed.us/atlas/tree%3e.

39. Cornell Cooperative Extension Species Program. Hemlock woolly adelgid (*Adelges tsugae*). New York Invasive Species Information. 2 April 2013. http:// www.nyis.info/index.php?action=invasive_detail& id=24.

40. Farrell, M. 2013. The sugarmaker's companion: an integrated approach to producing syrup from maple, birch, and walnut trees. Chelsea Green Publishing, White River Junction, Vt.

41. Bisbing, S. 2013. Trees on the move? Debating assisted migration in climate change mitigation. Early Career Ecologists. 25 September 2013. http://earlycareere-cologists.wordpress.com/2013/01/16/trees-on-the -move-debating-assisted-migration-in-climate-change-mitigation/.

42. Vitt, P., K. Havens, A.T. Kramer, D. Sollenberger, and E. Yates. 2010. Assisted migration of plants: Changes in latitudes, changes in attitudes. Biological Conservation. 143.1:18–27.

CHAPTER FOUR

1. Jacke and Toensmeier. 2005. p. 87.

2. Nelson, J.C., R.E. Sparks, L. DeHaan, and L. Robinson. 1998. Presettlement and contemporary vegetation patterns along two navigation reaches of the Upper Mississippi River. Perspectives on the land-use history of North America: A context for understanding our changing environment. US Geological Survey, Biological Resources Division. USGS/BRD/ BSR-1998-0003:51–60.

3. Peterson, R.N. 2003. Pawpaw variety development: a history and future prospects. HortTechnology.13.3: 449-454.

4. Archbold, D.D., R. Koslanund, and K.W. Pomper. 2003. Ripening and postharvest storage of pawpaw. HortTechnology. 13.3:439–441.

5. Peterson, R.N. 1991. Pawpaw (*Asimina*). Acta Hort. 290:567–600.

6. McLaughlin, J.L. 2008. Paw Paw and cancer: Annonaceous Acetogenins from discovery to commercial products. Journal of Natural Products. 71.7:1311–1321.

7. Mohebalian, P. 2011. U.S. consumer preference for elderberry products. Univer of Missouri, Columbia. MS Thesis.

8. Zakay-Rones, Z., E. Thom, T. Wollan, J. Wadstein. 2004. J. of Intern. Med. Res. 32.2:132–40. http:// www.jimronline.net/content/full/2004/47/0445 .pdf. doi:10.1177/147323000403200204. PMID 15080016.

9. Leyel, C.F. 1979. A modern herbal. Jonathan Cape Ltd., London.

10. Charlebois, D. 2007. Elderberry as a medicinal plant. Issues in new crops and new uses. ASHS, Alexandria, Va. 284–292.

11. Wellnessmama, K. 2011. How to make elderberry syrup for flu prevention. Wellness Mama. 26 December 2013. http://wellnessmama.com/1888/ how-to-make-elderberry-syrup-for-flu-prevention/.

12. Byers, P.L., A.L. Thomas, M.M. Cernusca, L.D. Godsey, and M.A. Gold. 2012. Growing and marketing

elderberry. 7/10/14. http://extension.missouri.edu/p/AF1017.

13. Byers, P.L., A.L. Thomas, M.M. Cernusca, L.D. Godsey, and M.A. Gold. 2012. Growing and marketing elderberry. 7/10/14. http://extension.missouri.edu/p/AF1017.

14. Cornell Gardening Resources. 2013. Cornell Guide to Growing Fruit at Home. 7 October 2014. http://www.gardening.cornell.edu/fruit/homefruit.html. 14 June 2013. http://www.fruit.cornell.edu/.

15. Ramanujan, K. 2013. Mutated white pine rust threatens Northeast trees | Cornell Chronicle. 8 December 2013. http://www.news.cornell.edu/stories/2013/11/mutated-white-pine-rust-threatens-northeast-trees.

16. Steven D. Ehrlich, NMD Hawthorn. 2011. Univ. of Maryland Med. Cent. 12 December 2013. http://www.umm.edu/altmed/articles/hawthorn-000256.htm.

17. Pubmed.gov. Efficacy and safety of a herbal drug containing hawthorn berries and D-camphor in hypotension and orthostatic circulatory disorders/results of a retrospective epidemiologic cohort study, B. Hempel, et al. Arzneimittelforschung. 2004; 55(8):443–50. http://science.naturalnews.com/pubmed/16149711.html

18. Chang, Q., Z. Zuo, F. Harrison, and M.S.S. Chow. 2002. Hawthorn. J. of Clin. Pharm. 42.6:604–612.

19. Home Herbals. Hawthorn tincture—home herbals. Home herbals. 12 December 2013. http://www.homeherbals.com/hawthorn-tincture.html.

20. Plants for a future. Schisandra chinensis. 1996–2012. 12 December 2013. http://www.pfaf.org/user/Plant.aspx?LatinName=Schisandra+chinensis.

21. Jihong, S., et al. 1997. Analysis of nutrition composition of Schisandra fruits. Jour. of For. Res. 8.2:97–98; Zhao, Xing-Nan, et al. 2013. Gender variation in a monoecious woody vine Schisandra chinensis (*Schisandraceae*) in northeast China. Annales Botanici Fennici. 50.4. Finnish Zoological and Botanical Publishing Board.

22. Cornell Cooperative Extension. Juneberries / Saskatoons. A new fruit for the Northeast. 2011. 12 December 2013. http://www.juneberries.org/.

23. Ahmad, S. D., S. M. Sabir, and M. Zubair. 2006. Ecotypes diversity in autumn olive (*Elaeagnus umbellata Thunb*): A single plant with multiple micronutrient genes. Chemistry and Ecology. 22.6:509–521.

24. Fordham, I.M., et al. 200. Fruit of autumn olive: A rich source of lycopene. HortScience 36.6:1136–1137.

25. Vander, W., B. Stephen. 2001. The evolutionary ecology of nut dispersal. The Botanical Review. 67.1:74–117.

26. Wolff, J.O. 1996. Population fluctuations of mast-eating rodents are correlated with production of acorns. Journal of Mammalogy. 850–856.

27. Jimmy Mengel. Money does grow on trees. 2013. 23 December 2013. http://www.outsiderclub.com/money-does-grow-on-trees/404.

28. Scott, R., and W.C. Sullivan. 2007. A review of suitable companion crops for black walnut. Agroforestry Systems. 71.3:185–193.

29. Jose, S., and A.R. Gillespie. 1998. Allelopathy in black walnut (*Juglans nigra L.*) alley cropping. II. Effects of juglone on hydroponically grown corn (*Zea mays L.*) and soybean (*Glycine max L. Merr.*) growth and physiology. Plant and Soil. 203.2:199–206.

30. Campbell, G.E., and J.O. Dawson. 1989. Growth, yield, and value projections for black walnut interplantings with black alder and autumn olive. North. J. of Appl. For. 6.3:129–132.

31. Reid, W., M.V. Coggeshall, and K.L. Hunt. 2004. Cultivar evaluation and development for black walnut orchards. United States Depatment of Agriculture Forest Service General Technical Report. NC 243:18.

32. Schmidt, S. K. 1988. Degradation of juglone by soil bacteria. Journal of Chemical Ecology. 14.7:1561–1571; Williamson, G.B, and J.D. Weidenhamer. 1990. Bacterial degradation of juglone. Journal of Chemical Ecology. 16.5:1739–1742.

33. Reid, W., et all. 2009. Growing black walnut for nut production. AF1011. 14 December 2013. http://extension.missouri.edu/p/AF1011.

34. Freinkel, S. 2007. American chestnut: the life, death, and rebirth of a perfect tree. Univ. of California Press, Oakland, Calif.

35. Oak, S. 2004. Restoring American chestnut to forest lands. NPS Conference Proceedings. Forest Health Impacts of the Loss of American Chestnut. 18 December 2013. http://ecosystems.psu.edu/research/chestnut/information/conference-2004/conference.

36. Gold, M.A., M.M. Cernusca, and L.D. Godsey. 2006. Competitive market analysis: chestnut producers. HortTechnology. 16.2:360–369.

37. Bartram, W. 1973. Travels through North and South Carolina, Georgia, east and west Florida: A facsimile of the 1792 London edition embellished with its nine original plates. University of Virginia Press, Charlottesville, Va.

38. Rutter, P.A. 1994. The potential of hybrid hazelnuts in agroforestry and woody agriculture systems, Paper presented at the North American Conference On Enterprise Development Through Agroforestry: Farming the Agroforest for Specialty Products, Minneapolis, MN, October 4-7, 1998, pp. 133-134.

39. Hampson, C.R., A.N. Azarenko, and J.R. Potter. 1996. Photosynthetic rate, flowering, and yield component alteration in hazelnut in response to different light environments, J. Amer. Soc. Hort. Sci. 121:1103–1111.

40. Farrell, M. 2013. The sugarmaker's companion: an integrated approach to producing syrup from maple, birch, and walnut trees. Chelsea Green Publishing, White River Junction, Vt.

41. University of Vermont. 2007. Comparison of the "small" spout with the traditional 7/16 spout. Proctor Maple Research Center. Univ. of Vermont, Burlington, Vermont. 2 January 2014. http://www.uvm.edu/~pmrc/?Page=publications.html.

42. Farrell, Michael., 2013.

43. Matta, Z., E. Chambers, and G. Naughton. 2004. Consumer and descriptive sensory analysis of black walnut syrup. Journal of Food Science. 70. 9: S610–S613; Naughton, G. G., W. A. Geyer, and E. Chambers IV. 2006. Making syrup from black walnut sap. Transactions of the Kansas Academy of Science. 109. 3:214–220.

44. Lord, B. 2013. Black birch tea: a delicate winter brew. Northern Woodlands. Winter 2013.

45. Shackford, S. 2013. Researchers tap potential of walnut and birch trees. Cornell Chronicle. 1 January 2014. http://www.news.cornell.edu/stories/2013/02/researchers-tap-potential-walnut-and-birch-trees.

46. Sen, I. 2011. When digging for ramps goes too deep. New York Times. 14 Sept 2013. http://www.nytimes.com/2011/04/20/dining/20forage.html?pagewanted=all&_r=0.

47. Rock, J.H., B. Beckage, and L.J. Gross. 2004. Population recovery following differential harvesting of *Allium tricoccum* Ait. in the southern Appalachians. Biological Conservation. 116.2:227–234.

48. Nault, A., and D. Gagnon.1993. Ramet demography of Allium tricoccum, a spring ephemeral, perennial forest herb. Journal of Ecology pp.101–119.

49. Barry, E., et all. Plugging the leak on wild leeks: The threat of over-harvesting wild leek populations in northern New York. St Lawrence University. 10 December 2013. http://web.stlawu.edu/sites/default/files/resource/wild_leek_conservation.pdf.

50. Pickowicz, N. 2011. Precious and dwindling: Quebec's wild leeks. The Gazette. http://www2.canada.com/montrealgazette/news/arts/story.html?id=4abb0399-2510-4cc7-8dde-4fdb139e854b

51. Davis, J.M., and J. Greenfield. 2002. Cultivating ramps: Wild leeks of Appalachia. Trends in new crops and new uses. ASHS Press, Alexandria, Va.: 449–452.

52. Carlisi, J. 2004. History, Culture, and Nutrition of *Apios americana*. Journal of Nutraceuticals, Functional & Medical Foods. 1089–4179. 4 .3-4:85–92.

53. Blackmon, W.J. and B.D. Reynolds. 1986. The crop potential of *Apios americana*—Preliminary evaluations. HortScience. 21:1334–1336.

CHAPTER FIVE

1. Mudge, K.W., A. Matthews, B. Waterman, and B. Hilshey. 2014. Best management practices for log-based shiitake cultivation in the northeastern United States. NE SARE research and education project. LNE 10-298.

2. Mushrooms, USDA National Agricultural Statistical Service, ISSN: 1949–1530, August 20, 2013. http://usda01.library.cornell.edu/usda/current/Mush/Mush-08-20-2013.pdf.

3. Hall, I.R., G. Brown, and A. Zambonelli. 2007. Taming the truffle: The history, lore, and science of the ultimate mushroom. Timber Press, Portland, Oregon: p. 304.

4. Ochterski, J. 2010. Selling forest grown mushrooms: customer, qualities, and opportunities. Cornell University Cooperative Extension, Ontario County.

5 Mudge, K.W., A. Matthews, B. Waterman, and B. Hilshey. 2014. Best management practices for log-based shiitake cultivation in the northeastern United States. NE SARE research and education project LNE 10-298.

6. University of Maryland Extension. 2013. Forest Stewardship Education. Shiitake mushroom production and marketing. 3 November 2013. http://www.naturalresources.umd.edu/Publications/html/shiitake.html.

7. Sanders, L. 2011. A red scare. New York Times. 1 January 2013. http://www.nytimes.com/2011/02/20/magazine/20fob-diagnosis-t.html?_r=0.

8. Code, 2009, Food and Drug Administration. Report no. PB2009112613. 1 January 2013. http://www.fda.gov/downloads/Food/GuidanceRegulation/UCM189448.pdf.

9. Gold, M.A., M.M. Cernusca, and L.D. Godsey. 2008. A competitive market analysis of the United States shiitake mushroom marketplace. HortTechnology. 18. 3:489–499.

10. Leatham, G.F. 1982. Cultivation of shiitake, the Japanese forest mushroom on logs: A potential industry for the United States. Forest Products Journal. 32:29–35.

11. Hobbs, C. 1995. Medicinal mushrooms: an exploration of tradition, healing & culture. No. Ed. 2. Botanica Press, Santa Cruz, Calif.

12. USDA SR-21. Conde Nast, 2014 Nutrition facts and analysis for mushrooms, shiitake, cooked, without salt. 5 January 2014. http://nutritiondata.self.com/facts/vegetables-and-vegetable-products/2488/2.

13. Bruhn, J.N., J.D. Mihail, and J.B. Pickens, 2009. Forest farming of shiitake mushrooms: An integrated evaluation of management practices. Bioresource Technology. 100:6472–6480.

14. Kalaras, M.D., and R.B. Beelman. 2008. Vitamin D2 enrichment in fresh mushrooms using pulsed UV light. Penn State University. 15 May 2014. foodscience. psu . edu/directory/rbb6/VitaminD Enrichment.pdf.

15. Keisuke Tokimoto. Aloha Medicinals, Inc. 2005. Shiitake log cultivation. 1 January 2013. http://www .alohamedicinals.com/book2/chapter-3-01.pdf.

16. Stamets, P. 3 June 2010. The petroleum problem. 6 January 2014. http://www.fungi.com/blog/items /the-petroleum-problem.html.

CHAPTER SIX

1. Dietary Supplements, United States Food and Drug Administration. 2014. 05/14/14. http://www.fda .gov/food/dietarysupplements/.

2. US Food and Drug Administration. 2013. Dietary Supplement Health and Education Act. 29 December 2013. http://www.fda.gov/RegulatoryInformation /Legislation/FederalFoodDrugandCosmeticActFDCAct /SignificantAmendmentstotheFDCAct/ucm148003 .htm.

3. National Center for Complementary and Alternative Medicine. 2013. National Institutes of Health. 29 December 2013. http://nccam.nih.gov/.

4. DeAngelis, C., and P. Fontanarosa. September 2003. Drugs alias dietary supplements. JAMA. 11:1519–1520.

5. Chrisp . Species At-Risk.05/08/2012. United Plant Savers. 5 January 2014. https://www.unitedplantsavers .org/content.php/121-species-at-risk.

6. Murphy, L.L., R.S. Cadena, D. Chávez, and J.S. Ferraro. 1998. Effect of American ginseng (*Panax quinquefolium*) on male copulatory behavior in the rat. Physiology & Behavior. 64.4:445–450.

7. National Health Interview Survey. 2008. National health statistics report 12. 29 December 2013. http:// www.cdc.gov/nchs/data/nhsr/nhsr012.pdf.

8. Onamson, B.J. 2010. Daniel Boone and the boatload of ginseng: Further considerations. 3 January 2014. http://prickettsfort.wordpress.com/2010/08/31 /daniel-boone-the-boatload-of-ginseng-further- considerations.

9. McGraw J.B., A.E. Lubbers, M. Van der Voort, E.H Mooney, M.A. Furedi, S. Souther, J.B. Turner, and J. Chandler. 2013. Ecology and conservation of ginseng (*Panax quinquefolius*) in a changing world. Annals of the New York Academy of Sciences. Issue: The year in ecology and conservation, 1–30.

10. US Fish and Wildlife Service. CITES in the United States. 4 January 2014. http://www.fws.gov /international/cites/.

11. Division of Management Authority, USFWS. 2012. US exports of American ginseng. 14 May 2014. http:// www.fws.gov/international/pdf/report-us-exports-of -american-ginseng-1992-2011.pdf.

12. American Herbal Products Association. 2013. US Department of the Interior Service Issues. CITES non-detrimental finding. 4 January 2014. www.ahpa .cog/Fefault.aspx?tabid=69&ald=949.

13. Persons, S. 1994. American ginseng: Green gold, a growers guide including ginseng's history and use. Bright Mountain Books, Fairview, N.C.

14. Agroforestry Resource Center. 2014. Great Northern Catskills, Greene County. 14 May 2014. http://www .cornell.edu/outreach/programs/program_view .cfm?ProgramID=1663.

15. R.C. Vaughan, J.L. Chamberlain, and J.F. 01/13/11. Munsell. Growing American ginseng (*Panax quinquefolius*) in forest lands. Virginia Cooperative Extension, Publication. No. 354-313.

16. Beyfuss, B. 2013. The practical guide to growing ginseng, New York ginseng specialist. Cornell University, (ret.). Cornell University Cooperative Extension, Agroforestry Resource Center, Acra, N.Y.

17. American Botanicals, 2013 Spring Price List. Eolia, Mo., www.american botanicals.com.

18. Persons, W.S., and J.M. Davis. 2007. Growing & marketing ginseng, goldenseal & other woodland medicinals. Bright Mountain Books, Fairview, N.C.

19. Division of Management Authority. 2012. US Exports of American Ginseng, USFWS. 8 July 2014. http:// www.fws.gov/international/pdf/report-us-exports-of -american-ginseng-1992-2011.pdf.

20. United Plant Savers. Species at Risk. 14 May 2014. https://www.unitedplantsavers.org/.

21. Burkhart, E.P., and M.G. Jacobson. 2009. Transitioning from wild collection to forest cultivation of indigenous medicinal forest plants in eastern North America is constrained by lack of profitability. Agroforestry Systems. 76.2:437–453.

22. American Botanicals, 2013 Spring Price List, www .americanbotanicals.com.

23. Stephen Foster. 2014. Black coshosh *Actaea racemosa* (*Cimicifuga racemosa*). 14 May 2014. http://www .stevenfoster.com/education/monograph/actaea.html.

24. Greenfield, J., and J. M. Davis. 2003. Analysis of the economic viability of cultivating selected botanicals in North Carolina. Strategic reports for the North Carolina consortium on natural medicinal products. https://sites.google.com/a/ncsu.edu/medicinal-herb

-natural-products-resources-for-educators/home
/links-to-natural-products-resources Medicinal Herb/
Natural Product Resources For Eductator: p. 242.

25. J.M. Davis, and A. Dressler, 2013. Black cohosh (*Actaea racemosa L.*). eXtension. 29 December 2013. http://
www.extension.org/pages/67832/black-cohosh
-actaea-racemosa-l.

26. Bureau of Consumer Protection. 2001. Dietary
supplements: an advertising guide for industry. 29
December 2013. http://business.ftc.gov/documents
/bus09-dietary-supplements-advertising-guide
-industry.

27. Cech, R.A. 2002. Balancing conservation with
utilization: restoring populations of commercially
valuable medicinal herbs in forests and agroforests.
Advances in Phytomedicine. 1:117–123.

28. Albrecht, M.A., and B.C. McCarthy. 2006.
Comparative analysis of goldenseal (*Hydrastis
canadensis L.*) population re-growth following human
harvest: implications for conservation. The American
Midland Naturalist. 156.2:229–236.

29. Schlag, E.M., and M.S. McIntosh. 2012. RAPD-
based assessment of genetic relationships among
and within American ginseng (*Panax quinquefolius
L.*) populations and their implications for a future
conservation strategy. Genetic Resources and Crop
Evolution. 59.7:1553–1568.

30. Naud, J., A. Olivier, A. Bélanger, and L. Lapointe.
2010. Medicinal understory herbaceous species
cultivated under different light and soil conditions
in maple forests in southern Québec, Canada.
Agroforestry Systems. 79.3:303–326.

CHAPTER SEVEN

1. Sharrock, R. 1660. The history of the propagation
& improvement of vegetables by the concurrence
of art and nature. Early English books online. 30
December 2013. http://gateway.proquest.com/
openurl?ctx_ver=Z39.88-2003&res_id=xri:eebo&rft_
id=xri:eebo:image:113671.

2. Mudge, K., J. Janick, S. Scofield, and E.E. Goldschmidt.
2009. A history of grafting. Horticultural Reviews.
35:437–493.

3. Northern Nut Growers Association Report of the
Proceedings at the 43rd Annual Meeting, Rockport,
Indiana, August 25, 26 and 27, 1952. 5 January 2014.
http://archive.org/stream/northernnutgrowe25935
gut/25935.txt.

4. Schopmeyer, C.S. 1974. Seeds of woody plants in the
United States. Agriculture handbook, US department
of agriculture 450. 30 December 2013. http://www
.nsl.fs.fed.us/nsl_wpsm.html.

5. Bailey, L.H. 1928. The standard cyclopedia of horticulture. MacMillan, Atlanta, Ga.: p. 1200.

6. Bir, R.E. Bilderback, and TG Ranney, Grafting and
Budding Nursery Crops Plants, AG 396. N.C. State
Univ. 5 January 2014. http://www.ces.ncsu.edu/depts
/hort/hil/grafting.html.

7. Vanek, S. University of Kentucky Cooperative
Extension Service. UK College of Agriculture. 2013.
Pot-in-pot nursery production. http://www.uky.edu
/Ag/CCD/introsheets/potinpot.pdf.

8. Pollan, M. 2001. The botany of desire: A plant's
eye-view of the world. Random House Digital, Inc.
New York.

CHAPTER EIGHT

1. Sloane, E. 1965. A reverence for wood. Funk &
Wagnalls, New York.

2. Odum, E.P. 1997. Ecology: A bridge between science
and society. Sinauer Associates Inc., Sunderland, Mass..

3. U.S. Energy Information Administration—EIA—
Independent Statistics and Analysis. 2009. Residential
energy consumption survey (RECS). 18 December
2013. http://www.eia.gov/consumption/residential
/data/2009/.

4. Bloom, I. 2012. New study: Frack fluids can migrate
to aquifers within years. Protecting our waters.
12 December 2013. http://protectingourwaters.
wordpress.com/2012/05/02/new-study-frack-fluids
-can-migrate-to-aquifers-within-years/.

5. Agroforestry Systems Journal. 1982–2004. Springer. 31
December 2013. http://link.springer.com/journal
/10457.

6. Ackerly, J. 2011. Facts and analysis on 2010 census
heating fuel data heated up! Heated up! 26 December
2013. http://forgreenheat.blogspot.com/2011/10
/facts-and-analysis-on-2010-census.html.

7. Ames, D. 2011. Your home: calculating
potential energy savings with the home heating
index. Intelligentutility. 7 December 2013.
http://intelligentutility.com/blog/11/11/
calculating-potential-energy-savings?quicktabs_4=1.

8. Smallidge, P. 2012. Estimating the volume of firewood
in a tree. Cornell Cooperative Extension Factsheet,
Ithaca, N.Y.

9. Fox, J. Winter 2013. How many cords of firewood do
you need?. Northern Woodlands: p. 4.

10. Gevorkiantz, S.R., and.L.P. Olsen. 1955. Composite
volume tables for timber and their application in the
lake states. US Dept. of Agriculture, No. 1104.

11. The Wood Heat Organization Inc. 2014. Outdoor
boilers. Woodheat.org. 28 December 2013. http://
woodheat.org/outdoor-boilers.html.

12. Hagan, J.M., and S.L. Grove. 1999. Coarse woody debris: Humans and nature competing for trees. Journal of Forestry. 97.1:6–11.

13. Utah State Forestry. 2014. Wood heating. 26 December 2013. http://forestry.usu.edu/htm/forest-products/wood-heating.

14. Evans, I., and L. Jackson. 2006. Rocket mass heaters. A Cob Cottage Company Publication, Coquille, Ore. 2006.

15. Bealer, A.W. 1972. Old ways of working wood. Barre Publishers, Wilkes Barre, Pa.

16. Krawczyk, M. Spring 2012. The polewood economy. Permaculture Activist. p. 4.

17. Long, J. 1998. Making bentwood trellises, arbors, gates & fences. Storey Books, North Adams, MA.

18. Pooktre tree shapers. Pooktre Tree Shapers. 12 November 2013. http://pooktre.com.

19. Lewis, L. 2000. Soil bioengineering: An alternative for roadside management: a practical guide. San Dimas Technology & Development Center, San Dimas, Calif.

20. Wikimedia Foundation. 2014. Fascine. Wikipedia. 14 December 2013. http://en.wikipedia.org/wiki/Fascine.

21. Remarks by Kevin Finney, Landscape Architect, at the Eleventh Annual California Salmon Restoration Federation Conference in Eureka, California, March 20, 1993.

22. Lehmann, J. Biochar projects. Cornell Univ. 12 October 2013. http://www.css.cornell.edu/faculty/lehmann/research/biochar/biocharproject.html.

23. UK Biochar Research Centre. UKBRC. 12 October 2013. http://www.biochar.ac.uk/.

24. International biochar initiative. 13 October 2013. http://www.biochar-international.org/.

25. Smolker, R. Biochar: Black gold or just another snake oil scheme? Earth Island Journal, Earth Island Institute. 23 November 2013. http://www.earthisland.org/journal/index.php/elist/eListRead/biochar_black_gold_or_just_another_snake_oil_scheme/.

26. Biofuelwatch. 2011. A critical review of biochar science and policy. Biofuelwatch. 1 December 2013. http://www.biofuelwatch.org.uk/2011/a-critical-review-of-biochar-science-and-policy/.

27. Dungait, Jennifer A.J., D.W. Hopkins, A.S. Gregory, and A. P. Whitmore. 2012. Soil organic matter turnover is governed by accessibility not recalcitrance. Global Change Biology. 18.6:1781–1796.

CHAPTER NINE

1. USDA National Agroforestry Center. Working trees for livestock. 16 December 2013. http://nac.unl.edu/documents/workingtrees/brochures/wts.pdf.

2. Baertsche, S. R., M. T. Yokoyama, and J. W. Hanover. 1986. Short rotation, hardwood tree biomass as potential ruminant feed: Chemical composition, nylon bag ruminal degradation and ensilement of selected species. Journal of Animal Science. 63.6:2028–2043.

3. Wilson, A. A. 1991. Browse agroforestry using honeylocust. The Forestry Chronicle. 67.3:232–235.

4. Chedzoy, B., and P. Smallidge. 2011. Silvopasturing in the northeast. Cornell ForestConnect. 12 December 2013. http://www2.dnr.cornell.edu/ext/info/pubs/MapleAgrofor/Silvopasturing3-3-2011.pdf.

5. Godsey, L. 8 November 2011. Silvopasture economics: three case studies. Lecture, Northeast Silvopasture Conference from Cornell Univer Cooperative Extension, Watkins Glen, N.Y.

6. Lin, C. H., R. L. McGraw, M. F. George, and H. E. Garrett. 1998. Shade effects on forage crops with potential in temperate agroforestry practices. Agroforestry Systems. 44, 2-3:109–119.

7. Smallidge, Peter. 2003. Enhancing goat meat production from controlled woodland browsing. NE SARE Research. 23 December 2013. http://mysare.sare.org/MySare/ProjectReport.aspx?do=viewProj&pn=LNE01-148.

8. Hall, M. 2009. Pine straw: A new mulch for Missouri. Green Horizons Newsletter. 31 December 2013. http://agebb.missouri.edu/agforest/archives/v13n2/gh6.htm.

9. Duryea, M.L., and J.C. Edwards. 1989. Pine-straw management in Florida's forests. Florida Cooperative Extension Service, Institute of Food and Agricultural Sciences, University of Florida, Vol. 831:p. 1.

10. PineStraw.com. Pine straw to your door. 31 December 2013. http://pinestraw.com/

11. Pote, D.H., B.C. Grigg, C.A. Blanche, and T. C. Daniel. 2004. Effects of pine straw harvesting on quantity and quality of surface runoff. Journal of Soil and Water Conservation. 59.5:197–203.

CHAPTER TEN

1. Mollison, B., and R.M. Slay. 1991. Introduction to permaculture. Tagari Publications, Tyalgum, NSW.

2. Yeomans, P.A. 1958. The challenge of landscape: The development and practice of keyline. Keyline Pub./Pty., Sydney, Australia

3. Yeomans, P.A. 1958. The challenge of landscape: The development and practice of keyline. Keyline Pub./Pty., Sydney, Australia.

4. Jacke, D., and E. Toensmeier. 2005. Edible forest gardens, vol. 2: Ecological design and practice for temperate-climate permaculture. Chelsea Green Publishing, White River Junction, Vt.

5. Whitefield, P. 2004. Earth care manual: a permaculture handbook for Britain and other temperate countries. Permanent Publications, East Meon, UK.

6. Cornell plantations. 2014. Ecological Communities. Cornell Plantations. 23 December 2013. http://www.cornellplantations.org/our-gardens/natural-areas/upper.cascadilla/ecological.communities.

7. Lancaster, B., and J. Marshall. 2008. Rainwater harvesting for drylands and beyond. Rainsource Press, Tucson, Ariz.

8. Land Trust Alliance. 31 December 2013. http://www.landtrustalliance.org/.

9. Jeffrey S.W., T.E. Worthley, P.J. Smallidge, and K.P. Bennett. 2006. Northeastern forest regeneration handbook. 4 March 2013. http://www.fws.gov/newengland/pdfs/forest_regn_hndbk06.pdf.

10. Jemison, G.M., and G.H. Hepting 1949. Timber stand improvement in the Southern Appalachian region. US Department of Agriculture, Forest Service. No. 693.

11. Perkey, A.W., and B.L. Wilkins. 1993. Crop tree management in eastern hardwoods. NA-TP USDA, Morgantown, W.Va.

12. Buckley, G. Peter, ed. 1992. Ecology and management of coppice woodlands. Springer, New York.

13. Petit, S., and C. Watkins. 2003. Pollarding trees: Changing attitudes to a traditional land management practice in Britain 1600–1900. Rural history. 14.2:157–176.

14. Admin. 2010. Stump culture: Coppice for conifers. Mast Tree Network. 5 January 2014. http://www.mast-producing-trees.org/2010/11/stump-culture-coppice-for-conifers/.

15. Loehle, Craig. 1987. Average and maximum lifespan of virginia trees. Tree Age. 5 January 2014. http://bigtree.cnre.vt.edu/TreeAge.htm.

16. The game of logging. Game of Logging. 5 January 2014. http://www.gameoflogging.com/.

17. Cornell Cooperative Extension. 2004. Chainsaw BMPs. Chainsaw. BMPs. 5 January 2014. http://www2.dnr.cornell.edu/ext/bmp/contents/diy/diy_chainsaw.htm.

Index

Note: page numbers followed by f or t refer to Figures/Photographs or Tables

Ken Mudge, associate professor at Cornell University, has been involved in agroforestry research, teaching, and extension for over twenty years. His domestic and international research has focused on nontimber forest products including nitrogen-fixing trees, American ginseng, forest-cultivated mushrooms, and others. He teaches both on-campus and online courses in forest farming, plant propagation, and grafting. He is director of the MacDaniels Nut Grove, which is the foremost center for forest-farming education in the country. Mudge was the principal investigator on a Northeast SARE-funded extension project in collaboration with the University of Vermont and Chatham University that trained forest owners in shiitake mushroom farming as a business enterprise, and has recently published the guidebook *Best Management Practices for Log-Based Shiitake Cultivation in the Northeastern United States*. He is the coordinator of the Northeast Forest Mushroom Growers Network (blogs.cornell.edu/mushrooms).

Steve Gabriel is an ecologist, educator, author, and forest farmer who has lived most of his life in the Finger Lakes region of New York. His personal mission is to reconnect people of all ages with the natural world and to provide the tools for good management of forests and other landscapes. He cofounded the Finger Lakes Permaculture Institute and currently works for the cooperative extension in the department of horticulture at Cornell, where he focuses on permaculture and agroforestry research and education. Along with his fiancée, Liz, he operates Wellspring Forest Farm (www.wellspringforestfarm.com) in Mecklenburg, New York, which produces shiitake mushrooms, duck eggs, pastured lamb, nursery trees, and maple syrup.

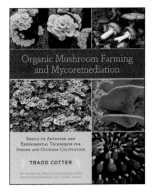